Applied Deep Learning with TensorFlow 2

Learn to Implement Advanced Deep Learning Techniques with Python

Second Edition

Umberto Michelucci

Apress®

Applied Deep Learning with TensorFlow 2: Learn to Implement Advanced Deep Learning Techniques with Python

Umberto Michelucci
Dübendorf, Switzerland

ISBN-13 (pbk): 978-1-4842-8019-5   ISBN-13 (electronic): 978-1-4842-8020-1
https://doi.org/10.1007/978-1-4842-8020-1

Managing Director, Apress Media LLC: Welmoed Spahr
Acquisitions Editor: Celestin Suresh John
Development Editor: James Markham
Coordinating Editor: Aditee Mirashi
Copy Editor: Kezia Endsley

Cover designed by eStudioCalamar

Cover image designed by Freepik (www.freepik.com)

Distributed to the book trade worldwide by Springer Science+Business Media New York, 1 New York Plaza, Suite 4600, New York, NY 10004-1562, USA. Phone 1-800-SPRINGER, fax (201) 348-4505, e-mail orders-ny@springer-sbm.com, or visit www.springeronline.com. Apress Media, LLC is a California LLC and the sole member (owner) is Springer Science + Business Media Finance Inc (SSBM Finance Inc). SSBM Finance Inc is a **Delaware** corporation.

For information on translations, please e-mail booktranslations@springernature.com; for reprint, paperback, or audio rights, please e-mail bookpermissions@springernature.com.

Apress titles may be purchased in bulk for academic, corporate, or promotional use. eBook versions and licenses are also available for most titles. For more information, reference our Print and eBook Bulk Sales web page at http://www.apress.com/bulk-sales.

Any source code or other supplementary material referenced by the author in this book is available to readers on GitHub via the book's product page, located at www.apress.com/978-1-4842-8019-5. For more detailed information, please visit http://www.apress.com/source-code.

Printed on acid-free paper

To my daughter Caterina and my wife Francesca.
You are the reason I do what I do.

Table of Contents

About the Author

Umberto Michelucci is the founder and the chief AI scientist of TOELT – Advanced AI LAB LLC, a company aiming to develop new and modern teaching, coaching, and research methods for AI, to make AI technologies and research accessible to every company and everyone. He's an expert in numerical simulation, statistics, data science, and machine learning. In addition to several years of research experience at the George Washington University (USA) and the University of Augsburg (DE), he has 15 years of practical experience in the fields of data warehouse, data science, and machine learning. His first book, *Applied Deep Learning—A Case-Based Approach to Understanding Deep Neural Networks,* was published by Springer in 2018. He published a second book, *Convolutional and Recurrent Neural Networks Theory and Applications,* in 2019. He's very active in artificial intelligence research. He publishes his research results regularly in leading journals and gives regular talks at international conferences. He also gives regular lectures on machine learning and statistics at various international universities. Umberto studied physics and mathematics. He holds a PhD in machine learning and physics, and he is also a Google Developer Expert in Machine Learning based in Switzerland.

About the Contributing Author

Michela Sperti is responsible for most of the code upgrades from TensorFlow 1 to TensorFlow 2 in this book. She is a second year PhD student at Politecnico di Torino, Bioengineering department, with Prof. M. A. Deriu. She graduated in Biomedical Engineering at Politecnico di Torino in 2019 with a thesis on machine learning techniques for cardiovascular risk prediction in rheumatic patients. She worked for one year as a research assistant under the European-funded MSCA VIRTUOUS project (which aims to apply machine learning techniques to investigate taste and food properties). Currently, she is studying explainability techniques for machine learning and deep learning models applied in various fields (from cardiovascular risk to food organoleptic properties prediction), with the final aim of understanding complex mechanisms that underlie physiological processes. She is very passionate about teaching and is committed to communicating her results. She is the author of eight articles published in peer-review journals and took part in three international workshops as both a teaching assistant and a speaker.

About the Technical Reviewer

Jojo Moolayil is an artificial intelligence professional and a published author of three books on machine learning, deep learning, and IoT. He is currently working with Amazon Web Services as a research scientist – A.I. in their Vancouver, BC office.

He was born and raised in Pune, India and graduated from the University of Pune with a major in Information Technology Engineering. His passion for problem-solving and data-driven decision-making led him to start a career with Mu Sigma Inc., the world's largest pure-play analytics provider. There, he was responsible for developing machine learning and decision science solutions to large, complex problems for healthcare and telecom giants. He later worked with Flutura (an IoT Analytics startup) and General Electric with a focus on industrial A.I in Bangalore, India.

In his current role with AWS, he works on researching and developing large-scale A.I. solutions for combating fraud and enriching the customer's payment experience in the cloud. He is also actively involved as a technical reviewer and AI consultant with leading publishers and has reviewed over a dozen books on machine learning, deep learning, and business analytics.

You can reach out to Jojo at

- `https://www.jojomoolayil.com/`
- `https://www.linkedin.com/in/jojo62000`
- `https://twitter.com/jojo62000`

Acknowledgments

This book would not have been possible without the help of many people who read drafts and gave me feedback. Prof. Marco Deriu helped greatly with many projects, ideas, and discussions. Dr. Piga read drafts and gave me feedback and ideas about how to make the chapters better. In particular, I am deeply indebted to Michela Sperti. She worked without pause and updated almost all of the book's code to TensorFlow 2. Not only that, but she also read all the chapters and gave me important feedback that made the book much better. Without her, the book would not be as good as it is. Of course, all the mistakes that are in the book are completely my fault.

I am also incredibly grateful to Aditee Mirashi, an untiring editor, Jojo John Moolayil, a wonderful technical editor, and Celestin John Suresh, the most wonderful acquisition editor one may want. Many thanks to a wonderful Apress editing team.

But more importantly, I am infinitely indebted to my daughter Caterina and my wife Francesca, who supported me during the entire process and had infinite patience with me while I was writing and updating this book. You are the reason I do what I do.

A last big thank you goes to all the readers of the first edition. I thank you for your trust and interest in what I wrote. You are my main motivation for updating the book to this second edition.

Foreword

Without even realizing it, we have been buried by the data we generate every day, in all areas of technology applications, social life, and health. Buried in this huge amount of data that we have been storing for years in the most disparate formats and ways lies knowledge that we yearn for and that we have not yet uncovered.

After an initial phase of caution that lasted a few years, today we all agree that artificial intelligence is a very powerful way to extract this knowledge from data.

For example, in my daily activity as a professor of biomedical engineering at the Politecnico di Torino, I often find myself dealing with topics that have to do with the world of health. I realize how much the clinical world is fascinated by the potential of these new technologies and sometimes looks at them as if they were something mystical. In healthcare, artificial intelligence, machine learning, and deep learning are in the spotlight for their ability to predict disease risk and for their efficiency in automating several steps that make up the investigation phase of biomedical images or signals that support clinical decision-making.

However, the road to making these technologies a definitive support in clinical practice is still long and tortuous and includes a better understanding of the functioning of biological systems, much of which has yet to be clarified despite the countless discoveries in medicine, biology, biochemistry, and biophysics. Despite the incredible impact of these methods in everyday life, most users of AI-driven technologies have no idea how these technologies work, and this is still true even for many scientific disciplines.

Therefore, there is a great need to educate society about AI technologies at different levels of depth in order to make sure that the professionals of the future, at least those involved in scientific disciplines, can actively use these methods. In other words, machine learning techniques, or deep learning, should not be considered a solution to a specific problem, but a tool or set of tools to achieve a solution to a specific problem.

In this context, this book proposes an approach that helps readers interface with complex methodologies by providing original application examples that are gradually more complex and resemble real problems while maintaining a scholastic character. Dr. Michelucci puts in this book all his skills as an exceptional trainer, combining his ability to explain very complex concepts in a clear and understandable way, while maintaining a good degree of mathematical formalism, and his ability to stimulate critical thinking through the development of practical problems. This book also provides a quick guide to using working environments such as Jupyter Notebooks to create and share documents that contain equations, code, and text.

A journey that starts with the single neuron teaches readers how to build neural networks, training techniques, testing, and validation through appropriate metrics, tuning hyperparameters, and much more.

All these topics are covered by providing code examples that help the readers make concepts and methods their own so that they can customize them to specific problems.

Therefore, this book is a valuable aid for those who want to not only learn about deep learning, but also want to make it part of their methodological background.

I believe that Dr. Michelucci's work will be useful to engineers, physicists, and mathematicians who are interested in making their own concepts and methods related to deep neural networks. The collaboration I undertook with Umberto Michelucci for several years was fundamental for my research group's professional growth, and I am sure that this book will be of support to many other passionate scientists.

Sincerely,
Marco A. Deriu, PhD

Introduction

This is the second edition of *Applied Deep Learning* and it has been updated for TensorFlow 2.X and expanded to cover additional advanced material, such as autoencoders and generative adversarial networks (GANs). The goal of this book is to teach you the necessary fundamentals of how neural networks work, how to train them, and how to implement them with Keras. We start by discussing what a neuron is and what you can achieve with just one, then move to multiple layers in feed-forward neural networks. You learn what regularization is and how to use it, how advanced optimizers (such as Adam) work, and how to do hyperparameter tuning. At the end of the book, we look at some advanced topics, such as autoencoders, metric analysis, and GANs.

If you are new to this subject, I suggest you read the chapters in order, but if you already have some experience and you want to learn about a specific topic, you can jump directly to the relevant chapter. The chapters are mostly self-contained, although each refers to concepts explained in previous chapters, so if you don't know what a specific symbol or concept means, you can refer to the previous chapters. I worked hard to keep the mathematical notation and programming style as consistent as possible to make following the book easier. I only discuss very short code snippets (the ones I consider relevant), so you will not find complete code to copy and use, but don't worry. This book has an online site where you can find lots of Jupyter Notebooks that will be updated regularly with new examples and topics. You can find them at `https://adl.toelt.ai`.

Anytime you want to see the complete code in action, go to that site and you will find complete examples that you can download or open in Google Colab to try. TensorFlow is updated often, so providing code examples in the book would make the book age very quickly! My suggestion is to study the concepts here in the book, and then go to the online site and try the complete code to see how what you learned works in practice.

At the end of each chapter there are some exercises that have the goal of making you think about what you learned and give you interesting insights.

Who This Book Is For

To benefit from this book, you should have intermediate Python programming experience. It's helpful if you understand how the NumPy library works, since it is used extensively with TensorFlow. You should also have a basic understanding of algebra and calculus. You should understand at least the following concepts:

- What is a matrix.
- How to do basic operations on matrixes, such as multiplying them, inverting them, and so on.
- What is a derivative (and what is a partial derivative).
- How to calculate easy derivatives.
- What a function is and what it means to minimize one.

If you understand those concepts, you should be able to follow the explanations in the book. I always give many practical hints in the book to make clear what the implications of the theoretical concepts are in practice. I hope this will help you with your real-life projects.

Do You Need to Know TensorFlow/Keras?

This is a tricky question. The more you know, the more you will be able to benefit from the book. The main goal of this book is not to teach you Keras, but to teach you how neural networks work and give you implementation examples in Keras. Let me stress it again: The focus is on understanding how neural networks work, not on how Keras works. This is not a book on Keras. The best way to learn all the particularities of Keras is to look at the official documentation (`https://www.tensorflow.org/learn`). It is always up-to-date and contains many examples. This book covers the necessary skills you need to understand basic examples, but if you want to understand all the subtleties, you should study the official documentation.

Note You will probably be able to understand most of the concepts even without knowing how Keras works, but the more experience you have with Keras, the easier it will be for you to follow the explanations.

Which Version of TensorFlow Is Used in this Book?

The code developed in this book has been tested on TensorFlow 2.5. I try to use only the fundamental Keras features to make it as compatible as possible with older and future versions. If you are using a different TensorFlow version, you may find that some of the code will not work. If you are running the code from `https://adl.toelt.ai` locally and you encounter this problem, I suggest you create a virtual environment[1] with TensorFlow 2.5. Versions of other packages, such as NumPy or Pandas, should not matter much. Any relatively modern (let's say from 2020 or 2021) version should work just fine.

How to Try the Code in the Book

There are several ways to try the code discussed in this book. I worked very hard to make sure that you can run all the examples in the book in Google Colab (`https://colab.research.google.com/`), so that you don't have to install anything on your personal laptop or PC. If you go to `https://adl.toelt.ai`, you can open all the examples directly in Google Colab. If you are on a page at `https://adl.toelt.ai`, simply hover the mouse over the small rocket icon on the top-right side of the examples (see Figure I-1). You have several options to open the notebook in an environment to try it.

[1] You can find instructions on how to do this on the official Python web page at `https://docs.python.org/3/tutorial/venv.html`. If you are using Anaconda, you can find a large amount of information on their official documentation at `https://bit.ly/3kjQDeL`.

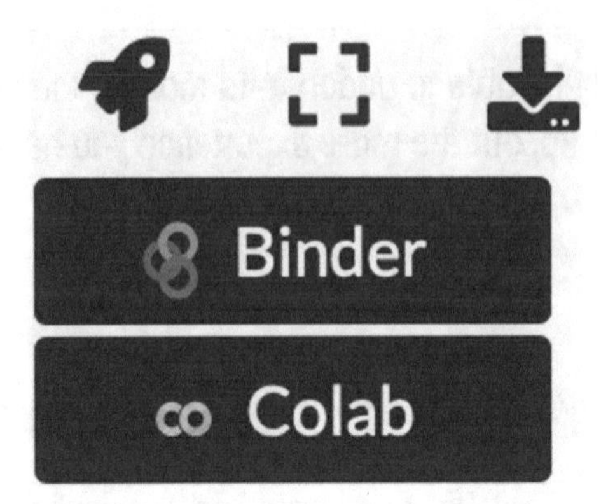

Figure I-1. *By hovering with the mouse over the small rocket icon on the top-right side of the pages at `https://adl.toelt.ai`, you can open the notebook directly in Google Colab without having to install anything on your machine*

You can simply choose Google Colab from the drop down list. The notebook will open in a browser in Google Colab so that you can test the code directly. Additionally, by clicking the icon with the arrow pointing down shown on the right of Figure I-1, you can also download the code on your laptop and run it locally[2].

Note You can run all the examples discussed in this book in Google Colab. You can find a direct link to open the notebooks online by going to `https://adl.toelt.ai` and hovering the mouse over the small rocket icon on the top-right side of the page.

If you don't know how Google Colab works, I suggest you watch the very short introductory video at `https://www.youtube.com/watch?v=inN8seMm7UI`. Basically, Google Colab is an online Jupyter Notebook with the Python engine running on Google servers. If you have worked with Jupyter Notebooks before you should be fine. If not, I suggest you go to the official project page at `https://jupyter.org` and study the many available tutorials. The Jupyter Notebook environment is widely used to do data science and is something that every practitioner should know about.

[2] This will of course require a local Python installation with the necessary packages installed.

Contents of the Book

The first chapter discusses the problem of optimization in general and how it relates to neural networks. We look at how the most important minimization algorithm, gradient descent, is working and how the mini-batch and stochastic variations of it function.

Chapter 2 looks at how one neuron is structured and at the most commonly used activation functions. Then it covers how to implement a neural network with one single neuron in Keras and how to do linear regression and logistic regression (classification) with it. We discuss the three fundamental components of any neural network model: the network architecture, the loss function, and the optimizer.

Chapter 3 moves to neural networks with multiple layers and many neurons. We discuss the concept of overfitting and how to do basic error analysis. Then we look at how to implement neural networks with multiple layers with Keras. Additionally, we look at weight initialization and discuss the various ways of doing it. Finally, we discuss how to estimate how much memory a neural network model implemented with Keras will need.

In Chapter 4, the concept of regularization is discussed. We look at the l_p norm and at l_2 and l_1 regularization. Then we discuss how dropout works and how to implement it in Keras. Finally, we look at how early stopping works.

In Chapter 5, the Adam, momentum, and RMSProp optimizers are discussed, starting with what exponentially weighted averages are, a concept necessary for understanding advanced optimizers.

In Chapter 6, we discuss hyper-parameter tuning. We discuss grid and random search and coarse-to-fine optimization. Then Bayesian optimization is explained and discussed at length. Finally, we discuss sampling on a logarithmic scale. In Chapter 7, we discuss convolutional neural networks and how to implement them in Keras.

Chapter 8 is a very short chapter with a very basic introduction to recurrent neural networks. In Chapter 9, we discuss autoencoders and their applications. Chapter 10 contains a discussion of metric and error analysis.

Chapter 11 is a brief introduction to generative adversarial networks. In Appendixes A and B, an introduction to Keras and a discussion about how to customize it are briefly discussed.

Final Words

I hope that this book gives you a clear curriculum to follow in order to study neural networks in the most structured and easy way. The topics are not easy and require effort and time. Thus, you should not be discouraged. Unfortunately, real machine learning projects involve much more than simply copy and pasting from blogs on the Internet. Programming is only a part of it, and without knowing how the algorithms work, writing the code will be useless, and in the worst case will give you the wrong results.

I hope that you will find this book useful and that you will profit from it for your career and research projects.

Dübendorf, 1st January 2022

CHAPTER 1

Optimization and Neural Networks

This chapter contains a basic introduction to the most important concepts of optimization and explains how they are related to neural networks. This chapter doesn't go into detail, leaving longer discussions to the following chapters. But at the end of this chapter, you should have a basic understanding of the most important concepts and challenges related to neural networks. This chapter covers the problem of learning, constrained and unconstrained optimization problems, what optimization algorithms are, and the gradient descent algorithm and its variations (mini-batch and stochastic gradient descent).

A Basic Understanding of Neural Networks

It is very useful to have a basic understanding of what is a neural network (NN) and how it learns. For this introductory section, we consider only what is called supervised learning.[1] Suppose you have a dataset of M tuples (x_i, y_i) with $i = 1, ..., M$. The x_i, called input observations or simply inputs, can be anything from images to multidimensional arrays, one-dimensional arrays or even simple numbers. The y_i are outputs (also called target variables or sometimes labels) and can be multidimensional arrays, numbers (for example, the probability of the input observation x_i being of a specific class), or even images. In the most basic formulation, an NN is a mathematical function (sometimes called a *network function*) that takes some kind of input (typically multi-dimensional) called x_i and generate some output. The subscript i indicates one of the inputs from a

[1] Supervised learning (SL) is the machine learning task of learning a function that maps an input to an output based on example input-output pairs (Source: Wikipedia).

U. Michelucci, *Applied Deep Learning with TensorFlow 2*, https://doi.org/10.1007/978-1-4842-8020-1_1

dataset of observations that is at your disposal. The output generated by the network function is called $\hat{y}_i$. The network function normally depends on a certain number N of parameters, which we will indicate with θ_i. We can write this mathematically as

$$\hat{y}_i \equiv f(\theta, x_i)$$

Where we have indicated the parameters in vector form: $\boldsymbol{\theta} = (\theta_1, \ldots, \theta_N) \in R^N$. Figure 1-1 shows a diagram of this idea. The blob in the middle represents the network function f that maps the input x_i to the output. Naturally, the output will depend on the parameters. The idea behind learning is to change the parameters until $\hat{y}_i$ is as close to y_i as possible. There are two very important undefined concepts in the last sentence: first, what "close" means, and second, how you update the parameters in an intelligent way to make $\hat{y}_i$ and y_i "close." We answer those exact two questions in depth in this book.

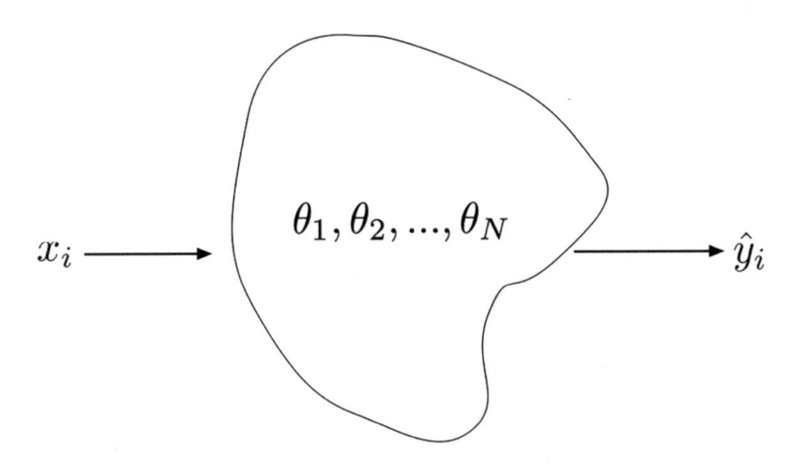

Figure 1-1. *A intuitive diagram of a neural network. x_i are the inputs (for i = 1, …, M), θ_i are the parameters (for i = 1, …, N), and $\hat{y}$ is the output of the network. The network function is depicted by the irregular shape in the middle*

To summarize, a neural network is nothing more than a mathematical function that depends on a set of parameters that are tuned, hopefully in some smart way, to make the network output as close as possible to some expected output. The concept of "close" is not defined here, but for the purposes of this section, a basic understanding is good enough. By the end of this book, you will have a more complete understanding of its meaning.

Note A neural network is nothing more than a mathematical function that depends on a set of parameters that are tuned, hopefully in some smart way, to make the network output as close as possible to some expected output.

The Problem of Learning

A First Definition of Learning

Let's now look at a more mathematical formulation of "learning" in the context of neural networks. For notation's simplicity, let's assume that each input is a mono-dimensional array, $x_i \in R^n$ with $i = 1, \ldots, M$. By the same token, we will assume that the output is a mono-dimensional array, $\hat{y}_i \in \mathbb{R}^k$, with k being some integer. We will assume we have a set of M input observations with the expected target variables $y_i \in R^k$. We also assume that we have a mathematical function $L(\hat{y},y) = L(f(\theta,\hat{y}),y)$, called a *loss function,* where we have used the vector notation $y = (y_1, \ldots, y_M)$, $\hat{y} = (\hat{y}_1,\ldots,\hat{y}_M)$ and $\theta = (\theta_1, \ldots, \theta_N)$. This function measures how "close" the expected (y) and predicted ($\hat{y}$) values are, given specific values of the parameters θ_i. We will not yet define how this function looks, as this is not relevant for this discussion. Let's summarize the notation we have defined so far:

- $x_i \in R^n$: Input observations (for this discussion, we assume that they are a mono-dimensional array of dimension $n \in N$). Examples could be age, weight, and height of a person, gray level values of pixels in an image, and so on.
- $y_i \in R^k$: Target variables (what we want the neural network to predict). Examples could be the class of an image, what movie to suggest to a specific viewer, the translated version of a sentence in a different language, and so on.
- $f(\theta, x_i)$: Network function. This function is built with neural networks and depends on the specific architecture used (feed-forward, convolutional, recurrent, etc.).

- $\theta = (\theta_1, \ldots, \theta_N)$: A set of real numbers, also called parameters or weights.
- $L(\hat{y},y) = L(f(\theta,\hat{y}),y)$: The loss or cost function. This function is a measure of how "close" y and $\hat{y}$ are. Or, in other words, how good the neural network's predictions are.

Those are the fundamentals elements that we need in order to understand neural networks.

[Advanced Section] Assumption in the Formulation

If you already have some experience with neural networks, it is important to discuss one important assumption that we have silently made. Note that skipping this short section in a first reading of this book will not impact the understanding of the rest. If you don't understand the points discussed here, feel free to skip this part and come back to it later.

The most important assumption here can be found in how the loss function $L(\hat{y},y)$ is written. In fact, as written, it is a function of all the M components of the two vectors $\hat{y}$ and y and this translates in *not* using a mini-batch during the training. The assumption here is that we will measure how good the network's predictions are by considering *all* the inputs and outputs simultaneously. This assumption will be lifted in the following sections and chapters and discussed at length. The experienced reader may notice that this will lead to advanced optimization techniques—stochastic gradient descent and the concept of mini-batch. Using all the M components of the two vectors $\hat{y}$ and y makes the learning process generally slower, although in some situations it's more stable.

A Definition of Learning for Neural Networks

With the notation defined previously, we can now formally define learning in the context of neural networks.

Definition Given a set of tuples (x_i, y_i) with $i = 1, \ldots M$, a mathematical function $f(\theta,\hat{y})$ (the network function), and a function (the loss function) $L(\hat{y},y) = L(f(\theta,\hat{y}),y)$, the process of *learning* is equivalent to minimizing the loss function with respect to the parameters $\boldsymbol{\theta}$. Or in mathematical notation

$$\min_{\theta \in \mathbb{R}^N} L(f(\theta,\hat{y}),y)$$

Note *Learning* is equivalent to minimizing the loss function with respect to the parameters θ, given a set of tuples (x_i, y_i) with $i = 1, \ldots M$.

The typical term for learning is *training* and that is the one we will use in this book. Basically, training a neural network is nothing more than minimizing a very complicated function that depends on a very large number of parameters (sometimes billions). This presents very difficult technical and mathematical challenges that we discuss at length in this book. But for now, this understanding is sufficient to start understanding how to tackle this problem.

The following sections discuss how to solve the problem of minimizing a function in general and explain the fundamentals theoretical concepts that are necessary to understand more advanced topics. Note that the problem of minimizing a function is called an *optimization problem.*

Constrained vs. Unconstrained Optimization

The problem of minimizing a function as described in the previous section is called an *unconstrained optimization problem.*

The problem of minimizing a function can be generalized by adding constraints to the problem. This can be formulated in the following way: we want to minimize a generic function $g(x)$ subject to a set of constraints

$$\begin{cases} c_i(x)=0, & i=1,\ldots,C_1 \text{ with } C_1 \in \mathbb{N} \\ q_i(x)\geq 0 & i=1,\ldots,C_2 \text{ with } C_2 \in \mathbb{N} \end{cases}$$

Where $c_i(x)$ and $q_i(x)$ are constraint functions that define some equations and inequalities that need to be satisfied. In the context of neural networks, you may have the constraint that the output (suppose for a moment that $\hat{y}_i$ is simply a number) must lie in the interval $[0, 1]$. Or maybe that it must be always greater than zero or smaller than a certain value. Or another typical constraint that you will encounter is when you want the network to output only a finite number of outputs, for example, in a classification problem.

Let's consider an example. Suppose that we want our network output to be $\hat{y}_i \in [0,1]$. Our learning problem could be formulated as follows

$$\min_{\theta \in \mathbb{R}^N} L(f(\theta,x),y) \text{ subject to } f(\theta,x_i) \in [0,1] \; i=1,\ldots,M$$

Or even more generally

$$\min_{\theta \in \mathbb{R}^N} L(f(\theta,x),y) \text{ subject to } f(\theta,x) \in [0,1] \; \forall \theta, x$$

This is clearly a *constrained optimization* problem. When dealing with neural networks, this problem is typically reformulated by designing the neural network in such a way that the constraint is automatically satisfied, and learning is brought back to an unconstrained optimization problem.

[Advanced Section] Reducing a Constrained Problem to an Unconstrained Optimization Problem

You may be confused by the previous section and wonder how constraints can be integrated into the network architecture design. This happens typically in the output layer of the network. For example, in the examples discussed in the previous section, to ensure that $f(\theta,\hat{y}_i) \in [0,1] \; i=1,\ldots,M$, it is enough to use the sigmoid function $\sigma(s)$ as an activation function for the output neuron. This will guarantee that the network output will always be between 0 and 1 since the sigmoid function maps any real number to the open interval (0, 1). If the output of the neural network should always be 0 or greater, you could use the ReLU activation function for the output neuron.

Note When dealing with neural networks, constraints are typically built into the network architecture, thus reframing the original constrained optimization problem into an unconstrained one.

Building constraints into the network architecture is extremely useful and it typically makes the learning process much more efficient. Constraints often come from a deep knowledge of the data and the problem you are trying to solve. It pays off to find as many constraints as possible and to build them into the network architecture.

Another example of a constrained optimization problem is when you have a classification problem with k classes. Typically, you want your network to output k real numbers p_i with $i = 1, \ldots, k$, where each p_i could be interpreted as the probability of the input observation being in a specific class. If we want to interpret p_i as the probability, the following equation must be satisfied

$$\sum_{i=1}^{k} p_i = 1$$

This is realized by having k neurons in the output layer and using for them the *softmax* activation function. This step reframes the problem into an unconstrained optimization problem, since the previous equation will be satisfied by the network architecture. If you don't know what the softmax activation function is, don't fret. We discuss it in the following chapters. Keep this example in mind, as it is the key to any classification problem with neural networks.

Absolute and Local Minima of a Function

Many algorithms that minimize a function are, by design, only able to find what is called a "local" minimum, or in other words, a point x_0 at which the function to minimize is smaller than at all other points in any *close* vicinity of x_0. Mathematically speaking x_0 is a local minimum of f if the following is satisfied (in a one-dimensional case)

$$\exists \eta \in \mathbb{R} \text{ such that } f(x) \leq f(x_0) \ \forall x \in [x_0 - \eta, x_0 + \eta]$$

In principle, we want to find the *global minimum* or, in other words, the point for which the function value is the smallest between all possible points. In the case of neural networks, identifying if the minimum is a local or a global minimum is impossible, due to the network function complexity. This is one (albeit not the only one) of the reasons that training large neural networks is such a challenging numerical problem. In the next chapters, we discuss at length what factors[2] may make finding the global minimum easier or more challenging.

[2] Factors include things like weight initialization, optimizer algorithm, optimizer parameters (as the learning rate) and so on.

Optimization Algorithms

So far, we have discussed the idea that learning is nothing less than minimizing a specific function, but we have not touched the issue of how this "minimizing" happens. This is achieved by what is called an "optimization algorithm," whose goal it is to find the location of the (hopefully) absolute minimum. Practically speaking, all unconstrained minimization algorithms require the choice of a starting point, which you denote by x_0. In the example of neural networks, this initial point would be the initial values of the weights. Typically starting from x_0, the optimization algorithms will generate a sequence of iterates $\{x_k\}_{k=0}^{\infty}$ that will converge toward the global minimum.

In all practical applications, only a finite number of terms will be generated, since we cannot generate an infinite number of x_k of course. The sequence will stop when progress can no longer be made (the value of x_k will not change any more[3]) or a specific solution has been reached with sufficient accuracy. Usually, the rule to generate a new x_k will use information about the function f to be minimized and one or more previous values (often properly weighted) of the x_k. In general, there are two main strategies for optimization algorithms: line search and trust regions. Optimizers for neural networks use all a line search approach.

Line Search and Trust Region

In the *line search* approach, the algorithm chooses a direction p_k and searches along this direction for a new value x_{k+1} when trying to minimize a generic function $L(x)$. In general, this approach, once a direction p_k has been chosen, consists in solving

$$\min_{\alpha>0} L(x_k + \alpha p_k)$$

for each iteration. In other words, you would need to choose the optimal α along the direction p_k. In general, this cannot be solved exactly, thus in a practical application (as you will see later), this approach is used by choosing a fixed α, or by reducing it in a way that is easy to calculate (independently of L). α is what is known as the *learning rate* when you deal with neural networks and is one of the most important hyper-parameters[4] when training networks. After you decide on a value for α, the new x_{k+1} is determined with the equation

[3] Don't be annoyed by the basic formulation. We will discuss it in more detail later.

[4] A *hyper-parameter* is a parameter that does not change during training and is not related to the training data. Even if the learning rate will change according to some fixed strategy, is still called a hyper-parameter since it does not change due to the training data.

$$x_{k+1} = x_k + \alpha p_k$$

In the *trust region* approach, the information available on L is used to build a model function m_k (typically quadratic in nature) that approximates f in a sufficiently small region around x_k. Then this approximation is used to choose a new x_{k+1}. This book does not cover trust region approaches, but the interested reader can find a very complete introduction in *Numerical Optimization, 2nd edition* by J. Nocedal and S.J. Wright, published by Springer.

Steepest Descent

The most obvious, and the most used search direction for line search methods is the steepest direction $p_k = -\nabla L(x_k)$. After all, this is the direction along which the function f decreases more rapidly. To prove it, we can use Taylor expansion[5] for $L(x_k + \alpha p)$ and try to determine along which direction the function decreases the most rapidly. We will stop at the first order and write

$$L(x_k + \alpha p) \approx L(x_k) + \alpha p \cdot \nabla L(x_k)$$

assuming that α is small enough. Our question (along which direction does the function L decrease more rapidly?) can be formulated as solving

$$\min_p L(x_k + \alpha p) \quad \text{subject to} \quad \|p\| = 1$$

Where $\|p\| = 1$ is the norm of the vector p (or in other words $\|p\|^2 = (p_1^2 + \ldots + p_n^2)$). Using the Taylor expansion and noting that $f(x_k)$ is a constant, we simply must solve for

$$\min_p p \cdot \nabla L(x_k)$$

Always subject to $\|p\| = 1$. Now, indicating with θ the angle between the direction p and $\nabla L(x_k)$, we can write

$$p \cdot \nabla L(x_k) = \|p\| \|\nabla L(x_k)\| \cos\theta$$

[5] If you don't know what the Taylor expansion is, you can visit Wikipedia to get an idea at `https://en.wikipedia.org/wiki/Taylor_series`. This is a fundamental tool that is used in calculus for various application.

It's easy to see that this is minimized when $\cos\theta = -1$, or in other words, by choosing the search direction parallel to the gradient of the loss function but pointing in the opposite direction. In other words

$$p = -\frac{\nabla L(x_k)}{\|\nabla L(x_k)\|}$$

as we claimed at the beginning.

Note The *steepest descent* method is a line-search method that searches for a better approximation of the minimum along the direction, minus the gradient of the function for every step. This method is at the basis of the *gradient descent optimizer*.

There are of course other directions that may be used, but for neural networks, those can be neglected. Just to cite an example, possibly the most important is the Newton direction, which can be derived from the second-order Taylor expansion of $f(x_k + p)$, but it requires you to know the Hessian $\nabla^2 L(x)$.

The Gradient Descent Algorithm

The gradient descent (GD) optimizer finds x_{k+1} by using the gradient of the function L according to the formula

$$x_{k+1} = x_k - \alpha \nabla L(x_k).$$

Thus, the GD algorithm is simply a line-search algorithm that searches for better approximations along the steepest descent direction. We can create a simple one-dimensional example ($x \in R$) and try the algorithm. Let's suppose we want to minimize the function

$$L(x) = x^2$$

This has a clear minimum at $x = 0$ as this is a simple quadratic form. If you know how to find the minimum of a function with calculus, it's easy to see that

$$\frac{dL(x)}{dx} = 0$$

implies that $2x = 0 \rightarrow x = 0$. This is indeed a minimum since

$$\frac{d^2L(x)}{dx^2} = 2 > 0$$

How does the GD algorithm work in this case? The algorithm will generate a sequence of x_k by using the formula $x_{k+1} = x_k - 2\alpha x_k$ (remember we are trying to minimize $f(x) = x^2$). We need of course to choose an initial value x_0 and a step α. For a first try, let's choose $x_0 = 1$ and $\alpha = 0.1$. The sequence can be seen in Table 1-1.

Table 1-1. *The Sequence x_k Generated for the Function $L(x) = x^2$ with the Parameters $x_0 = 1$ and $\alpha = 0.1$*

k	**x_k $x_0 = 1$**
0	1
1	0.8
2	0.64
3	0.512
...	...
40	0.00013
...	...
500	$3.5 \cdot 10^{-49}$

From Table 1-1 it should be evident how, albeit slowly, the GD algorithm converges toward the right answer, $x = 0$. That sounds good, right? What could go wrong? Not everything is so easy, and in the GD there is a marvelous hidden complexity. Let's rewrite the formula that is used to generate the sequence x_k:

$$x_{k+1} = x_k - 2\alpha x_k = x_k(1 - 2\alpha)$$

Consider for example the value $\alpha = 1$. In this case $x_{k+1} = -x_k$. It is easy to see that this generates an oscillating sequence that never converges. In fact, it is easy to calculate that $x_1 = -1$, $x_2 = 1$ and so on. An oscillating sequence will always be generated for all values of $1 - 2\alpha < 0$, or for $\alpha > \frac{1}{2}$. In Figure 1-2, you can see a plot of the sequence x_k for various values of the parameter α.

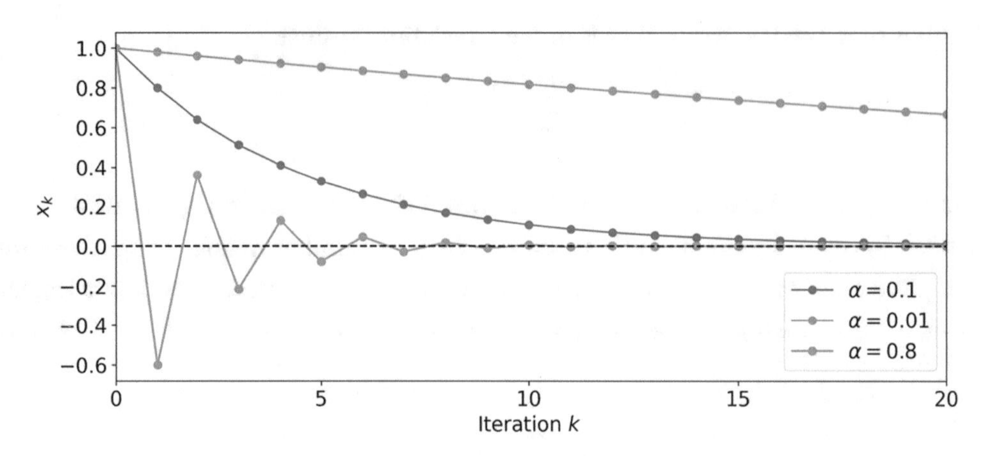

***Figure 1-2.** The sequence x_k generated for the function $L(x) = x^2$ for various values of α*

From Figure 1-2, when α is small, the convergence is very small (orange line), and as we discussed, for a value $\alpha > \frac{1}{2}$ (green line), an oscillating sequence is generated. It is interesting to note how this oscillating sequence converges quite faster than the others. The value $\alpha = 1$ generates a sequence that does not converge. But what happens for $\alpha > 1$? This is a very interesting case, as it turns out that the sequence diverges (albeit oscillating from positive to negative values). In Figure 1-3, you can see the plot of the sequence for $\alpha = 1.01$.

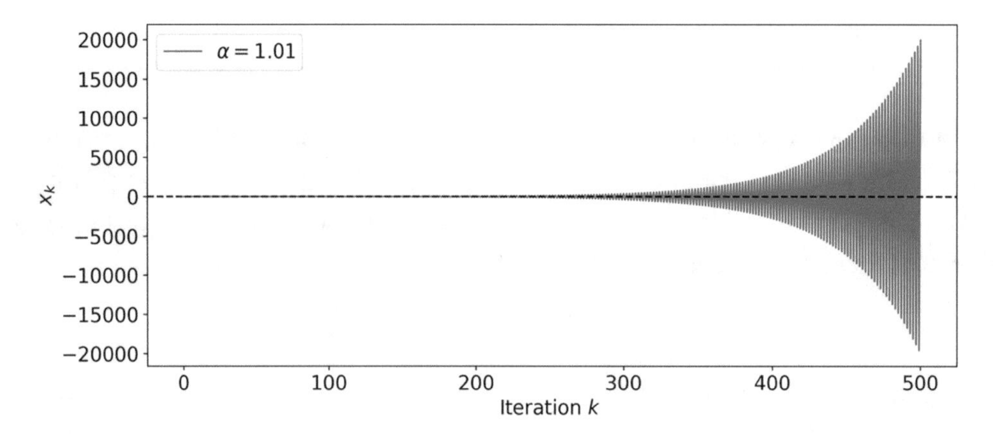

***Figure 1-3.** The sequence x_k for the parameter $\alpha = 1.01$. The sequence oscillates from positive to negative values while diverging in absolute value*

You can clearly see how it diverges. Note that from a numerical point of view, it is easy to get NaN (if you are using Python) or errors. If you are trying neural networks and you get NaN for your loss function (for example), one possible reason may be a learning rate that is too big.

Note The learning rate is possibly one of the most important hyper-parameters that you have to decide on when training neural networks. Choose one that's too small, and the training will be very slow, but choose one that's too big, and the training will not converge! More advanced optimizers (Adam for example) try to compensate this shortcoming by effectively varying the learning rate[6] dynamically, but the initial value is still important.

Now I must admit this is a very trivial case. In fact, the formula for x_k can also be written as

$$x_k = x_0 (1 - 2\alpha)^k$$

Therefore, it's easy to see that this sequence converges for $|1 - 2\alpha| < 1$ and diverges for $|1 - 2\alpha| > 1$. For $|1 - 2\alpha| = 1$, it stays at 1 and if $1 - 2\alpha = -1$, it oscillates between 1 and -1. Still, it is instructive to see how important the role of the learning rate is when using the GD.

Choosing the Right Learning Rate

You may be wondering how to choose the right α at this point. This is a good question but unfortunately there is no real precise answer, and some clarifications are in order. In all practical cases, you will not use the plain GD algorithm. Consider that, for example, in TensorFlow 2.X, the GD is not even available out of the box, due to its inefficiency. But in general, to check if the (in some cases only the initial) learning rate is optimal, you can follow these steps, assuming you are trying to minimize a function $L(x)$:

---

[6] To be precise, as we discuss next chapters, the Adam optimizer does not change the learning rate but uses a different algorithm that is very similar to, but not the same as, the GD. It is said that Adam updates the learning rate dynamically.

1. Choose an initial learning rate. Typical values[7] are 10^{-2} or 10^{-3}.
2. Let your optimizer run for a certain number of iterations, saving the $L(x_k)$ each time.
3. Plot the sequence $L(x_k)$. This sequence should show a convergent behavior. From the plot, you can get an idea whether the learning is too small (slow convergence) or too big (divergence). For example, Figure 1-4 shows the sequence $L(x_k)$ for the example we discussed in the previous section. The figure tells us that using $\alpha = 0.01$ (orange line) is very slow. Trying larger values for α shows how convergence can be faster (blue and green). With $\alpha = 0.1$ after 12-13 iterations, you already have a good approximation of the minimum, while for $\alpha = 0.01$ you are still very far from a solution.

Note When training neural networks, always check the behavior of your loss function. This will give you important information about how the training process is going.

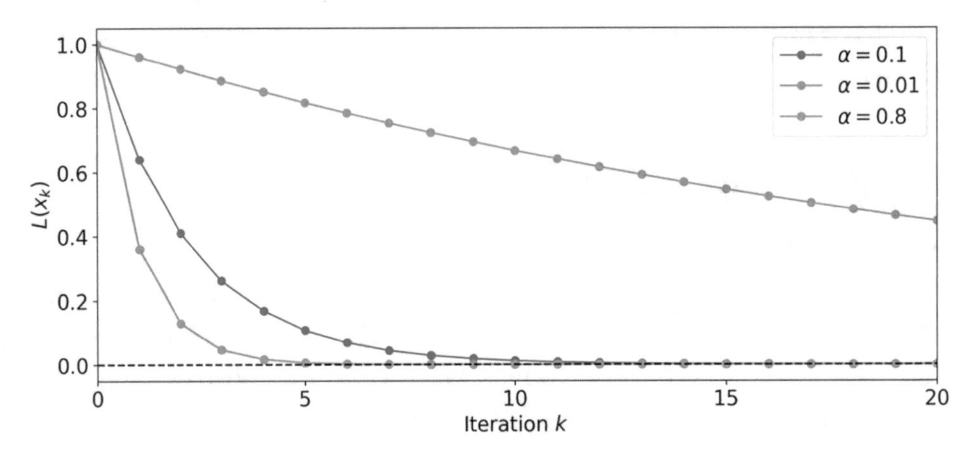

Figure 1-4. *The sequence $L(x_k)$ for the function $L(x) = x^2$ for various values of α*

[7] For example, the Adam optimizer in TensorFlow 2.X uses 0.001 as the standard learning rate, unless you specify otherwise.

This is why, when training neural networks, it is important to always check the behavior[8] of the loss function that you are trying to minimize. Never assume that your model is converging without checking the sequence $L(x_k)$.

Variations of GD

To understand variations of GD, the easiest way is to start with the loss function. As mentioned at the beginning of the chapter, in the "The Problem of Learning" section, the goal is to minimize the loss function $L\left(f\left(\theta,\hat{y}\right),y\right)$, where we have used the vector notation $y = (y_1, \ldots, y_M)$, $\hat{y}=\left(\hat{y}_1,\ldots,\hat{y}_M\right)$ and $\theta = (\theta_1, \ldots, \theta_N)$. In other words, we have at our disposal M input tuples that we can use. In the plain version of GD, the loss function is written as

$$L\left(f\left(\theta,\hat{y}\right),y\right)=\frac{1}{M}\sum_{i=1}^{M}l_i\left(f\left(\theta,\hat{y}_i\right),y_i\right)$$

Where l_i is the loss function evaluated over a single observation. For example, we could have a one-dimensional regression problem where our loss function is the mean square error (MSE). In this case we would have

$$l_i\left(f\left(\theta,\hat{y}_i\right),y_i\right)=\left|f\left(\theta,\hat{y}_i\right)-y_i\right|^2$$

And therefore

$$L\left(f\left(\theta,\hat{y}\right),y\right)=\frac{1}{M}\sum_{i=1}^{M}\left|f\left(\theta,\hat{y}_i\right)-y_i\right|^2$$

That is the classical formula for the MSE that you may have already seen. In plain GD, we would use this formula to evaluate the gradient that we need to minimize L. Using all M observations has pros and cons.

Advantages:

- Plain GD shows a stable convergence behavior

[8] Tools such as TensorBoard (from TensorFlow) have been built with exactly this problem in mind, to provide a real-time check about how the training is going.

Disadvantages:

- Usually, this algorithm is implemented in such a way that all the dataset must be in memory; therefore, it is computationally intensive.
- This algorithm is typically very slow with very big datasets.

Variations of the GD are based on the idea of considering only some of the observations in the sum in the previous equation instead of all M. The two most important variations are called *mini-batch GD* (MBGD, where you consider a small number of observations $m < M$) and *stochastic GD* (SGD, where you consider only one observation at a time). Let's look at both in detail, starting with the mini-batch GD.

Mini-Batch GD

To clarify the idea behind the method, we can write the loss function as

$$L_m\left(f(\theta,\hat{y}),y\right)=\frac{1}{M}\sum_{i=1}^{m}\left|f(\theta,\hat{y}_i)-y_i\right|^2$$

Where we have introduced $m \in \mathbb{N}$ with $m < M$, called *batch size.* L_m is defined by summing over m observations sampled from the initial dataset.

The mini-batch GD is implemented according to the following algorithm:

1. A mini-batch size m is chosen. Typical values are 32, 64, 128, or 256 (note that the mini-batch size m does not have to be a power of 2, and it can be any number, such as 137 or 17).
2. $N_b = \left\lfloor \frac{M}{m} \right\rfloor + 1$ subsets of observations are created[9] by sampling each time m observation from the initial dataset S without repetition. We indicate them with $S_1, S_2, \ldots, S_{N_b}$. Note that in general if M is not a multiple of m the last batch, S_{N_b} may have a number of observations smaller than m.
3. The parameters θ are updated N_b times using the GD algorithm with the gradient of L_m evaluated over the observations in S_i for $i = 1, \ldots, N_b$.
4. Repeat Step 3 until the desired result is achieved (for example, the loss function does not vary much anymore).

[9] The symbol $\lfloor x \rfloor$ indicates the integer part of x.

When training neural networks, you may have heard the term "epoch" instead of iteration. An epoch is *finished* after all the data has been used in the previous algorithm. Let's look at an example. Suppose we have $M = 1000$ and we choose $m = 100$. The parameters θ will be updated using 100 input observations each time. After ten iterations ($\frac{M}{m}$) the network will have used all M observations for its training. At this point, it is said that one epoch is finished. One epoch in this example consists of ten times the parameters being updated (or ten iterations). Here are the advantages and disadvantages.

Advantages:

- The parameters update frequency is higher than with plain gradient descent but lower than SGD, therefore allowing for a more robust convergence than SGD
- This method is computationally much more efficient than plain gradient descent or Stochastic GD since fewer calculations (as in SGD) and resources (as in Plain GD) are needed.
- This variation is by far the fastest of the three and the most commonly used.

Disadvantages:

- The use of this variation introduces a new hyper-parameter that needs to be tuned: the batch size (the number of observations in the mini-batch).

Note An epoch is *finished* after all the input data has been used to update the parameters of the neural network. Remember that in one epoch, the parameters of the network may be updated many times.

Stochastic GD

SGD is a very common version of the GD, and it simply is the mini-batch version with $m = 1$. This involves updating the parameters of the network by using one observation at a time for the loss function. This also has advantages and disadvantages.

Advantages:

- The frequent updates provide an easy way to check how the model learning is going (you don't need to wait until all the dataset has been considered).
- In a few problems this algorithm may be faster than plain gradient descent.
- The model is intrinsically noisy and that may help the model avoid local minima when trying to find the absolute minimum of the cost function.

Disadvantages:

- On large datasets, this method is quite slow, as it's very computationally intensive due to the continuous updates.
- The fact that the algorithm is noisy can make it hard for the algorithm to settle on a minimum for the cost function, and the convergence may not be as stable as expected.

How to Choose the Right Mini-Batch Size

So, what is the right mini-batch size m? Typical values used by practitioners are in the order of 100 or less. For example, the TensorFlow standard value (if you don't specify otherwise) is 32. Why is this value so special? To understand why, you need to study the behavior of MBGD for various choices of m. To make it resemble real cases, consider the MNIST dataset. You may have already seen it. It is a dataset that contains 70,000 hand-written digits from 0 to 9. The images are gray-level 28x28 pixel images. We will build a classifier with a neural network with 16 neurons using the ReLU activation function and use the Adam optimizer. Note that if you don't know what I am talking about, you can skip those details. The following discussion can be followed even without understanding the details of how the network is designed.

Secondly, using Adam is only for practical reasons, as in TensorFlow the MBGD is not available out of the box. But the conclusions continue to be valid. We have trained the network for ten epochs on 60,000 training images and then measured the running time[10]

[10] I have run these tests on Google Colab, and at the time of this writing, that meant an Intel Xeon CPU @ 2.20GHz and a GPU Tesla T4 with 15 GB memory.

needed, the reached value of the loss function, and the accuracy at the end of the training. We used the following values for the mini-batch size m: 60000 (effectively using all the data, so no mini-batches), 20000, 5000, 500, 50, 10, and 1. Note that while for $m = 10$ the required time is 2.34 min, when using $m = 1$, 19.18 minutes are needed for ten epochs!

Figure 1-4 shows the results of this study.

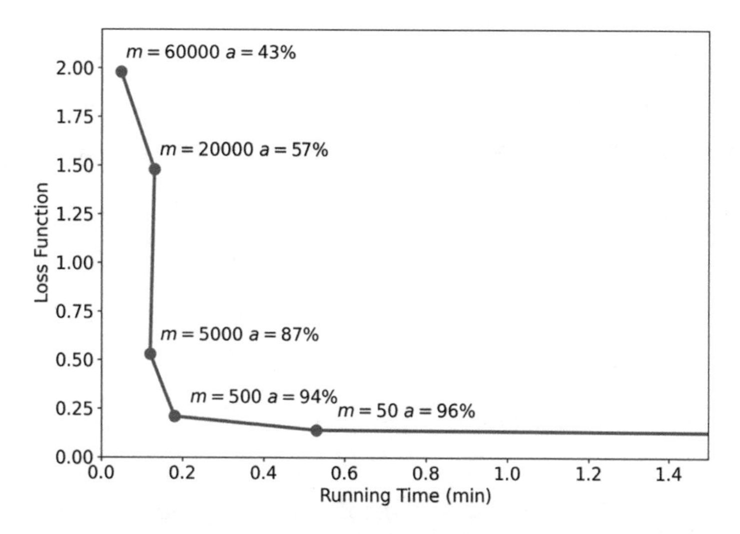

***Figure 1-5.** A plot of the loss function reached after ten epochs on the MNIST dataset plotted vs. the running time needed. a indicates the accuracy reached*

Let's see what Figure 1-5 is telling us. When we use $m = 60000$, the running time needed for ten epochs is the lowest, but the accuracy is quite low. Decreasing m increases the accuracy quite rapidly, until we reach the "elbow." Between $m = 500$ and $m = 50$, the behavior changes. Decreasing m does not increase the accuracy much, but the running time becomes larger and larger. So, when we reach the elbow, a decrease in m is no longer advantageous. As you will notice, around the elbow, m is in the order of 100. Figure 1-5 shows why typical values for m are in the order of 100. Of course, the optimal value is dependent on the data and some testing is required, but in most cases a value around 100 is a very good starting point.

Note A good starting point for the mini-batch size is in the order of 100. The optimal value is dependent on the data you use and on the neural network architecture you train, and testing is required to find the optimal value.

[Advanced Section] SGD and Fractals

In the previous sections, we discussed how choosing the wrong learning rate can make the convergence slow or even diverge. But the discussion was for a one-dimensional case and thus was very simple. This section shows you how much complexity is hidden when using SGD. You'll see how specific ranges of the learning rate make the convergence chaotic (in the mathematical sense of the word). This is one of the many hidden gems that you can find when dealing with optimization problems. Let's consider a problem[11] in which the M inputs $x^{[i]}$ are bi-dimensional, in other words $x^{[i]} = \left(x_1^{[i]}, x_2^{[i]}\right) \in \mathbb{R}^2$. We call the target variables $y^{[i]}$. The optimization problem we are trying to solve involves minimizing this function

$$L = \frac{1}{2}\sum_{i=1}^{M}\left(f\left(x_1^{[i]}, x_2^{[i]}\right) - y^{[i]}\right)^2$$

with

$$f\left(x_1^{[i]}, x_2^{[i]}\right) = w_1 x_1^{[i]} + w_2\, x_2^{[i]}$$

This is a simple linear combination of the inputs. The problem is simple enough right? We minimize the MSE (Mean Square Error) and try to find the best parameters w_1 and w_2 that minimize L. Consider $M = 3$ inputs as well. In particular, to make it more concrete, consider the following input matrix[12]

$$X = \begin{pmatrix} 0 & 1 \\ 1 & \frac{1}{2} \\ 1 & -\frac{1}{2} \end{pmatrix}$$

We write our labels in matrix form as well

$$Y = \begin{pmatrix} 0 \\ 4 \\ 0 \end{pmatrix}$$

[11] This problem is an adaptation of the one described in *Rojas, R. (2013). Neural Networks: A Systematic Introduction.* Springer Science & Business Media.

[12] Note that all the inputs can be written in matrix form for simplicity.

Note that what we show you here is not dependent on the numerical values. You can reproduce the results with different values without a problem. Let's first find the minimum of L exactly (since in this easy case, we can do that). To do that, we need simply to derive L and solve the two equations

$$\frac{\partial L}{\partial w_1} = 0; \quad \frac{\partial L}{\partial w_2} = 0$$

The calculations are boring but not overly complex. By solving these two equations, you will find that the minimum is at $x^* = \left(2, \frac{4}{3}\right)$.

Exercises

EXERCISE 1

Solve these two equations

$$\frac{\partial L}{\partial w_1} = 0; \quad \frac{\partial L}{\partial w_2} = 0$$

And prove that L has its global minimum at $x^* = \left(2, \frac{4}{3}\right)$.

To implement an SGD optimizer, the following algorithm can be followed:

1. Choose a learning rate α.
2. Choose a random value between {1,2,3} and assign it to i.
3. Update the parameters w_1, w_2 by using $l_i = \frac{1}{2}\left(f\left(x_1^{[i]}, x_2^{[i]}\right) - y^{[i]}\right)^2$; in other words, use l_i to calculate the derivatives to update the weights according to the gradient descent rule $w_j \rightarrow w_j - \alpha \partial l_i / \partial w_j$ for $j = 1, 2$. Each time, save the values w_1, w_2, for example in a Python list.
4. Repeat Steps 2 and 3 a certain number of times, N.

By following the previous algorithm, you can plot in the (w_1, w_2) space all the points you obtained and saved in Step 3. Those are the all the values that the two parameters w_1 and w_2 will assume during the optimization procedure. Figure 1-6 shows the result for $\alpha = 0.65$. The result is nothing short of amazing.

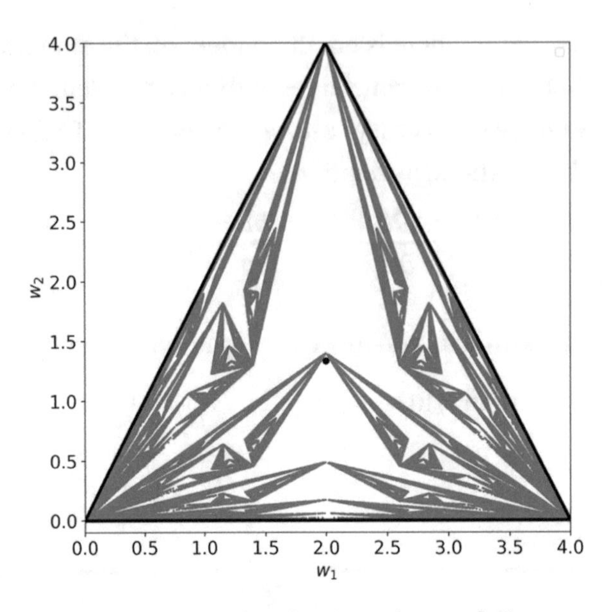

Figure 1-6. *Each blue point is a tuple of values (w_1, w_2) that are generated by using SGD as described in the section for a value of the learning rate $\alpha = 0.65$. The plot has been obtained using $4 \cdot 10^5$ iterations*

EXERCISE 2 (LEVEL: DIFFICULT)

Can you derive the equations of the lines delimiting the triangle from the input matrix X?

EXERCISE 3

Try to reproduce the image by implementing the SGD algorithm as described in this section from scratch. If you are stuck, you can find a complete implementation in the online version of the book.

It can be shown that what you see in Figure 1-6 is indeed a fractal. The mathematical proof is way beyond the scope of this book, but if you are interested, you can consult the book *Fractals Everywhere,* by M.F. Barnsley, published by Dover. One of the main property of fractals is that when you zoom in, you will find the same structure that you observe at a larger scale. To convince you that this is happening, Figure 1-7 shows you a detail of Figure 1-6. In the zoomed area you can observe the same kind of structure that you see at a larger scale.

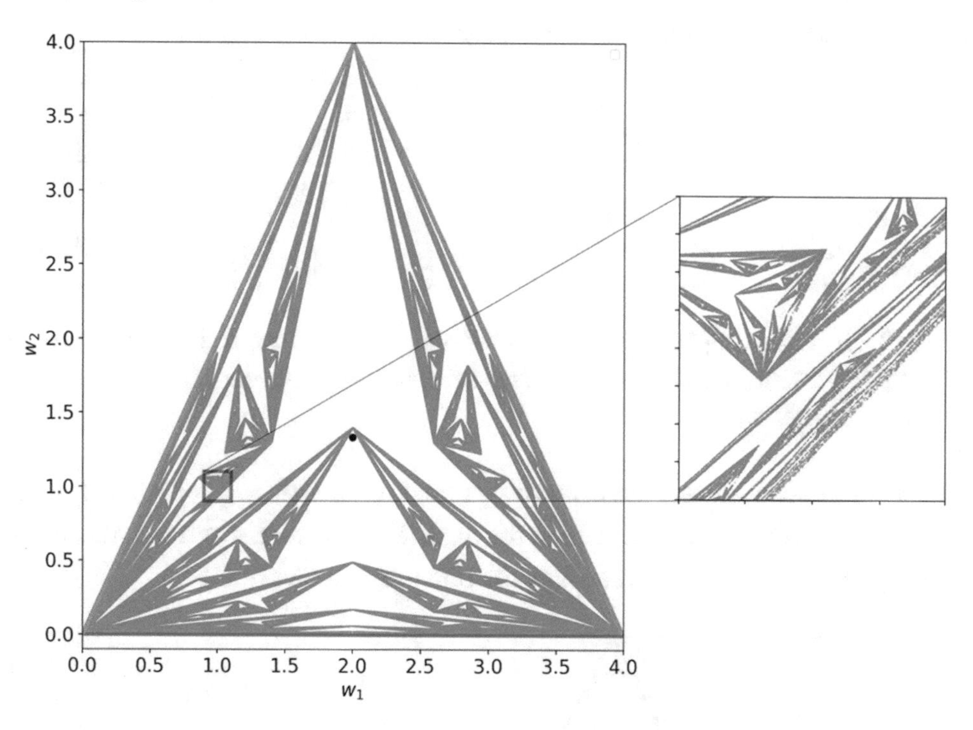

Figure 1-7. *A zoomed region that shows the fractal nature that the SGD algorithm can generate. This picture has been generated with a learning rate of $\alpha = 0.65$ and with 10^7 iterations. In the zoomed region, you can clearly see the same kind of structure that you observe at a larger scale on the left. The zoomed region is less sharp than the one on the left since only a fraction of the 10^7 points happen to be in the small regioned zoomed in*

The particular structure of the fractal depends on the learning rate. In Figure 1-8, you can see the fractal structure for different learning rates, from 0.65 to 1.0. It is fascinating to see how the structure changes, showing the great complexity that is hidden in using SGD.

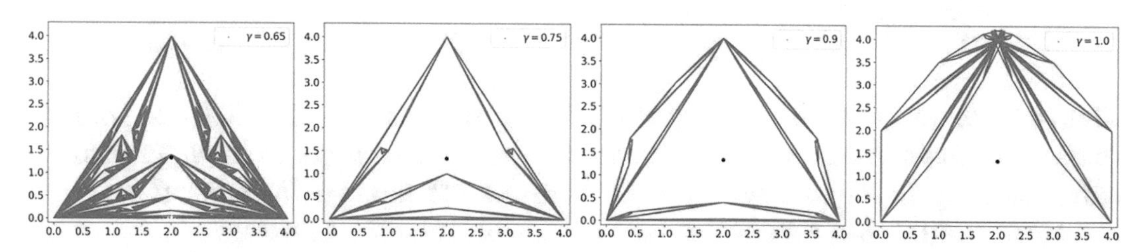

Figure 1-8. *Fractal shapes obtained by SGD for different learning rates. Note in the figure the learning rate is indicated with γ.*

When using smaller learning rates, at a certain point the fractal structure completely disappears quite suddenly, leaving an unstructured cloud of points, as you can see in Figure 1-9. The smaller the learning rate, the smaller the cloud of points.

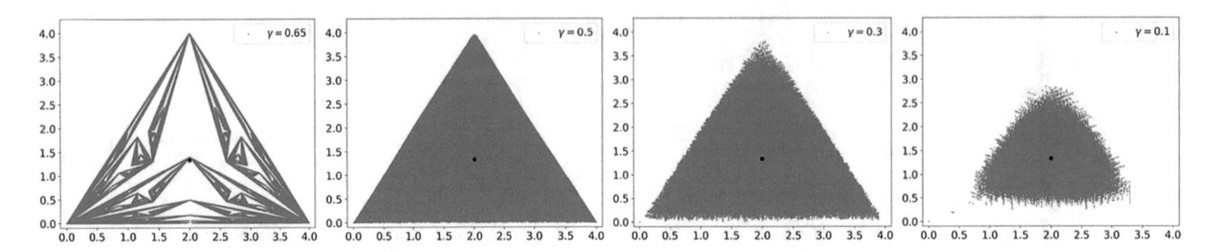

Figure 1-9. *By choosing smaller and smaller learning rates, the fractal structures completely disappear, leaving an unstructured cloud of points centered on the global minimum x^* of L.*

Finally, by choosing a very small learning rate, for example $\alpha = 5 \cdot 10^{-4}$, SGD delivers the behavior you would expect, meaning the algorithm converges and remains close to the expected minimum. You can see how the plot looks in Figure 1-10.

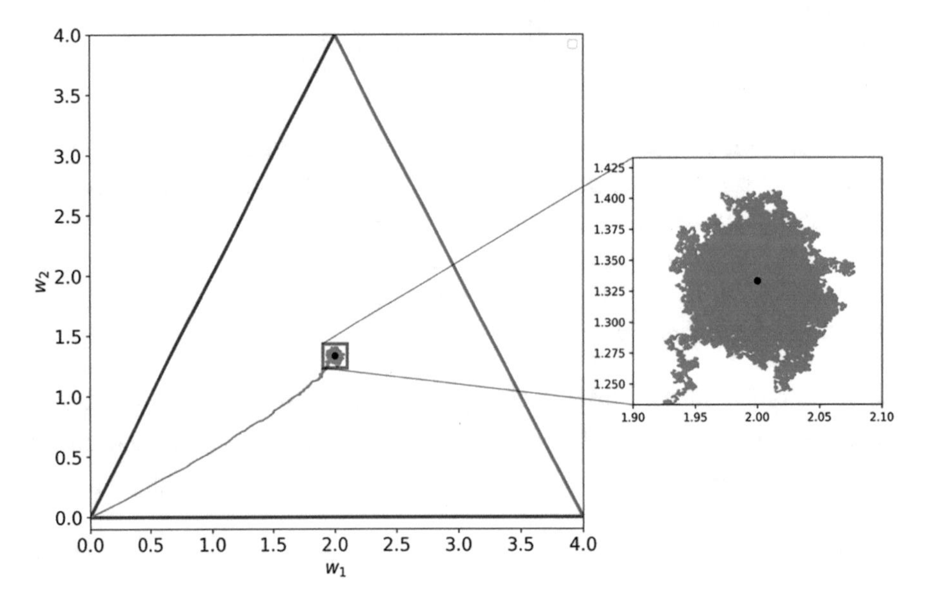

Figure 1-10. *By choosing a very small learning rate of $\alpha = 5 \cdot 10^{-4}$, SGD will move in the direction of the expected global minimum x^* and remain in its vicinity, as can be seen in the zoomed-in region. Still, SGD continues to deliver points that remain around x^*, but never converge to it. The smaller the learning rate, the smaller the cloud of points around x^**

EXERCISE 4 (LEVEL: DIFFICULT)

Prove that each of the updates done with SGD, as described in the previous section, moves the point in parameter space in the direction perpendicular to one of the three lines that describe the triangle in the previous figures. In other words, the direction between two subsequent updates of the parameters (w_1, w_2) and $\left(w_1 - \gamma\frac{\partial L}{\partial w_1}, w_2 - \frac{\partial L}{\partial w_2}\right)$ is perpendicular to one of the three lines that delimitate the triangle in the previous figures. Note that this is not an easy exercise, and you can find some tips in the online version of the book if you get stuck.

Note The gradient descent algorithm, especially in its stochastic version, has incredible hidden complexity, even for a trivial case, as the one described in the previous section. This is the reason that training neural networks can be so difficult and tricky, and why choosing the right learning rate and optimizer is so important.

Conclusion

You should now have all the ingredients to (at least at a basic level) understand what it means for neural networks to learn. We have not yet covered how to build a neural network, except for the fact that it is a very complicated function f of the inputs that depend on a large number of parameters. We go into a lot more detail about how neurons work, how non-linearity is introduced, and so much more.

Let's now summarize what we discussed in this chapter. To train neural networks, you need the following major ingredients:

- A neural network architecture, namely a way of getting from an input x to an answer $\hat{y}$ (remember the function f?) that can be tuned by changing a large number of parameters.
- A set of input observations x_k (possibly a large number of them) with the expected values we want to predict (we are dealing with supervised learning).

- An optimizer, or in other words, an algorithm that can find the best parameters of the network to get the outputs as close as possible to what you expect.

This chapter discussed these points in a basic way. It is important that you get the main idea behind what training neural networks means. The next chapters discuss each of the three points in more detail and include a lot of examples to make the discussion as clear as possible. We build on what's discussed here to bring you to a point where you can use these more advanced techniques for your projects.

CHAPTER 2

Hands-on with a Single Neuron

After reading about learning with neural networks in the previous chapter, you are now ready to explore their most fundamental component, the neuron. In this chapter, you learn the main components of the neuron. You also learn how to solve two classical statistical problems (i.e., linear regression and logistic regression) by using a neural network with just one neuron. To make things a bit more fun, you do that using real datasets. We discuss the two models and explain how to implement the two algorithms in Keras.

First, we briefly explain what a neuron is, including its typical structure and its main characteristics (e.g., the activation function, weights, etc.). Then we look at how you can formally express it in a matrix (this step is fundamental to obtaining optimized codes and exploiting all TensorFlow and NumPy functionalities). Finally, we look at some code examples in Keras. You can find the complete Jupyter Notebooks discussed in this chapter at `https://adl.toelt.ai`.

A Short Overview of a Neuron's Structure

Deep learning is based on large and complex networks made up of a large number of simple computational units. Companies at the forefront of research are dealing with networks with 160 billion parameters [1]. To put things in perspective, this number is half of the number of stars in our galaxy or 1.5 times the number of people that ever lived. On a basic level, neural networks are large sets of differently interconnected units,[1]

[1] More advanced architecture, such as convolutional or recurrent neural networks, have a structure that is more complex than what is described here. In this chapter, we describe the neurons used to build so-called Feed-Forward Neural Networks (FFNN).

U. Michelucci, *Applied Deep Learning with TensorFlow 2*, https://doi.org/10.1007/978-1-4842-8020-1_2

each performing a specific (and usually relatively easy) computation. They remind me of Lego building blocks, where you can build very complicated things using elementary and fundamental units.

Due to the biological parallel with the brain [2], these basic units are known as *neurons*. Each neuron (at least the ones most commonly used and the ones we use in this book) does a straightforward operation: it takes a certain number of inputs (real numbers) and calculates an output (also a real number). Remember that the inputs are indicated in this book with $x_i \in \mathbb{R}$ (real numbers) where $i = 1, 2, \ldots, n_x$, $i \in \mathbb{N}$ is an integer, and n_x is the number of input attributes (often called features). As an example of input features, you can imagine the age and weight of a person (so we would have $n_x = 2$). x_1 could be the age and x_2 could be the weight. In real life, the number of features can be easily very large (of the order of $10^2 - 10^3$ or higher).

There are several kinds of neurons that have been extensively studied. In this book we concentrate on the most commonly used one. The neuron we are interested in simply applies a function to a linear combination of all the inputs. In a more mathematical form, given n_x real parameters $w_i \in \mathbb{R}$ (with $i = 1, 2, \ldots, n_x$) and a constant $b \in \mathbb{R}$ (usually called bias), the neuron will first calculate what is traditionally indicated in literature with z:

$$z = w_1 x_1 + w_2 x_2 + \ldots + w_{n_x} x_{n_x} + b$$

It will then apply a function f (normally non-linear) to z, giving the output $\hat{y}$

$$\hat{y} = f(z) = f\left(w_1 x_1 + w_2 x_2 + \ldots + w_{n_x} x_{n_x} + b\right)$$

Note Practitioners generally use the following terminology: w_i are called *weights*, b is the *bias*, x_i are the *input features*, and f is the *activation function*.

Let's summarize the neuron computational steps again.

1. Linearly combine all inputs x_i, calculating $z = w_1 x_1 + w_2 x_2 + \ldots + w_{n_x} x_{n_x} + b$.
2. Apply f to z giving the output $\hat{y} = f(z) = f\left(w_1 x_1 + w_2 x_2 + \ldots + w_{n_x} x_{n_x} + b\right)$.

In the literature, you can find numerous representations for neurons. Figure 2-1 shows a graphical representation of the mathematical operations we just discussed to obtain the output $\hat{y}$ from the inputs.

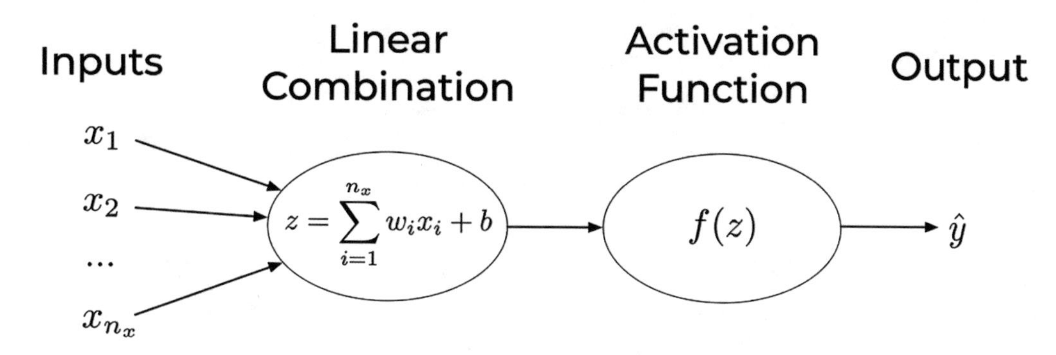

***Figure 2-1.** A representation of a single neuron with the operation highlighted. This is also called the computational graph of a single neuron, or in other words, a graphical representation of the operations needed to calculate $\hat{y}$ from the inputs*

Figure 2-1 must be interpreted in this way:

- The inputs are not placed in a bubble, simply to distinguish them from nodes that perform an actual calculation.
- The weight's names are typically not written. The expected behavior is that before passing the inputs to the central bubble (or node), the inputs will be multiplied by the relative weight. The first input x_1 will be multiplied by w_1, x_2 by w_2, and so on.
- The first bubble (or node) will sum the inputs multiplied by the weights (the $x_i w_i$ for $i = 1, 2, ..., n_x$) and then sum the result to the bias b.
- The last bubble (or node) will finally apply to the resulting value the activation function f.

All the neurons we deal with in this book have exactly this structure. Very often an even simpler representation is used, as in Figure 2-2. In such a case, unless otherwise stated, it's understood that the output is as follows

$$\hat{y} = f(z) = f\left(w_1 x_1 + w_2 x_2 + \ldots + w_{n_x} x_{n_x} + b\right)$$

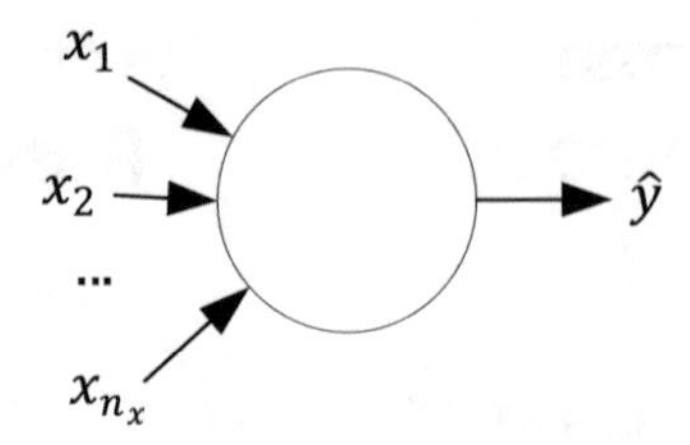

Figure 2-2. *A simplified version of Figure 2-1. Unless otherwise stated, it is usually understood that the output is* $\hat{y}=f(z)=f\left(w_1x_1+w_2x_2+\ldots+w_{n_x}x_{n_x}+b\right)$. *The weights are often not explicitly reported in the neuron representation*

A Short Introduction to Matrix Notation

When dealing with big datasets, the number of features is typically large (n_x will be large) and so it is better to use a vector notation for the features and weights. The inputs can be indicated with

$$\boldsymbol{x}=\begin{pmatrix} x_1 \\ \vdots \\ x_{n_x} \end{pmatrix}$$

where we have indicated the vector with a bold-faced $\boldsymbol{x}$. For the weights, we use the same notation

$$\boldsymbol{w}=\begin{pmatrix} w_1 \\ \vdots \\ w_{n_x} \end{pmatrix}$$

For consistency with formulas that we will use later, to multiply $\boldsymbol{x}$ and $\boldsymbol{w}$, we use matrix multiplication notation and therefore we write

$$\boldsymbol{w}^T\boldsymbol{x}=\left(w_1\ldots w_{n_x}\right)\begin{pmatrix} x_1 \\ \vdots \\ x_{n_x} \end{pmatrix}=w_1x_1+w_2x_2+\ldots+w_{n_x}x_{n_x}$$

Where $\boldsymbol{w}^T$ indicates the transpose of $\boldsymbol{w}$. z can then be written with this vector notation as

$$z=\boldsymbol{w}^T\boldsymbol{x}+b$$

and the neuron output $\hat{y}$ as

$$\hat{y} = f(z) = f\left(\boldsymbol{w}^T \boldsymbol{x} + b\right)$$

Let's now summarize the different components that define this neuron and the notation we use in this book:

- $\hat{y}$: Neuron (and later network) output
- $f(z)$: Activation function (sometimes called a transfer function) applied to z
- $\boldsymbol{w}$: Weights (vector with n_x components)
- b: Bias

An Overview of the Most Common Activation Functions

There are many activation functions at your disposal to change the output of a neuron. Remember, an activation function is simply a mathematical function that transforms z in the output $\hat{y}$. Let's look at the most common ones.

Identity Function

This is the most basic function that you can use. It's typically indicated with $I(z)$. It merely returns the input value, unchanged (see Figure 2-3). Mathematically we can write it as

$$f(z) = I(z) = z$$

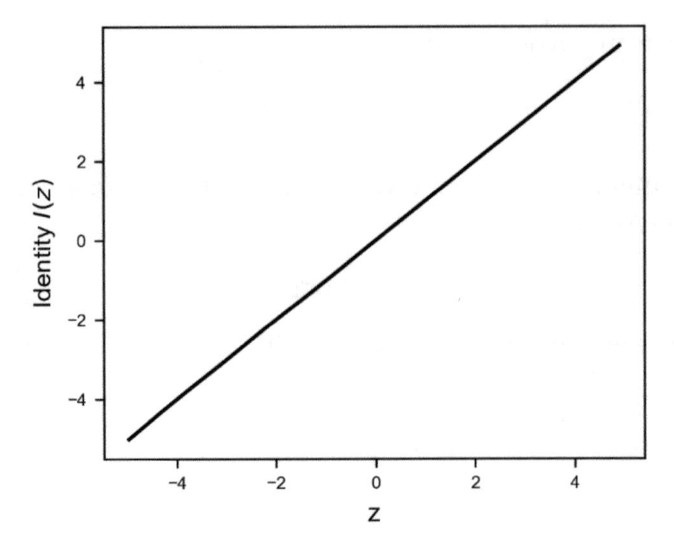

Figure 2-3. *The identity function*

This simple function will come in handy when discussing linear regression with one neuron later in the chapter. Implementing it[2] in Python with NumPy is incredibly trivial:

```
def identity(z):
    return z
```

Sigmoid Function

This is a very commonly used function that gives only values between 0 and 1. It is usually indicated with $\sigma(z)$

$$f(z)=\sigma(z)=\frac{1}{1+e^{-z}}$$

It is primarily used for classification models, where we want to predict the probability as an output (remember that a probability may only assume values between 0 and 1).

The calculation can be written in this form using NumPy functions

```
s = np.divide(1.0, np.add(1.0, np.exp(-z)))
```

[2] Note that you will not have to implement it yourself in Python. Keras offers this function, and many more, out of the box.

Note It is very useful to know that if we have two NumPy arrays, `A` and `B`, the following are equivalent: `A/B` is equivalent to `np.divide(A,B)`, `A+B` is equivalent to `np.add(A,B)`, `A-B` is equivalent to `np.subtract(A,B)`, and `A*B` is equivalent to `np.multiply(A,B)`. If you know object-oriented programming, we say that in NumPy basic operations like `/`, `*`, `+` and `-` are *overloaded*. Note also that these four basic operations in NumPy act element-by-element (also called element-wise).

We can write the sigmoid function in a more readable (at least for humans) form as

```
def sigmoid(z):
    s = 1.0 / (1.0 + np.exp(-z))
    return s
```

As stated, `1.0 + np.exp(-z)` is equivalent to `np.add(1.0, np.exp(-z))` and `1.0 / (np.add(1.0, np.exp(-z)))` to `np.divide(1.0, np.add(1.0, np.exp(-z)))`. We want to draw your attention to another point in the formula. `np.exp(-z)` will have the dimensions of `z` (usually a vector that will have a length equal to the number of observations), while `1.0` is a scalar (a one-dimensional entity). How can Python sum the two? What happens is called *broadcasting*. Python, subject to certain constraints, "broadcasts" the smaller array (in this case, the `1.0`) across the larger one so that, at the end, the two have the same dimensions. In this case, the `1.0` becomes an array of the same dimensions of `z`, all filled with `1.0`. This is an important concept to understand, as it is very useful. You do not have to transform numbers in arrays, for example. Python will take care of this process for you. The rules on how broadcasting works in other cases are rather complex and go beyond this book's scope. However, it is important to know that Python is doing something in the background.

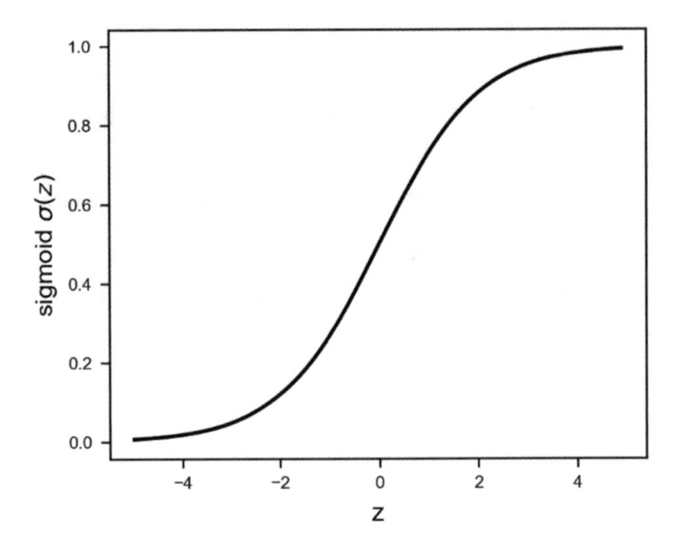

Figure 2-4. *The sigmoid activation function is a s-shaped function that goes from 0 to 1*

The sigmoid activation function (that you can see in Figure 2-4) is especially used for models where we must predict the probability as an output, as logistic regression (remember that a probability may only assume values between 0 and 1). Note that in Python, if z is big enough, it can happen that the function returns exactly 0 or 1 (depending on the sign of z) for rounding errors. In classification problems we will calculate $\log\sigma(z)$ or $\log(1 - \sigma(z))$ very often, and therefore this can be a source of errors in Python since it will try to calculate $\log 0$, which is not defined. For example, you can start seeing nan appearing while calculating the cost function (more on that later).

Note Although $\sigma(z)$ should never be exactly 0 or 1, while programming in Python the reality can be quite different. It may happen that, due to a very big z (positive or negative), Python will round the results to exactly 0 or 1. This may give you errors while calculating the cost function for classification, since you need to calculate $\log\sigma(z)$ and $\log(1 - \sigma(z))$. Therefore, Python will try to calculate $\log 0$, which is not defined. This may happen, for example, if you don't correctly normalize the input data or if you don't correctly initialize the weights. For the moment, it's important to remember that although everything seems under control mathematically, the reality while programming can be more difficult. It's good to keep this in mind while debugging models that give, for example, nan as a result for the cost function.

Tanh (Hyperbolic Tangent) Activation Function

The hyperbolic tangent is also an s-shaped curve that goes from -1 to 1, as shown in Figure 2-5.

$$f(z) = \tanh(z)$$

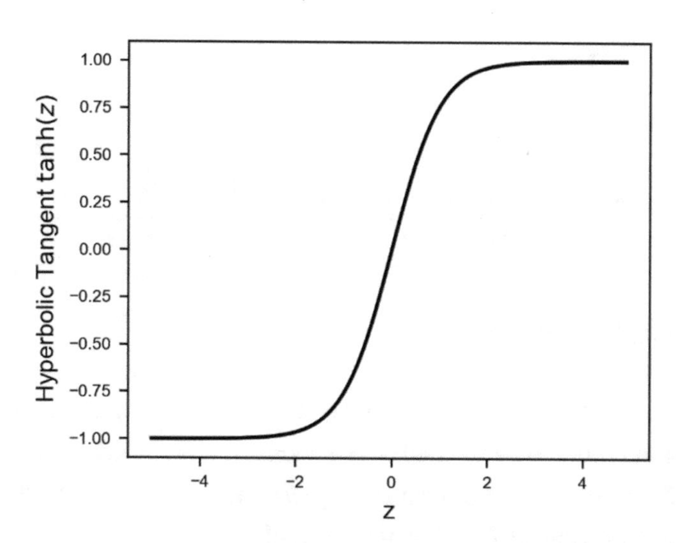

Figure 2-5. *The tanh (or hyperbolic function) is an s-shaped curve that goes from -1 to 1*

In Python, this can be easily implemented

```
def tanh(z):
    return np.tanh(z)
```

ReLU (Rectified Linear Unit) Activation Function

The ReLU has the following formula (see Figure 2-6):

$$f(z) = \max(0, z)$$

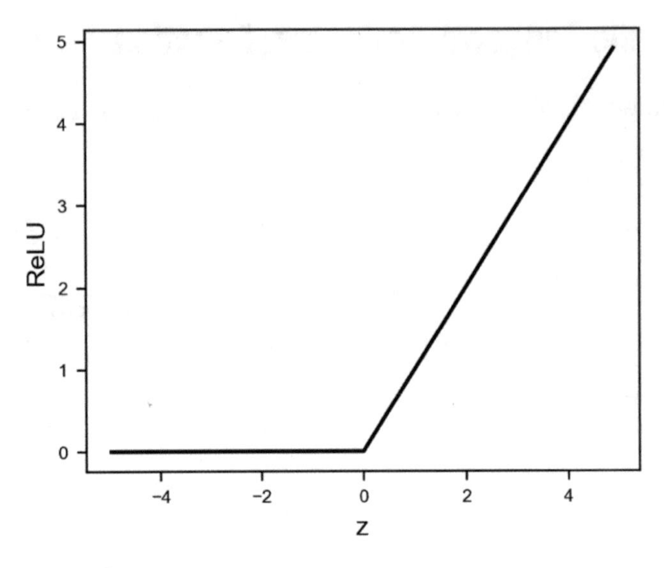

Figure 2-6. *The ReLU function*

It is worth it to spend a few moments to see how to implement ReLU in a smart way in Python. Note that when you start using TensorFlow, it is implemented for you, but it's still very instructive to see how different Python implementations can make a difference when implementing complex deep learning models.

In Python you can implement the ReLU function in several ways. Here are four different ways (try to understand why they work before going on):

1. `np.maximum(x, 0, x)`
2. `np.maximum(x, 0)`
3. `x * (x > 0)`
4. `(abs(x) + x) / 2`

The four methods have very different execution speeds. Let's generate a NumPy array with 10^8 elements:

```
x = np.random.random(10**8)
```

Now we measure the time needed by the four different versions of the ReLU function when applied to it. Let the following code run

```
x = np.random.random(10**8)
print("Method 1:")
%timeit -n10 np.maximum(x, 0, x)

print("Method 2:")
%timeit -n10 np.maximum(x, 0)

print("Method 3:")
%timeit -n10 x * (x > 0)

print("Method 4:")
%timeit -n10 (abs(x) + x) / 2
```

The results are

```
Method 1:
2.66 ms ± 500 µs per loop (mean ± std. dev. of 7 runs, 10 loops each)
Method 2:
6.35 ms ± 836 µs per loop (mean ± std. dev. of 7 runs, 10 loops each)
Method 3:
4.37 ms ± 780 µs per loop (mean ± std. dev. of 7 runs, 10 loops each)
Method 4:
8.33 ms ± 784 µs per loop (mean ± std. dev. of 7 runs, 10 loops each)
```

The difference is stunning. Method 1 is four times faster than method 4. The `numpy` library is highly optimized, with many routines written in C. But knowing how to code efficiently still makes a difference and can have a great impact. Why is `np.maximum(x, 0, x)` faster than `np.maximum(x, 0)`? The first version updates `x` in place, without creating a new array. This can save a lot of time, especially when the arrays are big. If you don't want to (or can't) update the input vector in place, you can still use the `np.maximum(x, 0)` version.

An implementation could look like this

```
def relu(z):
    return np.maximum(z, 0)
```

Note Remember that, when optimizing your code, even small changes can make a huge difference. In deep learning programs, the same chunk of code is repeated millions and billions of times, so even a small improvement can have a huge impact in the long run. Spending time optimizing your code is a necessary step that will pay off.

Leaky ReLU

The Leaky ReLU (also known as Parametric Rectified Linear Unit) is given by the formula

$$f(z) = \begin{cases} \alpha z & for \quad z < 0 \\ z & for \quad z \geq 0 \end{cases}$$

with α a parameter typically of the order of 0.01. See Figure 2-7.

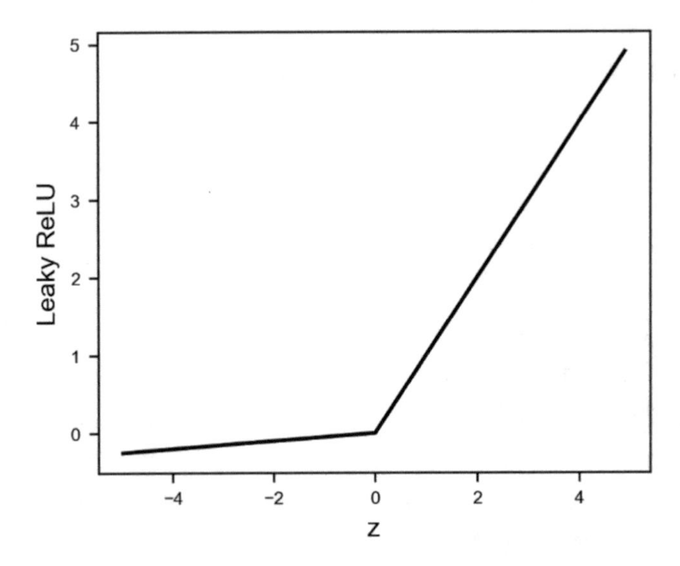

Figure 2-7. *The Leaky ReLU activation function with $\alpha = 0.05$. This value has been chosen to make the difference between $x > 0$ and $x < 0$ more marked. Smaller values for α are typically used. Testing your model is required to find the best value*

In Python, this can be implemented if the `relu(z)` function has been defined as

```
def lrelu(z, alpha):
  return relu(z) - alpha * relu(-z)
```

The Swish Activation Function

Ramachandran, Zopf, and Le from Google Brain [4] recently studied a new activation function that shows great promise in the deep learning world; they named it Swish. It is defined as

$$f(z) = z\sigma(\beta z)$$

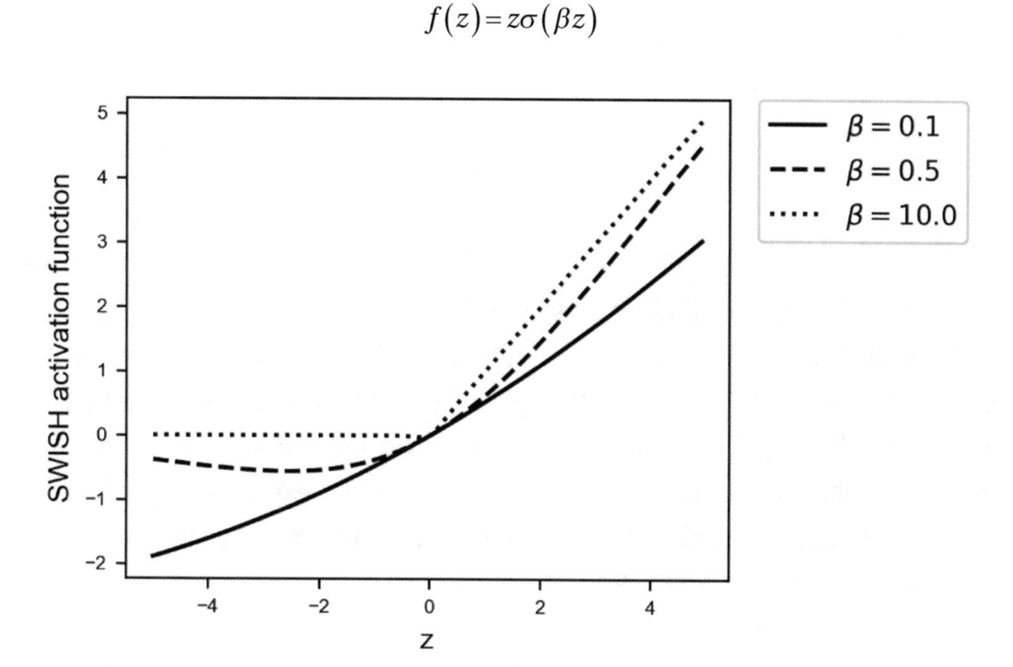

Figure 2-8. *The Swish activation function for three different values of the parameter β*

where β is a learnable parameter (see Figure 2-8). Their studies have shown that simply replacing ReLU activation functions with Swish improves classification accuracy on ImageNet by 0.9%, which in today's deep learning world is a lot. ImageNet is a large database of images that is often used to benchmark new network architectures or algorithms, as in this case, networks with a different activation functions. You can find more information about ImageNet at `http://www.image-net.org/`.

Other Activation Functions

There are many other activation functions, but those are rarely used. As a reference, here are some additional ones. The list is by no means comprehensive and should serve the purposes of giving you an idea of the variety of activation functions that can be used when developing neural networks.

- ArcTan

$$f(z) = \tan^{-1} z$$

- Exponential Linear unit (ELU)

$$f(z) = \begin{cases} \alpha(e^z - 1) & for \quad z < 0 \\ z & for \quad z \geq 0 \end{cases}$$

- Softplus

$$f(z) = \ln(1 + e^z)$$

Note Practitioners almost always use only two activation functions: the sigmoid and the ReLU (with probably the ReLU in the lead). With both you can achieve good results. Both can, given a complex enough network architecture, approximate any nonlinear function.[3] Remember that when using TensorFlow, you do not have to implement them by yourself. TensorFlow offers an efficient implementation for you to use. But is important to know how each activation function behaves to understand when to use which one.

Now that we briefly discussed all the necessary components, you can use the neuron on some real problems. Let's first see how to implement a neuron in Keras and then how to perform linear and logistic regression with it.

How to Implement a Neuron in Keras

Building a network with a single neuron in Keras is straightforward and can be done with

```
model = keras.Sequential([
   layers.Dense(1, input_shape = [...])
])
```

[3] Montufar G., Pascanu R., Cho K. and Bengio Y., "On the Number of Linear Regions of Deep Neural Networks," https://papers.nips.cc/paper/5422-on-the-number-of-linear-regions-of-deep-neural-networks.pdf, last accessed 10th Jan. 2018; Fortuner B., "Can neural networks solve any problem?," https://towardsdatascience.com/can-neural-networks-really-learn-any-function-65e106617fc6, last accessed 10th Jan. 2018.

The **Sequential** class groups a linear stack of layers into a `tf.keras.Model`. In this straightforward case, we need just one layer made by a single neuron, defined by the **layers.Dense** command, which specifies **1** unit (neuron) inside a layer and the shape of the input dataset. The `Dense` class implements densely connected neural networks' layers (more on that in the next chapters).

In the next paragraphs, you see two practical examples of how to use this simple approach—choosing the right activation function and the proper loss function—to solve two different problems, namely linear regression and logistic regression.

Python Implementation Tips: Loops and NumPy

As you have just seen, Keras does all the dirty work for you. Of course, you can also implement the neuron from scratch, using Python standard functionalities such as lists and loops, but those tend to become very slow as the number of variables and observations grows. A good rule of thumb is to avoid loops when possible and use NumPy (or TensorFlow) methods as often as possible.

It is easy to understand how fast NumPy can be (and how slow loops are). Let's start by creating two standard lists of random numbers in Python with 10^7 elements in each:

```
import random
lst1 = random.sample(range(1, 10**8), 10**7)
lst2 = random.sample(range(1, 10**8), 10**7)
```

The actual values are not relevant for our purposes. We are simply interested in how fast Python can multiply two lists, element by element. The time reported was measured on a 2017 Microsoft Surface laptop and will vary greatly depending on the hardware that the code runs on. We are not interested in the absolute values, but only in how much faster NumPy is in comparison to standard Python loops. If you are using a Jupyter Notebook, it is useful to know how to time Python code in a cell. To do this, you can use a "magic command." Those commands start (in a Jupyter Notebook) with %% or with %. It's a good idea to check the official documentation to better understand how they work (`http://ipython.readthedocs.io/en/stable/interactive/magics.html`).

Coming back to the test, let's measure how much time a standard laptop takes to multiply, element by element, the two lists with standard loops. Using the code

```
%%timeit
ab = [lst1[i]*lst2[i] for i in range(len(lst1))]
```

gives us the following result (note that on your computer, you will probably get different numbers):

```
2.06 s ± 326 ms per loop (mean ± std. dev. of 7 runs, 1 loop each)
```

The code needed roughly two seconds averaged over seven runs. Now let's try to do the same multiplication, but this time using NumPy

```
%%timeit
out2 = np.multiply(list1_np, list2_np)
```

We first converted the two lists to numpy arrays with the following code

```
import numpy
list1_np = np.array(lst1)
list2_np = np.array(lst2)
```

This time we get the following results

```
20.8 ms ± 2.5 ms per loop (mean ± std. dev. of 7 runs, 10 loops each)
```

The NumPy code needed only 21ms or, in other words, it was roughly 100 times faster than the code with standard loops. NumPy is faster for two reasons: the underlying routines are written in C, and it uses vectorized code as much as possible to speed up calculations on a large amount of data.

Note Vectorized code refers to operations performed on multiple components of a vector (or a matrix) at the same time (in one statement). Passing matrixes to NumPy functions is an excellent example of *vectorized code*. NumPy will perform operations on big chunks of data simultaneously, obtaining much better performance than standard Python loops since the latter must operate on one element at a time. Note that part of the excellent performance NumPy shows is also due to the underlying routines being written in C.

While training deep learning models, you will find yourself doing this kind of operation over and over. Such a speed gain will make the difference between having a model that can be trained and one that will never give you a result.

Linear Regression with a Single Neuron

This section explains how to build your first model in Keras and how to use it to solve a basic statistical problem. You can, of course, perform linear regression quickly by applying traditional math formulas or using dedicated functions such as those found in Scikit-learn. For example, you can find the complete implementation of linear regression from scratch with NumPy, using the analytical formulas (see `http://adl.toelt.ai/single_neuron/Linear_regression_with_numpy.html`) in the online version of the book. However, it is instructive to follow this example, since it gives a practical understanding of how the building blocks of deep learning architectures (i.e., neurons) work.

If you remember, we have said many times that NumPy is highly optimized to perform several parallel operations simultaneously. To get the best performance possible, it's essential to write your equations in matrix form and feed the matrixes to NumPy. In this way, the code will be as efficient as possible. Remember: avoid loops at all costs whenever possible.

The Dataset for the Real-World Example

To make things a bit more interesting, let's use an instructive dataset. We will use the so-called radon dataset [3]. Radon is a radioactive gas that enters homes through contact points with the ground. It is a carcinogen that is the primary cause of lung cancer in non-smokers. Radon levels vary significantly from household to household. The dataset contains measured radon levels in U.S. homes by county and state. The `activity` label is the measured radon concentration in pCi/L (referred to as the *target* variable, i.e., the one we want to predict using a linear regression model). Important features are:

- Floor (the floor of the house in which the measurement was taken)
- County (the U.S. county in which the house is located)
- uppm (a measurement of the uranium level of the soil by county)

This dataset fits a classical regression problem well since it contains a continuous variable (radon activity) to predict. The model that we will build is made of one neuron and will fit a linear function to the different features.

You do not need to understand or study the features. The goal here is to understand how to build a linear regression model with what you have learned. Generally, in a machine learning project, you would first study your input data, check its distribution, quality, missing values, and so on. But we skip this part to concentrate on the implementation with Keras.

Note In machine learning, the variable we want to predict is usually called the *target variable*.

Now, let's look at the data. We will skip all the import and load details for simplicity and concentrate on the fundamental steps of the code, such as dataset preparation, model creation, and performance evaluation. Remember that you can find the complete code on the online version of the book. We start by checking how many observations we have

```
num_counties = len(county_name)
num_observations = len(radon_features)
print('Number of counties included in the dataset: ', num_counties)
print('Number of total samples: ', num_observations)
```

The code will give the following results

```
Number of counties included in the dataset:  85
Number of total samples:  919
```

So, we have 919 different measurements of radon activity in 85 distinct counties. The `radon_features.head()` command will give the following table as output (the first five lines of the pandas dataframe)

```
   floor  county  log_uranium_ppm  pcterr
0      1       0         0.502054     9.7
1      0       0         0.502054    14.5
2      0       0         0.502054     9.6
3      0       0         0.502054    24.3
4      0       1         0.428565    13.8
```

We have four features (`floor`, `county`, `log_uranium_ppm`, and `pcterr`) that we will use as predictors of radon activity.

As suggested, we have prepared the data in matrix form. Let's briefly review the notation, which will come in handy when building the neuron. Normally we have many observations (919 in this case). We use an upper index to indicate the different observations between parentheses. The ith observation is indicated with $\boldsymbol{x}^{(i)}$, and the jth feature of the ith observation is indicated with $\boldsymbol{x}_j^{(i)}$. We indicate the number of observations with m.

Note In this book, m is the *number of observations* and n_x is the *number of features.* Our j^{th} feature of the ith observation is indicated with $\boldsymbol{x}_j^{(i)}$. In deep learning projects, the bigger the m, the better. So be prepared to deal with a massive number of observations.

---

n_x in this example is equal to 4, while m is equal to 919. The entire set of inputs (features and observations) can be therefore written using the following notation

$$X = \begin{pmatrix} x_1^{(1)} & \dots & x_4^{(1)} \\ \vdots & \ddots & \vdots \\ x_1^{(m)} & \dots & x_4^{(m)} \end{pmatrix}$$

where each row is an observation, and each column represents a feature in the matrix X that has the dimensions $m \times 4$.

Dataset Splitting

In any machine learning project, to check how the model generalizes to unseen data, you need to split the dataset into different subsets.[4] When you build a machine learning model, you first need to train the model, and then you have to test it (i.e., verify the model's performances on novel data). The most common way to do this is to split the dataset into two subsets: 80% of the original dataset to train the model (the more data you have, the better your model will perform) and the remaining 20% to test it.[5]

---

[4] It is assumed here that you know why it is important.

[5] The ratio 80/20 is simply a convention; you can choose 75/25, 70/30, or anything in between. Normally 80/20 is chosen since you want to have as much training data as possible and enough test data to make your testing reasonable.

The following code splits the dataset randomly into two parts with the following proportions: 80%/20%.

```
np.random.seed(42)
rnd = np.random.rand(len(radon_features)) < 0.8

train_x = radon_features[rnd] # training dataset (features)
train_y = radon_labels[rnd] # training dataset (labels)
test_x = radon_features[~rnd] # testing dataset (features)
test_y = radon_labels[~rnd] # testing dataset (labels)

print('The training dataset dimensions are: ', train_x.shape)
print('The testing dataset dimensions are: ', test_x.shape)
```

This code will give as output the following lines

```
The training dataset dimensions are:  (733, 4)
The testing dataset dimensions are:  (186, 4)
```

We will use 733 observations to train our linear regression model, and we will then evaluate it on the remaining 186 observations.

Linear Regression Model

Keep in mind that using a single-neuron model is overkill for a linear regression task. We could solve linear regression exactly without using gradient descent or a similar optimization algorithm. As mentioned, you can find an exact regression solution example, implemented with NumPy library, in the book's online version.

Given that the dataset can be expressed as a matrix (X) and the label we want to predict as a column vector (y), when we employ one neuron to perform linear regression, we simply try to find the best weights (W) that appear in the following equation:

$$\hat{y} = WX + b$$

The weights and bias need to be chosen so that that the network output is as similar as possible to the expected target variable.

If you remember how a neuron is structured, you can easily see that, for the neuron to give as output a linear combination of the inputs, we need to use the *identity activation function*. How can we measure how close the neuron's outputs are to the target variables? The difference is measured by using the Mean Squared Error (MSE) function.

$$L(\boldsymbol{w},b)=\frac{1}{m}\sum_{i=1}^{m}\left(y^{(i)}-\boldsymbol{w}^T\boldsymbol{x}^{(i)}-b\right)^2$$

where the sum is over all m observations. This is the typical loss function chosen in a regression problem. By minimizing $L(\boldsymbol{w}, b)$ with respect to the weights and bias, we can find their optimal values.

Minimizing $L(\boldsymbol{w}, b)$ is done with an optimizer. If you remember from the last chapter, gradient descent is the most basic example of an optimizer and could be used to solve this problem. Since is not available out of the box in TensorFlow, we use the RMSProp optimizer for practical reasons. Don't worry if you don't know how it works. Just know that it is simply a more intelligent version of the GD algorithm. You learn how it works in detail in the following chapters.

Keras Implementation

If you have no experience with Keras, you can consult the appendixes of this book. There you will find an introduction to Keras that will give you enough information to be able to understand the following discussion.

Implementing what we discussed with Keras is straightforward. The following function builds the one-neuron model for linear regression.

```
def build_model():

  model = keras.Sequential([
    layers.Dense(1, input_shape = [len(train_x.columns)])
  ])

  optimizer = tf.keras.optimizers.RMSprop(
    learning_rate = 0.001)
```

```
  model.compile(loss = 'mse',
                optimizer = optimizer,
                metrics = ['mse'])

  return model
```

Let's analyze what this code does:

- First of all, we defined the network structure (also called network architecture) with the `keras.Sequential` class, adding one layer[6] made of one neuron (`layers.Dense`) and with `input dimensions` equal to the number of features used to build the model. The activation function is the one set by default, i.e. the identity function.
- Then, we defined the optimizer (`tf.keras.optimizers.RMSprop`), setting the learning rate to 0.001. The optimizer is the algorithm that Keras will use to minimize the cost function. We use the RMSprop algorithm in this example.
- Finally, we compile the model (i.e., we configure the model for training), setting its loss function (i.e., the cost function to be minimized), its optimizer, and the metric to be calculated during performance evaluation (`model.compile`). The function returns the built `model` as a single Python object.

The *learning rate* is a very important parameter for the optimizer. In fact, it strongly influences the convergence of the minimization process. It is a common and good behavior to try different learning rate values and observe how the model's convergence changes.

Now, let's apply the `build_model` function and look at the model summary

```
model = build_model()
model.summary()
```

[6] We discuss at length what layers are in the next chapter. For the moment, just ignore this if you don't understand what it means.

This code gives the following output

```
Model: "sequential"

_________________________________________________________________
Layer (type)                 Output Shape              Param #
=================================================================
dense (Dense)                (None, 1)                 5
=================================================================
Total params: 5
Trainable params: 5
Non-trainable params:

_________________________________________________________________
```

Thus, we have five parameters to be trained—the weights associated with the four features, plus the bias.

The Model's Learning Phase

Training our neuron means finding the weights and biases that minimize the chosen cost function (the MSE in this case). The *minimization process* is iterative; therefore, it is necessary to decide when to stop it. For this example, we simply set a fixed number of epochs. We train the model for 1000 epochs and then look at the results in terms of the MSE.

```
EPOCHS = 1000

history = model.fit(
  train_x, train_y,
  epochs = EPOCHS, verbose = 0
)
```

As you can see, training the model in Keras is straightforward. It is enough to apply the `fit` method to our model object. `fit` takes as inputs the training data and the number of epochs.

The cost function clearly decreases for a while and then stays almost constant. That is a good sign, indicating that the cost function has reached a minimum. That does not mean that our model is good or that it will give good predictions. This tells us only that learning has somehow worked. A very good way to immediately visualize the loss function decrease is by plotting the cost function vs. the number of epochs. This can be seen in Figure 2-9.

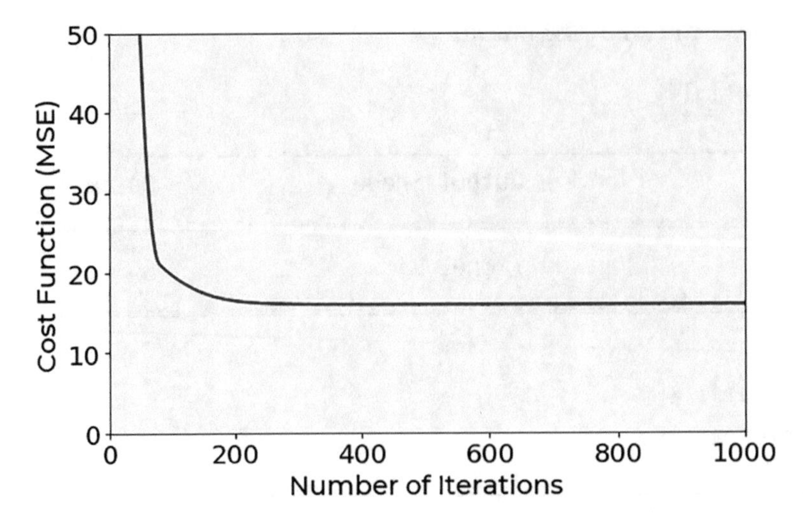

***Figure 2-9.** The cost function behavior during the model training, applied to the radon dataset with a learning rate of 0.001*

Looking at Figure 2-9, you can see that, after 400 epochs, the cost function remains almost constant in its value, indicating that a minimum has been reached.

Since we are doing a linear regression, we are interested in the coefficients of the linear equation. Those are the weights of our neuron. The first four are the linear regression coefficients, while the last one is the bias term. You can compare these numbers with the ones obtained by performing linear regression with traditional math formulas and using the NumPy library in the online version of the book. To get the weights in Keras, you simply use the `get_weights()` call.

```
weights = model.get_weights() # return a numpy list of weights
print(weights)
```

which returns

```
[array([[-6.6795307e-01],
       [ 2.7279984e-03],
       [ 2.8733387e+00],
       [-2.0828046e-01]], dtype=float32),
 array([4.2394686], dtype=float32)]
```

Those are the weights we were expecting.

Model's Performance Evaluation on Unseen Data

To determine if the model you just trained is suited for unseen data, you must check its performance over a dataset containing only unseen data (the test set). Then, predicted radon activity values are compared with real values (`test_y`) by simply plotting predictions vs. true values, as shown in Figure 2-10. A perfect model will show points distributed over the black line. The less precise the predictions are, the more spread out the points will be around the diagonal.

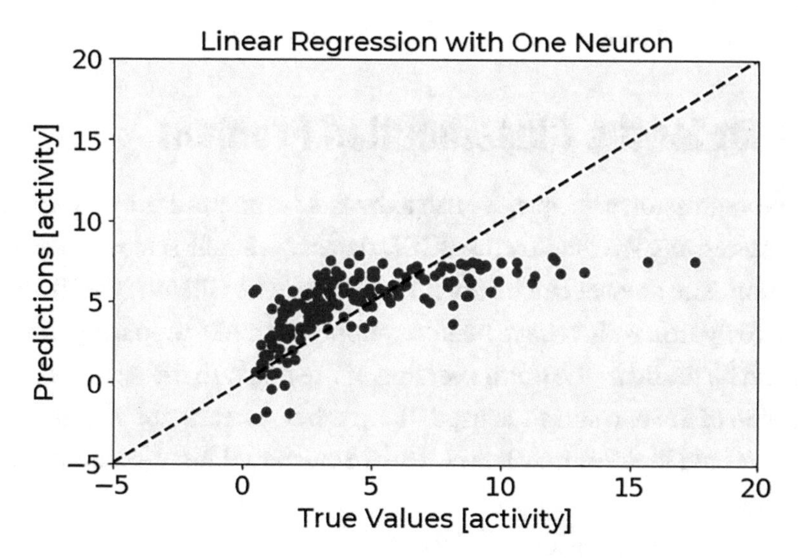

***Figure 2-10.** The predicted target value vs. measured target value for our model, applied to our testing data*

If you have followed so far, congratulations! You just built your very first neural network, with just one neuron, but still a neural network!

Logistic Regression with a Single Neuron

Now let's try to solve a classification problem with a single neuron. Logistic regression is a classical algorithm and probably the simplest classification. We will consider a binary classification problem: that means we will deal with the problem of recognizing two classes (that we label as 0 or 1). We need an activation function that's different from the one we used for linear regression, a different cost function to minimize, and a slight modification of the output of our neuron. The goal is to be able to build a model that can predict if a certain new observation is in one of two classes. The neuron should give as output the probability $P(y = 1|x)$ of the input x to be of class 1. We will then classify our observation as being in class 1 if $P(y = 1|x) > 0.5$ or in class 0 if $P(y = 1|x) < 0.5$.

It is instructive to compare this example with the one about linear regression, since they both are applications of the one-neuron model yet are used to solve different tasks. You should pay attention to the similarities and differences with the linear regression model discussed previously. You are going to see how simple using Keras is in this case and how, by changing a few things (such as the activation and loss functions), you can easily obtain a different model that can solve a different problem.

The Dataset for the Classification Problem

As in the linear regression example, we use a dataset taken from the real world, to make things more interesting. We employ the BCCD dataset, a small-scale dataset for blood cell classification. The dataset can be downloaded from its GitHub repository [8]. For this dataset, two Python scripts have been developed to make preparing the data easier. All the code can be found in the online version of the book. In the example, a slightly modified version of the two scripts is used. Remember, you are not interested at this point in how the data looks or how it is cleaned. You should focus on how to build the model with Keras.

The dataset contains three kinds of labels:

1. Red blood cells (RBC)
2. White blood cells (WBC)
3. Platelets

To make it a binary classification problem, we will consider only the RBC and WBC types. The model that will be trained has one neuron and will predict if an image contains an RBC or a WBC type by using the `xmin`, `xmax`, `ymin`, and `ymax` variables as features.

For simplicity, as in the linear regression example, we will skip all the import and load details and concentrate on the fundamental steps of the code, such as dataset preparation, model creation, and performance evaluation. You can find the complete code on the online version of the book. Note that the greatest differences between this case and the linear regression lie in the chosen activation and cost function.

Here is the data:

```
num_observations = len(bccd_features)
print('Number of total samples: ', num_observations)
```

This code returns

```
Number of total samples:  4527
```

Let's display the first lines of the data

```
bccd_features.head()
```

which prints to the screen

```
    xmin xmax ymin ymax
0   192  292  376  473
1   301  419  320  424
2   433  510  273  358
3   434  528  368  454
4   507  574  381  454
```

The dataset is made up of 4527 observations, one target column (cell_type), and four features (xmin, xmax, ymin, and ymax). When working with images, it is always useful to get an idea of how they look. You can see an example in Figure 2-11.

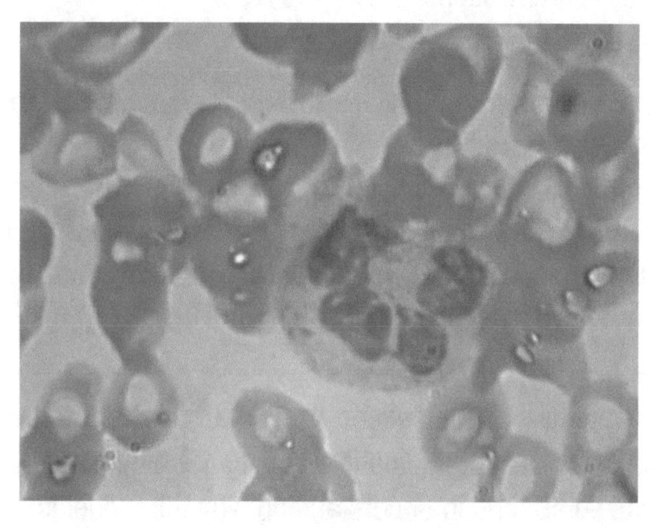

Figure 2-11. *A sample image from the BCCD dataset*

Note that the features we employ in this example are not the pixel values of the image, but the location of the edges of the bounding box of the cell. In fact, for each image, we have only four values (`xmin`, `xmax`, `ymin`, and `ymax`).

Dataset Splitting

As stated, in any machine learning project, you have to split the dataset in different subsets. Let's create a `train` and a `test` dataset by splitting the dataset randomly into two parts with a ratio of 80/20, as we did with the radon dataset.

```
np.random.seed(42)
rnd = np.random.rand(len(bccd_features)) < 0.8

train_x = bccd_features[rnd] # training dataset (features)
train_y = bccd_labels[rnd] # training dataset (labels)
test_x = bccd_features[~rnd] # testing dataset (features)
test_y = bccd_labels[~rnd] # testing dataset (labels)

print('The training dataset dimensions are: ', train_x.shape)
print('The testing dataset dimensions are: ', test_x.shape)
```

This code will give as output the following lines

```
The training dataset dimensions are:  (3631, 4)
The testing dataset dimensions are:  (896, 4)
```

So, we use 3631 observations to train our logistic regression model and then evaluate it on the remaining 896 observations.

Now comes a particularly important point. The labels in this dataset as imported will be `'WBC'` or `'RBC'` strings (they simply tell you if an image contains white or red blood cells). But we will build our cost function with the assumptions that our class labels are 0 and 1. That means we need to change our `train_y` and `test_y` arrays.

Note When doing binary classifications, remember to check the values of the labels you are using for training. Sometimes using the wrong labels (not 0 and 1) may cost you quite some time in understanding why the model is not learning.

```
train_y_bin = np.zeros(len(train_y))
train_y_bin[train_y == 'WBC'] = 1

test_y_bin = np.zeros(len(test_y))
test_y_bin[test_y == 'WBC'] = 1
```

Now all images containing RBC will have a label of `0`, and all images containing WBC will have a label of `1`.

The Logistic Regression Model

This model will be made up of one neuron and its goal will be to recognize two classes (labeled `0` or `1`, referring to RBC or WBC inside a cell image). This is a *binary classification problem.*

As opposed to linear regression, the activation function will be a *sigmoid function* (leading to a different neuron's output) and the cost function will be *cross-entropy* [7]. If you don't remember what a sigmoid function is, reread the beginning of this chapter. We use it because we want our neuron to output the probability of our observation being in class 1. Therefore, we need an activation function that can assume only values between 0 and 1; otherwise we cannot regard it as a probability. The cross-entropy for one observation is

$$L\left(\hat{y}^{(i)}, y^{(i)}\right) = -\left(y^{(i)} \log \hat{y}^{(i)} + \left(1 - y^{(i)}\right) \log\left(1 - \hat{y}^{(i)}\right)\right)$$

In presence of more than one observation, the cost function is the sum over all observations

$$L(\boldsymbol{w}, b) = \frac{1}{m} \sum_{i=1}^{m} L\left(\hat{y}^{(i)}, y^{(i)}\right)$$

Explaining the cross-entropy loss function goes beyond the scope of this book, but if you are interested, you can find it described in many books and websites. For example, you can check out [7].

Keras Implementation

The following function builds the one-neuron model for logistic regression. The implementation is remarkably similar to that of linear regression. The differences, as already mentioned, are the activation function, the cost function, and the metrics (accuracy in this case, which we analyze more in detail in the testing phase).

```
def build_model():

  model = keras.Sequential([
    layers.Dense(1, input_shape = [len(train_x.columns)], activation =
    'sigmoid')
  ])

  optimizer = tf.keras.optimizers.RMSprop(
              learning_rate = 0.001)

  model.compile(loss = 'binary_crossentropy',
                optimizer = optimizer,
                metrics =
               ['binary_crossentropy','binary_accuracy'])

  return model
```

Now, let's apply the build_model function and look at the model summary

```
model = build_model()
model.summary()
```

This code gives the following output

```
Model: "sequential"

_________________________________________________________________
Layer (type)                 Output Shape              Param #
=================================================================
dense (Dense)                (None, 1)                 5
=================================================================
Total params: 5
Trainable params: 5
Non-trainable params: 0

_________________________________________________________________
```

We have five parameters to be trained in this case as well—the weights associated with the four features plus the bias.

The Model's Learning Phase

As with the linear regression example, training our neuron means finding the weights and biases that minimize the cost function. The cost function we chose to minimize in our logistic regression task is the cross-entropy function, as discussed in the previous section.

We start training our model for 500 epochs and then look at the summary in terms of performances (accuracy).

```
EPOCHS = 500

history = model.fit(
  train_x, train_y_bin,
  epochs = EPOCHS, verbose = 1
)
```

In Figure 2-12, you can see the cost function vs. the number of iterations associated with the learning phase.

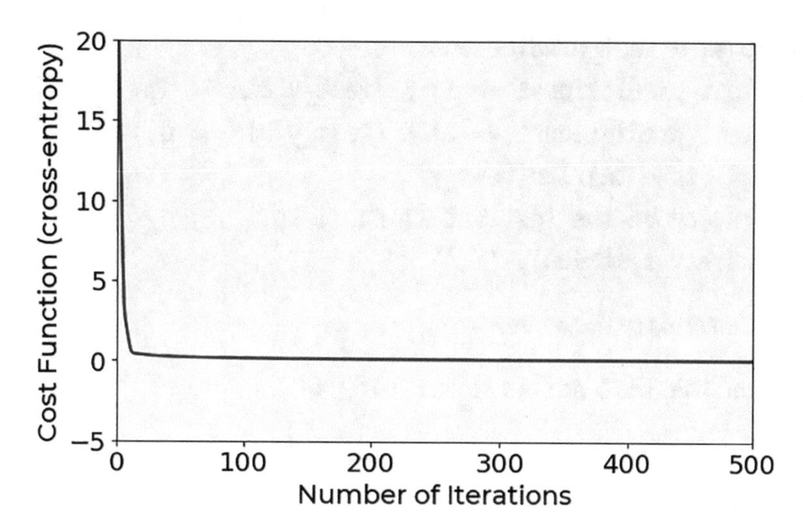

Figure 2-12. *The cost function resulting in our model applied to the BCCD dataset with a learning rate of 0.001*

Note that, after 100 epochs, the cost function remains almost constant, indicating that a minimum has been reached.

The Model's Performance Evaluation

Now, to determine if the model you have just trained is suited for unseen data, you must check its performance against the test set. Moreover, an *optimizing metric* must be chosen. For a binary classification problem, a classic metric is the *accuracy,* which can be understood as a measure of how well the classifier correctly identified the two classes of the dataset. Mathematically it can be calculated as

$$accuracy = \frac{number\ of\ cases\ correctly\ identified}{total\ number\ of\ cases}$$

where the number of cases correctly identified is the sum of all positive samples and negative samples (i.e., all 0s and 1s) that were correctly classified. These are usually called true positives and true negatives.

To get the accuracy, we can run this code (remember, we will classify an observation i to be in class 0 if $P(y^{(i)} = 1|\boldsymbol{x}^{(i)}) < 0.5$ or to be in class 1 if $P(y^{(i)} = 1|\boldsymbol{x}^{(i)}) > 0.5$)

```
test_predictions = model.predict(test_x).flatten()
test_predictions1 = test_predictions > 0.5
tp = np.sum((test_predictions1 == 1) & (test_y_bin == 1))
tn = np.sum((test_predictions1 == 0) & (test_y_bin == 0))
accuracy_test = (tp + tn)/len(test_y)
print('The accuracy on the test set is equal to: ',
      int(accuracy_test*100), '%.')
```

This code prints this to the screen

```
The accuracy on the test set is equal to:  98 %.
```

With this model, we reach an accuracy of 98%. Not bad for a network with just one neuron.

Conclusion

This chapter looked at many things. You learned how a neuron works and what its main components are. You also learned about the most common activation functions and saw how to implement a model with a single neuron in Keras to solve two problems: linear and logistic regression. In the next chapter, we look at how to build neural networks with a large number of neurons and how to train them.

Note Linear and logistic regression are two classical statistical models that can be implemented in many ways. This chapter used the neural network language to implement them and to show you how flexible neural networks are. When you understand their inner components, you can use them in many ways.

Exercises

EXERCISE 1 (LINEAR REGRESSION) (LEVEL: EASY)

Try using only one feature to predict radon activity and see how the results change.

EXERCISE 2 (LINEAR REGRESSION) (LEVEL: MEDIUM)

Try to change the `learning_rate` parameter and then observe how the model's convergence changes. Then try to reduce the `EPOCHS` parameter and observe when the model cannot reach convergence.

EXERCISE 3 (LINEAR REGRESSION) (LEVEL: MEDIUM)

Try to see how a model's results change based on the training dataset's size (reduce it and use different sizes, comparing the results).

EXERCISE 4 (LOGISTIC REGRESSION) (LEVEL: MEDIUM)

Try to change the `learning_rate` parameter and observe how the model's convergence changes. Then try to reduce the `EPOCHS` parameter and observe when the model cannot reach convergence.

EXERCISE 5 (LOGISTIC REGRESSION) (LEVEL: MEDIUM)

Try to see how a model's results change based on the training dataset's size (reduce it and use different sizes comparing the results).

EXERCISE 6 (LOGISTIC REGRESSION) (LEVEL: DIFFICULT)

Try to add to labels to `Platelets` samples and generalize the binary classification model to a multiclass one (three possible classes).

References

[1] `https://spectrum.ieee.org/tech-talk/computing/software/biggest-neural-network-ever-pushes-ai-deep-learning`, last accessed 27.12.2017.

[2] R. Rojas (1996), *Neural Networks: A Systematic Introduction,* Springer-Verlag Berlin Heidelberg.

[3] `https://www.tensorflow.org/datasets/catalog/radon`, last accessed 09.01.2021.

[4] Lever, Jake, Martin Krzywinski, and Naomi Altman. "Points of significance: model selection and overfitting." (2016): 703.

[5] Srivastava, Nitish, et al. "Dropout: a simple way to prevent neural networks from overfitting." The journal of machine learning research 15.1 (2014): 1929-1958.

[6] Bengio, Yoshua. "Practical recommendations for gradient-based training of deep architectures." Neural networks: Tricks of the trade. Springer, Berlin, Heidelberg, 2012. 437-478.

[7] `https://rdipietro.github.io/friendly-intro-to-cross-entropy-loss/`, last accessed 10.01.2021.

[8] `https://www.tensorflow.org/datasets/catalog/bccd`, last accessed 10.01.2021.

CHAPTER 3

Feed-Forward Neural Networks

In the last chapter we did some amazing things with one neuron, but that is hardly flexible enough to tackle more complex cases. The real power of neural networks comes into light when several (thousands, even millions) neurons interact with each other to solve a specific problem. The network architecture (how neurons are connected to each other, how they behave, and so on) plays a crucial role in how efficient the learning of a network is, how good its predictions are, and what kind of problems it can solve. There are many kinds of architectures that have been extensively studied and that are very complex, but from a learning perspective, it is important to start from the simplest kind of neural network with multiple neurons. It makes sense to start with a *feed-forward neural network*, where data enters at the input layer and passes through the network, layer by layer, until it arrives at the output layer (this gives the networks their name: feed-forward neural networks).

This chapter discusses networks where each neuron in a layer gets its input from all neurons from the preceding layer and feeds its output into each neuron of the next layer. As it is easy to imagine, with more complexity come more challenges. It is more difficult to get quick learning and good accuracy, since the number of hyper-parameters that are available grows due to the increase in network complexity. A simple gradient descent algorithm is not as efficient when dealing with big datasets. When developing models with many neurons, we need to have at our disposal an expanded set of tools that allow us to deal with all the challenges that those networks present.

This chapter starts looking at some more advanced methods and algorithms that will allow you to work efficiently with big datasets and big networks. These complex networks can do some interesting multiclass classifications, one of the tasks that big networks are most often required to do (for example, handwriting recognition, face recognition, image recognition, and so on), so we will use a dataset that will allow us to perform multiclass classification and study its difficulties.

U. Michelucci, *Applied Deep Learning with TensorFlow 2*, https://doi.org/10.1007/978-1-4842-8020-1_3

We start the chapter with the network architecture and the needed matrix formalism. Next is a short overview of the new hyper-parameters that come with this new type of networks. You learn how to do multiclass classifications using the softmax function and what kind of output layer is needed. Then, before starting with the Python code, we will go into a brief digression to explain what exactly overfitting is with a simple example, and how to do a basic error analysis with complex networks.

Then we will start using Keras to construct bigger networks, applying them to a MNIST-similar dataset based on images of clothing items (the Fashion-MNIST dataset, from Zalando). Then we will look at how to add many layers in an efficient way and how to initialize the weights and the biases in the best way possible to make training fast and stable. We will look at Xavier and He initialization for the sigmoid and ReLU activation functions, respectively. Finally, we describe a rule of thumb for comparing the complexity of networks going beyond only the number of neurons. This chapter concludes with some tips on how to choose the right networks and a method to estimate the memory footprint depending on the architecture.

A Short Review of Network's Architecture and Matrix Notation

The network architecture is quite easy to understand. It consists of an *input layer* (the inputs $x_{i,j}$), several layers (called *hidden* because they are sandwiched between the input and the output layers, so they are "invisible" from the outside so to speak), and then an *output layer*. In each layer you may have one to several neurons. The main property of such a network is that each neuron gets input from each neuron in the preceding layer and feeds its output to every neuron in the next layer. Figure 3-1 shows a graphical representation of such a network (in the inputs, we omitted the first index indicating the observation index for clarity).

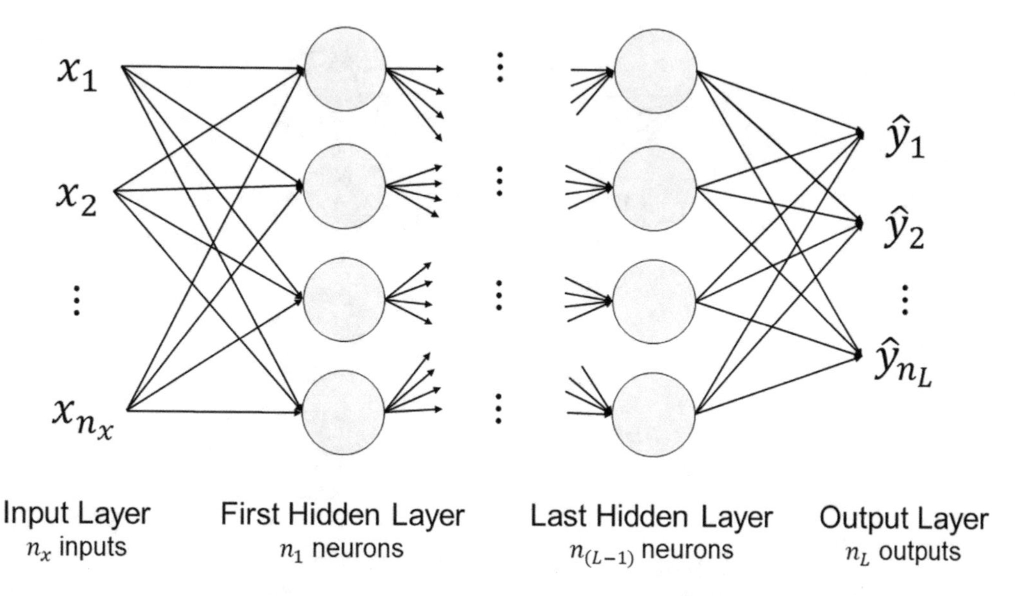

Figure 3-1. *The schematic representation of a deep feed-forward neural network with many hidden layers, where each neuron gets input from each neuron in the preceding layer and feeds its output to every neuron in the next layer*

To jump from one neuron to this is quite a big step. To build the model, we need to work with matrix formalism and therefore we need to get all the matrix dimension right. Let's first discuss some new notation:

- L is the number of hidden layers, excluding the input layer but including the output layer
- n_l is the number of neurons in layer l

In a network such as the one in Figure 3-1, we indicate the total number of neurons with $N_{neurons}$, which can be written as follows

$$N_{neurons} = n_x + \sum_{i=1}^{L} n_i = \sum_{i=0}^{L} n_i$$

Where, by convention, we define $n_0 = n_x$. Each connection between two neurons will have its own weight. Let's indicate the weight between neuron i in layer l and neuron j in layer $l - 1$ with $w_{ij}^{[l]}$. Figure 3-2 shows only the first two layers (input and layer 1) of the generic network from Figure 3-1, with the weights between the first neuron in the input layer and all the others in layer 1. All other neurons are grayed out for clarity.

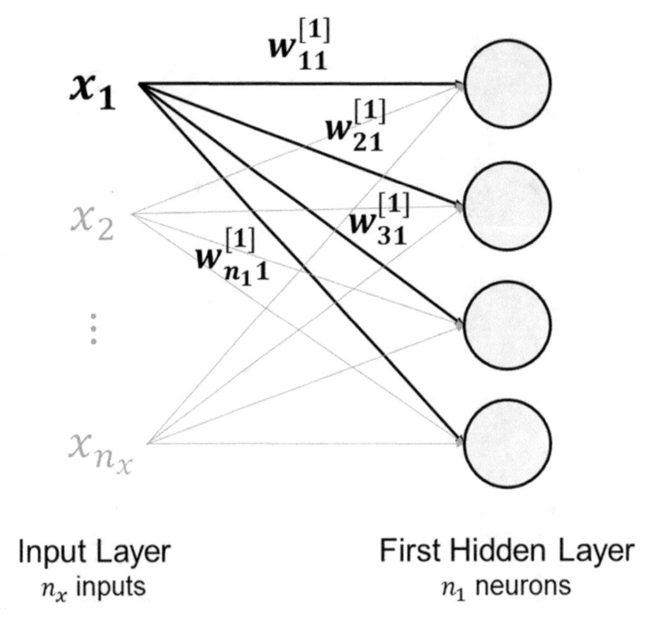

***Figure 3-2.** The first two layers of a generic neural network, with the weights of the connections between the first neuron in the input layers and the others in the second layer. All other neurons and connections are drawn in light gray to make the diagram clearer*

The weights between the input layer and layer 1 can be written as a matrix as follows

$$W^{[1]} = \begin{pmatrix} w_{11}^{[1]} & \cdots & w_{1n_x}^{[1]} \\ \vdots & \ddots & \vdots \\ w_{n_1 1}^{[1]} & \cdots & w^{[1]}{}_{n_1 n_x} \end{pmatrix}$$

That means that the matrix $W^{[1]}$ has dimensions $n_1 \times n_x$. This of course can be generalized between any two layers l and $l - 1$. Meaning that the weight matrix between two adjacent layers l and $l - 1$, which we indicate with $W^{[l]}$, will have dimensions $n_l \times n_{l-1}$. By convention, $n_0 = n_x$ is the number of input features (not the number of observations, which we indicate with m).

Note The weight matrix between two adjacent layers l and $l - 1$, which we indicate with $W^{[l]}$, will have dimensions $n_l \times n_{l-1}$, where, by convention, $n_0 = n_x$ is the number of input features.

The bias (indicated with b) will be a matrix this time. Remember that each neuron that receives inputs will have its own bias, so when considering our two layers l and $l-1$ we will need n_l different values of b. We indicate this matrix with $b^{[l]}$ and it will have dimensions $n_l \times 1$.

Note The bias matrix for two adjacent layers l and $l-1$, which we indicate with $b^{[l]}$, will have dimensions $n_l \times 1$.

Output of Neurons

Now let's consider the output of our neurons. To begin, we will consider the ith neuron of the first layer (remember our input layer is by definition layer 0). Let's indicate its output with $\hat{y}_i^{[1]}$ and let's assume that all neurons in layer l use the same activation function that we indicate with $g^{[l]}$. Then we will have

$$\hat{y}_i^{[1]} = g^{[1]}\left(z_i^{[1]}\right) = g^{[i]}\left(\sum_{j=1}^{n_x}\left(w_{ij}^{[1]}\, x_{i,j} + b_i^{[1]}\right)\right)$$

where we have indicated z_i as

$$z_i^{[1]} = \sum_{j=1}^{n_x}\left(w_{ij}^{[1]}\, x_{i,j} + b_i^{[1]}\right)$$

As you can imagine, we want to have a matrix for all the output of layer 1, so we will use this notation

$$Z^{[1]} = W^{[1]}X + b^{[1]}$$

Where $Z^{[1]}$ will have dimensions $n_1 \times 1$, and where with X we have indicated our matrix with all our observations (rows for the features and columns for observations). We assume that all neurons in layer l will use the same activation function, which we will indicate with $g^{[l]}$.

We can easily generalize the previous equation for a layer l:

$$Z^{[l]} = W^{[l]}Z^{[l-1]} + b^{[l]}$$

Since layer l will get its input from layer $l-1$, we just need to substitute X with $Z^{[l-1]}$. $Z^{[l]}$ will have dimensions $n_l \times 1$. Our output in matrix form will then be

$$Y^{[l]} = g^{[l]}\left(Z^{[l]}\right)$$

where the activation function acts, as usual, element by element.

A Short Summary of Matrix Dimensions

Let's summarize the dimensions of all the matrixes we have described so far

- $W^{[l]}$ has dimensions $n_l \times n_{l-1}$ (where we have $n_0 = n_x$ by definition)
- $b^{[l]}$ has dimensions $n_l \times 1$
- $Z^{[l-1]}$ has dimensions $n_{l-1} \times 1$
- $Z^{[l]}$ has dimensions $n_l \times 1$
- $Y^{[l]}$ has dimensions $n_l \times 1$

In each case, l goes from 1 to L.

Example: Equations for a Network with Three Layers

To make all this discussion a bit more concrete, let's consider an example of a network with three layers (so $L = 3$) with $n_1 = 3$, $n_2 = 2$, and $n_3 = 1$, as depicted in Figure 3-3.

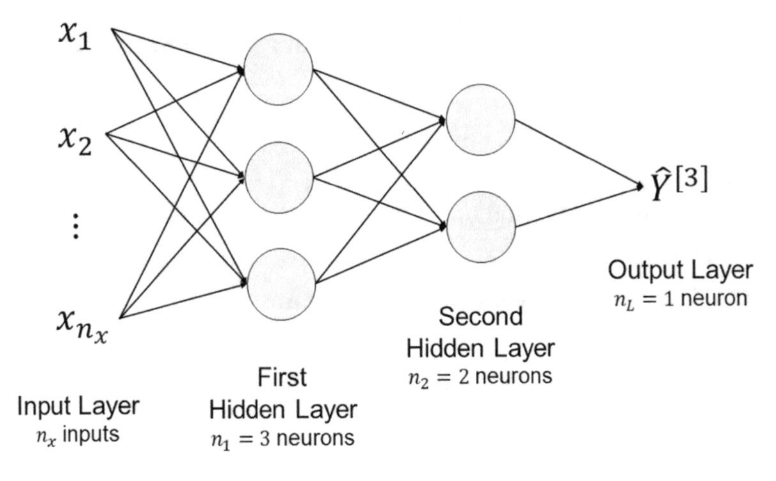

Figure 3-3. *A practical example of a feed-forward neural network*

In this case, we need to calculate the following quantities

- $\hat{Y}^{[1]} = g^{[1]}\left(W^{[1]}X + b^{[1]}\right)$, whereby $W^{[1]}$ has dimensions $3 \times n_x$, b has dimensions 3×1, and X has dimensions $n_x \times m$
- $\hat{Y}^{[2]} = g^{[2]}\left(W^{[2]}\hat{Y}^{[1]} + b^{[2]}\right)$, whereby $W^{[2]}$ has dimensions 2×3, b has dimensions 2×1, and $\hat{Y}^{[1]}$ has dimensions $3 \times m$
- $\hat{Y}^{[3]} = g^{[3]}\left(W^{[3]}\hat{Y}^{[2]} + b^{[3]}\right)$, whereby $W^{[3]}$ has dimensions 1×2, b has dimensions 1×1, and $\hat{Y}^{[2]}$ has dimensions $2 \times m$

The network output $\hat{Y}^{[3]}$ will have, as expected, dimensions $1 \times m$.

All this may seem rather abstract (and in fact it is). You will see later in the chapter how easy it is to implement in Keras simply building the right architecture, based on the steps just discussed.

Hyper-Parameters in Fully Connected Networks

In networks as the ones just discussed, there are quite a few parameters that you can tune to find the best model for your problem.

Note Parameters that you fix at the beginning and then do not change during the training phase are called hyper-parameters (like the number of epochs).

You need to tune the additional following hyper-parameters for feed-forward networks:

- The number of layers, L
- The number of neurons in each layer n_i for i, from 1 to L
- The choice of activation function for each layer $g^{[l]}$

Then of course you still have the following hyper-parameters:

- The number of iterations (or epochs)
- The learning rate

A Short Review of the Softmax Activation Function for Multiclass Classifications

You still need to suffer a bit more theory before getting to some TensorFlow code. These kinds of networks start to be complex enough to be able to perform multiclass classifications with reasonable results. To do this, we must introduce the softmax function.

Mathematically speaking, the softmax function S transforms a k dimensional vector into another k dimensional vector of real values, each between 0 and 1, that sum up to 1. Given k real values z_i for $i = 1, \dots, k$ we define the vector $\boldsymbol{z} = (z_1, \dots, z_k)$ and we define the softmax vector function $\boldsymbol{S}(\boldsymbol{z}) = (S(\boldsymbol{z})_1 \; S(\boldsymbol{z})_2 \dots \; S(\boldsymbol{z})_k)$ as follows

$$S(z)_i = \frac{e^{z_i}}{\sum_{j=1}^{k} e^{z_j}}$$

Since the denominator is always bigger than the nominator $S(\boldsymbol{z})_i < 1$. Additionally, we have

$$\sum_{i=1}^{k} S(z)_i = \sum_{i=1}^{k} \frac{e^{z_i}}{\sum_{j=1}^{k} e^{z_j}} = \frac{\sum_{i=1}^{k} e^{z_i}}{\sum_{j=1}^{k} e^{z_j}} = 1$$

So $S(\boldsymbol{z})_i$ behaves like a probability since its sum over i is 1 and its elements are all less than 1. We will look at $S(\boldsymbol{z})_i$ as a probability distribution over k possible outcomes. For this example, $S(\boldsymbol{z})_i$ will simply be the probability of our input observation of being of class i. Let's suppose we are trying to classify an observation into three classes. We may get the following output: $S(\boldsymbol{z})_1 = 0.1$, $S(\boldsymbol{z})_2 = 0.6$, and $S(\boldsymbol{z})_3 = 0.3$. That means that the observation has a 10% probability of being in class 1, a 60% probability of being in class 2, and a 30% probability of being in class 3. Normally, you would choose to classify the input observation into the class with the higher probability—in this example, class 2 with 60% probability.

Note We will look at $S(\boldsymbol{z})_i$ with $i = 1, \dots, k$ as a probability distribution over k possible outcomes. For this example, $S(\boldsymbol{z})_i$ will simply be the probability of our input observation being in class i.

To be able to use the softmax function for classification, we need to use a specific output layer. We need to use ten neurons (in the case of a ten-class multiclass classification problem, like the one we see later in the chapter), where each will give z_i as its output and then one neuron that will output $\boldsymbol{S}(\boldsymbol{z})$. This neuron will have the softmax function as the activation function and will have as inputs the ten outputs z_i of the last layer with ten neurons. In Keras, you use the `tf.keras.activations.softmax` function applied to the last layer with ten neurons. Remember that this Keras function will act element by element. You will a concrete example from start to finish on how to implement this practically later in this chapter.

A Brief Digression: Overfitting

One of the most common problems that you will encounter when training deep neural networks is *overfitting*. Your network may, due to its flexibility, learn patterns that are due to noise, errors, or simply wrong data. It is very important to understand what overfitting is, so now we will go through a practical example of what can happen, to get an decent understanding of it. To make it easier to visualize, we will work with a simple two-dimensional dataset created for this purpose.

A Practical Example of Overfitting

To understand overfitting, consider the following problem: find the best polynomial that approximates a given dataset. Given a set of two-dimensional points (x_i, y_i), we want to find the best polynomial of degree K in the form[1]

$$f(x_i) = \sum_{j=0}^{K} a_j x_i$$

That minimizes the mean square error

$$\frac{1}{m}\sum_{i=1}^{m}\left(y_i - f(x_i)\right)^2$$

[1] Note that in this example the input is one-dimensional, and therefore the only index we need is the one identifying the observation. The number of features is 1 in this example.

where, as usual, m indicates the number of data points we have. We do not only want to determine all the parameters a_j but also the value of K that best approximates the data. K in this case measures our model complexity. For example, for $K = 0$ we simply have $f(x_i) = a_0$ (a constant), the simplest polynomial we can think of. For higher K, we have higher order polynomials, meaning our function is more complex and has more parameters available for training. Here is an example of the function for $K = 3$

$$f(x_i) = \sum_{j=0}^{3} a_j x_i^j = a_0 + a_1 x_i + a_2 x_i^2 + a_3 x_i^3$$

Where we have four parameters that can be tuned during the model's training. Let's generate some data, starting from a second-order polynomial ($K = 2$)

$$1 + 2x_i + 3x_i^2$$

We are adding some random error (this will make overfitting visible). Let's first import our standard libraries with the addition of the `curve_fit` function, which will automatically minimize the standard error and find the best parameters. Do not worry too much about this function; the goal here is to show you what can happen when you use a model that is too complex.

```
import numpy as np
import matplotlib.pyplot as plt
from scipy.optimize import curve_fit
```

Let's define a function for a second-degree polynomial

```
def func_2(p, a, b, c):
    return a + b*p + c*p**2
```

Then let's generate the dataset

```
x = np.arange(-5.0, 5.0, 0.05, dtype = np.float64)
y = func_2(x, 1, 2, 3) + 18.0 * np.random.normal(0, 1, size = len(x))
```

To add some random noise to the function, we used the `np.random.normal(0, 1, size = len(x))` function, which generates a NumPy array of random values from a normal distribution of length `len(x)` with average 0 and standard deviation 1.

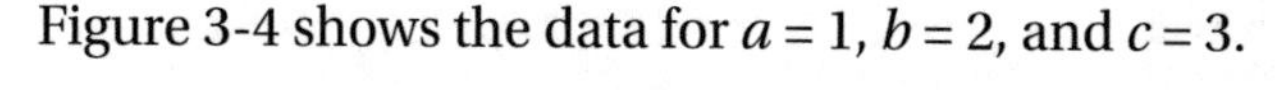

Figure 3-4 shows the data for $a = 1$, $b = 2$, and $c = 3$.

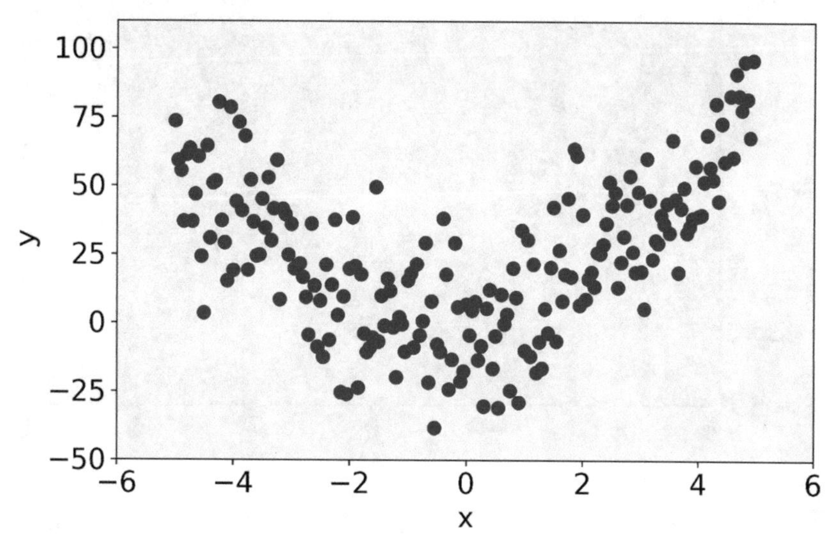

Figure 3-4. *The data we generated with a = 1, b = 2, and c = 3*

Now let's consider a model that is too simple to capture the feature of the data, meaning we will see what a model with high bias[2] can do. Consider a linear model ($K = 1$). The code will be

```
def func_1(p, a, b):
    return a + b*p
popt, pcov = curve_fit(func_1, x, y)
```

That will give the best values for a and b that minimize the standard error. In Figure 3-5, it's clear how this model completely misses the main feature of the data, being too simple.

[2] Bias is a measure of the error originating from models that are too simple to capture the real features of the data.

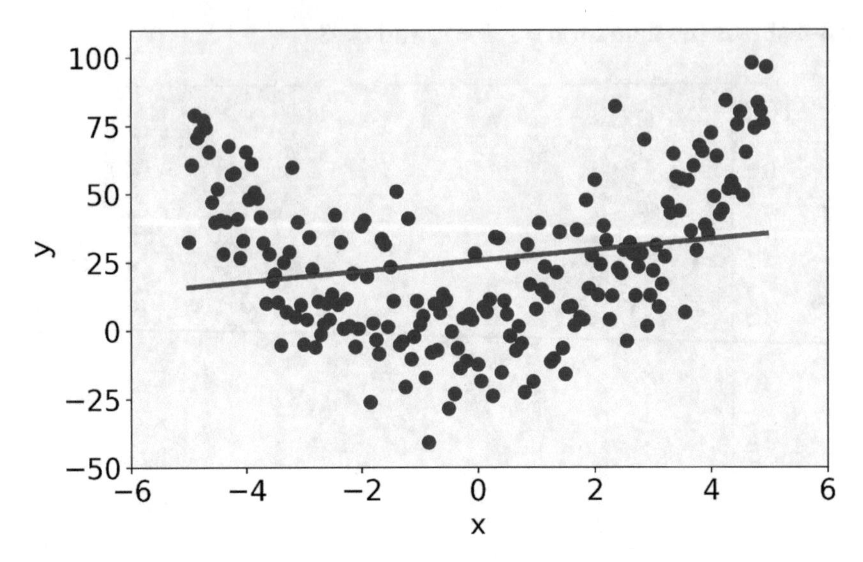

***Figure 3-5.** The linear model misses the main feature of the data being too simple. In this case, the model has high bias*

Let's try to fit a two-degree polynomial ($K = 2$). The results are shown in Figure 3-6.

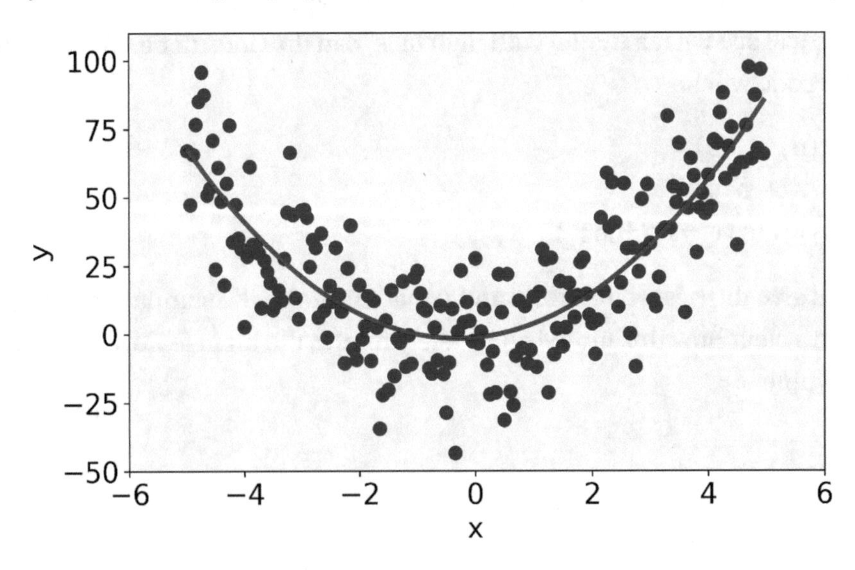

***Figure 3-6.** The result (red line) for a two-degree polynomial*

That is better. This model seems to capture the main features of the data, ignoring the random noise. Now let's try a very complex model, a 21-degree polynomial ($K = 21$). The results are shown in Figure 3-7.

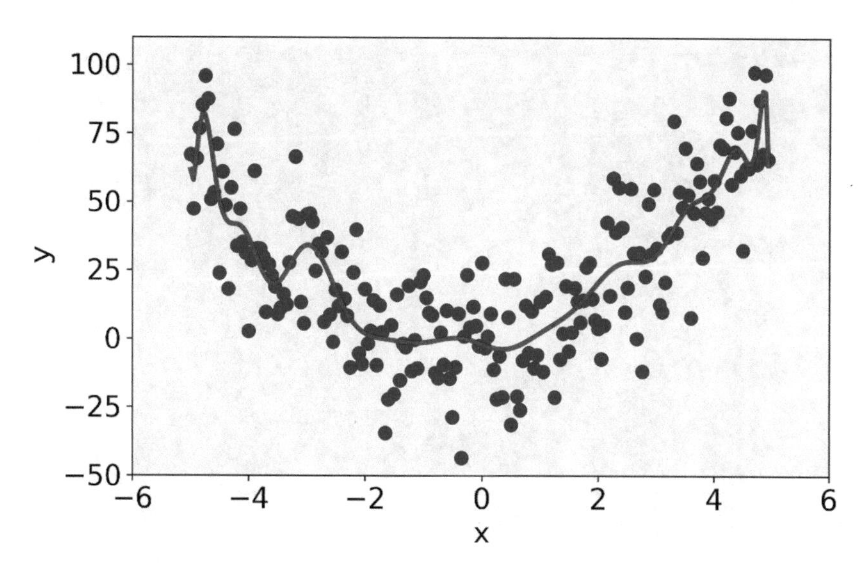

Figure 3-7. *The results for a 21-degree polynomial model*

This model shows features that we know are wrong (since we created our data). Those features are not present, but the model is so flexible that it captures the random variability that we introduced with noise. The oscillations that have appeared using this high-ordered polynomial are wrong and do not describe the data correctly.

In this case, we talk about *overfitting*, meaning we start capturing with our model features that are due to random error, for example. It is easy to understand that this generalizes quite badly. If we applied this 21-degree polynomial model to new data it would not work well, since the random noise would be different in the new data and so the oscillations (the ones represented in Figure 3-7) would make no sense.

Figure 3-8 shows the best 21-degree polynomial models obtained by fitting data generated with ten different random noise values . You can clearly see how much it varies. It is not stable and is strongly dependent on the random noise. The oscillations are always different! In this case, we talk about high variance.

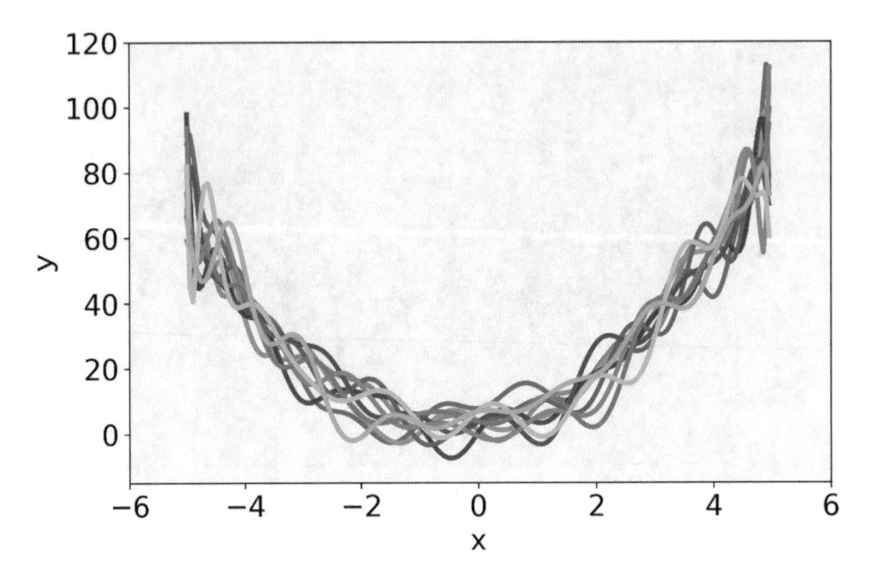

Figure 3-8. *The result of our model with a 21-degree polynomial fitted to ten different datasets generated with different random noise values*

Now let's run the same plot with this linear model, while varying the random noise as we did in Figure 3-8. You can see the results in Figure 3-9.

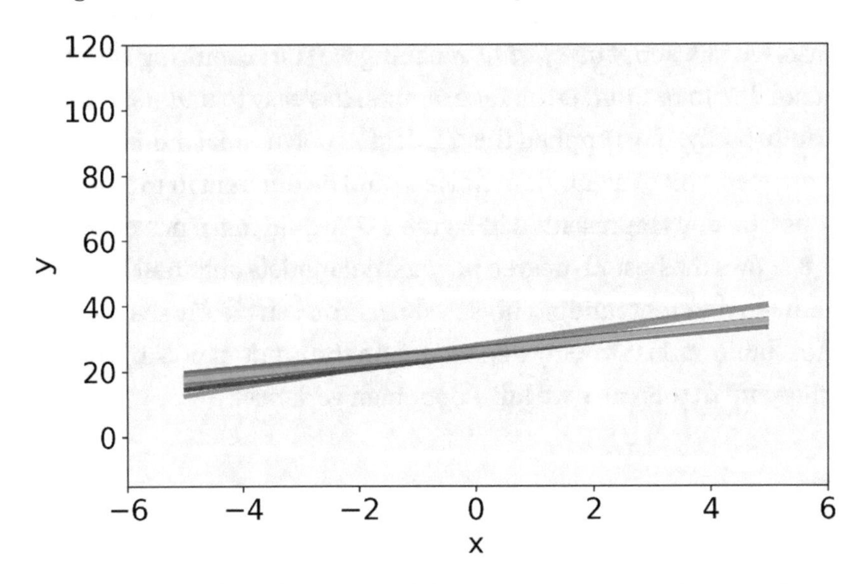

Figure 3-9. *The result of the linear model applied to data where we have randomly changed the random noise. For easier comparison with Figure 3-8, we used the same scale*

You can see that the model is much more stable. The linear model does not capture features that are dependent from the noise, but it misses the main features of the data (the concave nature). We talk here of high bias.

Figure 3-10 illustrates the concepts of bias and variance.

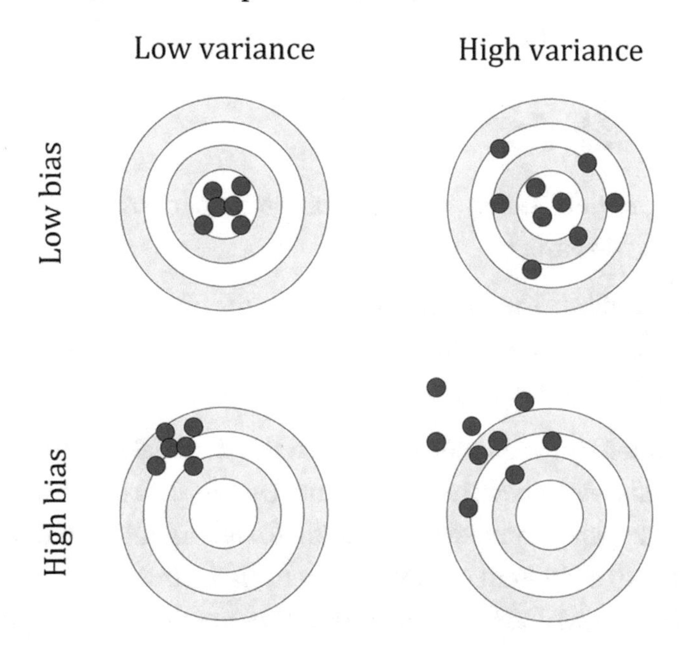

Figure 3-10. *Bias is a measure of how close the measurements are to the true values (the center in the figure) and variance is a measure of how spread the measurements are around the average (not necessarily the true value, as you can see on the left)*

In the case of neural networks, we have many hyper-parameters (number of layers, number of neurons in each layer, activation function, and so on) and it is very difficult to know in which regime we are. How can we tell if our model has a high variance or a high bias, for example? An entire chapter is dedicated to this subject, but the first step to performing this error analysis is to split the dataset in two different subsets. Let's see what that means and why we have to do it.

Note The essence of *overfitting* is to have unknowingly extracted some of the residual variation (i.e., the noise) as if that variation represented underlying model structure [1]. The opposite is called *underfitting,* when the model cannot capture the structure of the data.

The problem with overfitting and deep neural networks is that there is no way of easily visualizing the results. Therefore we need a different approach to determine if our model is overfitting, underfitting, or is just right. This can be achieved by splitting the dataset into different parts and comparing the metrics of the parts. You learn about the basic idea in the next section.

Basic Error Analysis

To check how your model is doing and to perform a proper error analysis, you need to split your dataset in two parts[3]:

- Training dataset: The model is trained on this dataset using the inputs and the relative labels and using an optimizer algorithm like gradient descent. Often this is called the training set.
- Development (or validation) set: The trained model will then be used on this dataset to check how it is doing. On this dataset we will test different hyper-parameters. For example, we can train two different models with a different number of layers on the training dataset and test them on this dataset to check how they are doing. Often this is called the dev set.

There is an entire chapter dedicated to error analysis, but it is a good idea to read an overview of why it is important to split the dataset. Let's suppose we are dealing with classification and the metric we use to judge how good our model is the accuracy minus one, or in other words the percentage of the cases that are wrongly classified.

Table 3-1. *Examples of the Difference Between Models with High Bias and Models with High Variance*

Error	Case A	Case B	Case C	Case D
Train set error	1%	15%	14%	0.3%
Dev set error	11%	16%	32%	1.1%

[3] To do a proper error analysis, you need at least three parts, sometimes four. But to get a basic understanding of the process, two parts are enough.

Let's consider the three cases reported in Table 3-1:

- **Case A**: We are overfitting (high variance), because we are doing very well on the training set, but our model generalizes very badly to our dev set (see Figure 3-8 again).
- **Case B**: We see a problem with high bias, meaning that our model is not doing very well generally, on both datasets (see Figure 3-9 again).
- **Case C**: We have a high bias (the model cannot predict the training set very well) and high variance (the model does not generalize on the dev set very well).
- **Case D**: Everything seems okay. There is good error on the training set and good data on the dev set. It is a good candidate for our best model.

We will explain much better all those concepts later in the book, where we provide recipes on how to solve problems of high bias, high variances, both, and even more complex cases.

To recap, to do a very basic error analysis, you need to split your dataset into at least two sets: train and dev. You should then calculate your metric on both sets and compare them. You want to have a model that has low error on the train set, low error on the dev set (as with Case D in the previous example), and the two values should be comparable.

Note Your main take away from this section should be two-fold: **1)** you need a set of recipes and guidelines for understanding how your model is doing: is it overfitting, underfitting, or is it just right? **2)** to answer question 1 and perform the analysis, you need to split the dataset into two parts (later in the book, you will also see what you can do with the dataset split in three or even four parts).

Implementing a Feed-Forward Neural Network in Keras

Building a feed-forward neural network in Keras is straightforward and is simply a generalization of the one neuron model you built in last chapter. Let's compare the two cases. Here you have the schematic one-neuron model Keras implementation

```
model = keras.Sequential([
   layers.Dense(1, input_shape = [...])
])
```

The following is a feed-forward network model with one hidden layer made of 15 neurons (the first model we will use for our multiclass classification task on the Zalando dataset):

```
model = keras.Sequential([
   layers.Dense(15, input_shape = [...])
   layers.Dense(10)
])
```

As you can see, we added more neurons (15) to the hidden layer (that in the one-neuron model was already the output one) and we added the output layer, made of ten neurons, since we have ten classes. As you can notice, you can easily create very complex models by simply adding to the stack one layer after another.

In the next paragraphs, you will see a practical example about how you use this model, choosing the right activation function and the right loss function (given as additional parameters) to solve a multiclass classification task.

Multiclass Classification with Feed-Forward Neural Networks

The task we are going to solve together is a multiclass classification problem on the Zalando dataset. It consists of predicting the corresponding label among ten possible cases (ten different types of clothing). To solve it, we will use a feed-forward network architecture and try different configurations (different optimizers and architectures), performing some error analysis to see which situation is better. Let's start by looking at the data.

The Zalando Dataset for the Real-World Example

Zalando SE is a German commerce company based in Berlin. The company maintains a cross-platform store that sells shoes, clothing, and other fashion items [2]. For a Kaggle competition, they prepared a MNIST-similar dataset of Zalando's clothing article images [4], where they provided 60000 training images and 10000 test images. (If you do not know what Kaggle is, check out their website [3], where you can participate in many competitions where the goal is to solve problems with data science.) As in MNIST, each image is 28x28 pixels in grayscale. They classified all images in ten different classes and provided the labels for each image. The dataset has 785 columns, where the first column is the class label (an integer going from 0 to 9) and the remaining 784 contain the pixel gray value of the image (you can calculate that 28x28=784).

Each training and test example is assigned to one of the following labels (as from the documentation):

- 0: T-shirt/top
- 1: Trouser
- 2: Pullover
- 3: Dress
- 4: Coat
- 5: Sandal
- 6: Shirt
- 7: Sneaker
- 8: Bag
- 9: Ankle boot

Figure 3-11 shows an example of each class chosen randomly from the dataset.

Figure 3-11. *One example from each of the ten classes in the Zalando dataset*

The dataset has been provided under the MIT License[4]. The datafile can be downloaded from Kaggle (`https://www.kaggle.com/zalando-research/fashionmnist/data`) or directly from GitHub (`https://github.com/zalandoresearch/fashion-mnist`). If you choose the second option, you need to prepare the data a bit

[4] `https://en.wikipedia.org/wiki/MIT_License`

(you can convert it to CSV with the script located at `https://pjreddie.com/projects/mnist-in-csv/`). If you download it from Kaggle, you have all the data in the right format. You will find two CSV files zipped on the Kaggle website. The ZIP file contains `fashion-mnist_train.csv`, with 60000 images (roughly 130 MB) and `fashion-mnist_test.csv`, with 10000 (roughly 21 MB).

In our example, we will retrieve the dataset in a third way: from the TensorFlow datasets catalog (`https://www.tensorflow.org/datasets/catalog/fashion_mnist`), since in this way we will not have to perform any preprocessing steps and we will automatically import the data inside our notebook. Now, let's code!

We will need the following imports in our code

```
# general libraries
import pandas as pd
import numpy as np
import matplotlib
import matplotlib.pyplot as plt
import matplotlib.font_manager as fm
from random import *
import time

# tensorflow libraries
from tensorflow.keras.datasets import fashion_mnist
import tensorflow as tf
from tensorflow import keras
from tensorflow.keras import layers
import tensorflow_docs as tfdocs
import tensorflow_docs.modeling
```

Then, to retrieve the dataset, we can simply run the following command

```
((trainX, trainY), (testX, testY)) = fashion_mnist.load_data()
```

Incredibly easy! Now we have two numpy matrices (`trainX` and `testX`) containing all the pixel values describing each of the training and test images and two NumPy arrays (`trainY` and `testY`) containing the associated labels.

Let's print the datasets' dimensions

```
print('Dimensions of the training dataset: ', trainX.shape)
print('Dimensions of the test dataset: ', testX.shape)
print('Dimensions of the training labels: ', trainY.shape)
print('Dimensions of the test labels: ', testY.shape)
```

which will return as output

```
Dimensions of the training dataset:  (60000, 28, 28)
Dimensions of the test dataset:  (10000, 28, 28)
Dimensions of the training labels:  (60000,)
Dimensions of the test labels:  (10000,)
```

Tip Remember that you should not focus on the Python implementation. Focus on the model, on the concepts behind the implementation. You can achieve the same results using pandas, NumPy, or even C. Try to concentrate on how to prepare the data, how to normalize it, how to check the training, and so on.

As you can see, we have a training dataset made of 60000 items, stored as images of 28x28 pixels each, and a test dataset made of 10000 items, stored in the same way. Then, an array of corresponding labels is associated with each dataset.

Now we need to modify our data to obtain a "flattened" version of image, meaning an array of 754 pixels, instead of a matrix of 28x28 pixels. This step is necessary, because, as we saw when we discussed the feed-forward network architecture, it receives as inputs all the features as separate values. Therefore, we need to have all the pixels stored in the same array. On the contrary, Convolutional Neural Networks (CNNs) do not work with flattened versions of the images, but that's a topic for Chapter 7. For now, keep this in mind.

The following lines reshape the matrix dimensions

```
data_train = trainX.reshape(60000, 784)
data_test = testX.reshape(10000, 784)
```

Let's summarize our data so far

- `labels`: Has dimensions (60000) and contains the class labels (integers from 0 to 9)
- `train`: Has dimensions $m \times n_x$ (60000x784) and contains the features, where each column contains the grayscale value of a single pixel in the image (remember 28x28=784)

See Figure 3-11 again to get an idea of how the images look. Finally let's normalize the input so that instead of having values from 0 to 255 (the grayscale values), it has only values between 0 and 1. This is very easy to do with the code

```
data_train_norm = np.array(data_train/255.0)
data_test_norm = np.array(data_test/255.0)
```

Before developing the network, we need to solve another problem. Labels must be provided in a different form when performing a multiclass classification task.

Modifying Labels for the Softmax Function: One-Hot Encoding

In classification, we use the following cost function (called cross-entropy)

$$L(\hat{y}_i, y_i) = -(y_i \log \hat{y}_i + (1 - y_i) \log(1 - \hat{y}_i))$$

where y_i contains our labels and $\hat{y}_i$ is the result of our network. So, the two elements must have the same dimensions. In our case we saw that our network will give as output a vector with ten elements, while a label in our dataset is simply a scalar. So, we have $\hat{y}_i$ that has dimensions (10,1) and y_i that has dimensions (1,1). This will not work if we do not do something smart. We need to transform our labels in a vector that have dimensions (10,1). A vector with a value for each class, but what value should we use?

What we need to do is what is called one-hot encoding[5]. Meaning we will transform our labels (integers from 0 to 9) to vectors with dimensions (1,10) with this algorithm: our one-hot encoded vector will have all zeros, except at the index of the label.

[5] As a side note, this technique is often used to feed categorical variables into machine learning algorithms.

For example, for a label 2 our 1x10 vector will have all zeros except at position with index 2, or in other words, it will be (0,0,1,0,0,0,0,0,0,0). Let's look at some other examples (see Table 3-2).

Table 3-2. *Examples of How One-Hot Encoding Works. Remember that Labels Go from 0 to 9, as Indexes*

Label	One-Hot Encoded Label
0	(1,0,0,0,0,0,0,0,0,0)
2	(0,0,1,0,0,0,0,0,0,0)
5	(0,0,0,0,0,1,0,0,0,0)
7	(0,0,0,0,0,0,0,1,0,0)

Figure 3-12 shows a graphical representation of the process of one-hot encoding a label.

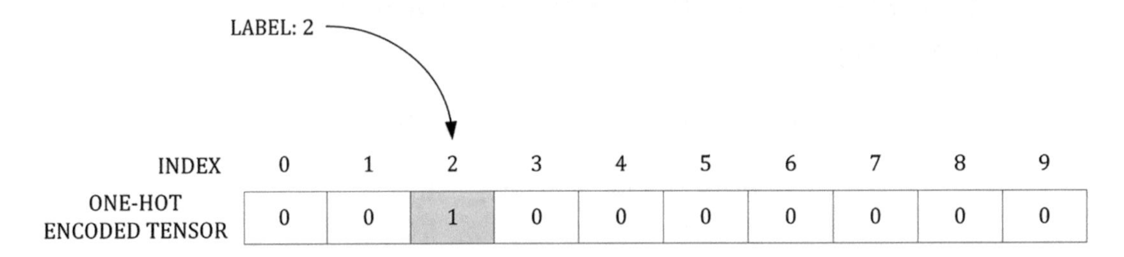

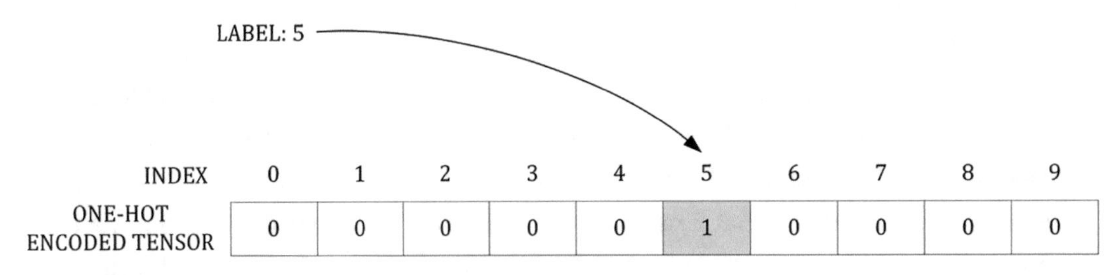

Figure 3-12. *A graphical representation of the process of one-hot encoding a label. Two labels (2 and 5) are one-hot encoded in two vectors. The grayed element of the vector is the one the becomes 1, while the white marked ones remain 0*

Sklearn has several ways of doing this automatically (check for example the function `OneHotEncoder()`). But I think it's instructive to do it manually to really see how it's done. Once you understand why you need it, and in which format you need it, you can use the function you like best. The Python code to do this is very simple:

```
labels_train = np.zeros((60000, 10))
labels_train[np.arange(60000), trainY] = 1

labels_test = np.zeros((10000, 10))
labels_test[np.arange(10000), testY] = 1
```

First you create a new array with the right dimensions: (60000,10), and then you fill it with zeros with the NumPy function `np.zeros((60000,10))`. Then you set to 1 only the columns related to the label itself, using pandas functionalities to slice dataframes with the line `labels_train[np.arange(60000), trainY] = 1` (the same of course is also performed in the case of the test dataset). In the end, you obtain the dimensions (60000,10), where each row indicates a different observation.

Now we can compare y_i and $\hat{y}_i$, since both have the dimensions (10,1) for one observation, or when considering the entire test dataset of (10000,10). The same can of course be asserted for the training dataset. Each column in $\hat{y}_i$ will represent the probability of our observation being of a specific class. At the very end when calculating the accuracy of our model, we will assign the class with the highest probability to each observation.

Note Our network will give us the ten probabilities for the observation of being of each of the ten classes. At the end, we will assign to the observation the class that has the highest probability.

The Feed-Forward Network Model

We will start with a network with just one hidden layer. We will have an input layer with 784 features, then a hidden layer (where we will vary the number of neurons), then an output layer of ten neurons that will feed their output into a neuron that will have as an activation function the softmax function. Figure 3-13 shows a graphical representation of the network. We will spend some time looking at the various parts, especially the output layers.

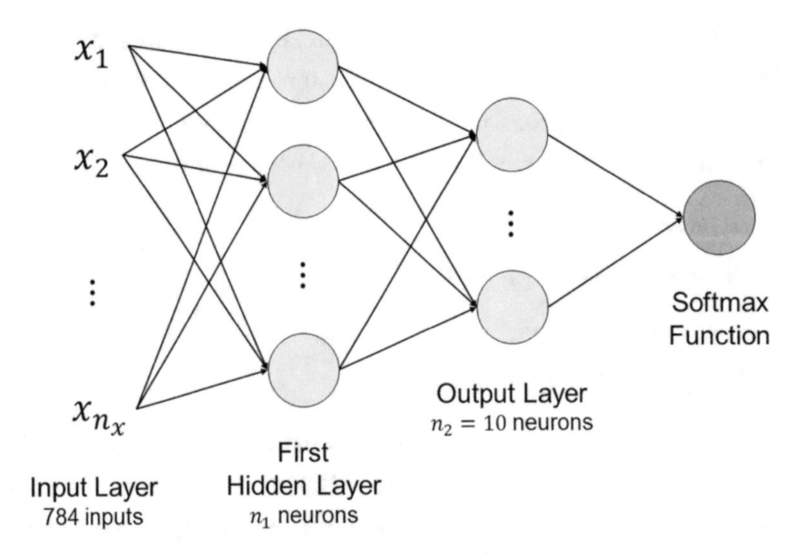

Figure 3-13. *The network architecture with a single hidden layer. We will vary the number of neurons n_1 in the hidden layer during our analysis*

Let's look at why this strange output layer has ten neurons and why we need an additional neuron for the softmax function. Remember, we want to be able to tell which class each image belongs to. To do this, as explained when discussing the softmax function, we need to get ten outputs for each observation: each being the probability of the image being in each of the classes. So given input $\boldsymbol{x}^{(i)}$, we need ten values: $P(y_i = 1|x_i)$, $P(y_i = 2|x_i)$, ..., $P(y_i = 10|x_i)$ (the probability of the observation class $y^{(i)}$ being one of the ten possibilities given the input x_i). Or, in other words, our output should be a vector of dimensions 1x10 in this form

$$\hat{\boldsymbol{y}} = \left(P(y_i = 1|x_i) \quad P(y_i = 2|x_i) \quad \dots \quad P(y_i = 10|x_i) \right)$$

And since the observation must be of one single class, this condition must be satisfied

$$\sum_{j=1}^{10} P(y_i = j|x_i) = 1$$

This can be understood as: the observation has a 100% probability of being one of the ten classes. Or, in other words, all the probabilities must add up to 1. We solve this problem in two steps:

- We create an output layer with ten neurons, in this way we will have our ten values as output.
- Then we feed the ten values into a new neuron (let's call it "softmax" neuron) that will take the ten inputs and give as output ten values that are all less than one, which adds to 1.

Calling z_i the output of the ith neuron in the last layer (with i going from 1 to 10), we will have

$$P(y_i = j|x_i) = \frac{e^{z_i}}{\sum_{j=1}^{10} e^{z_j}}$$

In Keras, this is straightforward. But it is instructive to know exactly what each line of code does. That is what the Keras function `model.add(Dense(10, activation = 'softmax'))` does. It takes a vector as the input and returns a vector with the same dimensions as the input, but "normalized," as discussed above. In other words, if we feed $\boldsymbol{z} = (z_1 \ \ z_2 \ \ \dots \ \ z_{10})$ into the function, it will return a vector with the same dimensions as $\boldsymbol{z}$, meaning 1x10, where each element added to the others gives 1.

Keras Implementation

Now is time to build our model with Keras. The following code will do the job

```
def build_model(opt):
  # create model
  model = Sequential()
  # add first hidden layer and set input dimensions
  model.add(Dense(15, input_dim = 784, activation = 'relu'))
  # add output layer
  model.add(Dense(10, activation = 'softmax'))
  # compile model
  model.compile(loss = 'categorical_crossentropy',
                optimizer = opt,
                metrics = ['categorical_accuracy'])
  return model
```

Now we will not go through each line of code, since you should understand by now how a basic Keras model is built (remember our simple one-neuron model in Chapter 2). But there are a few details of the code that need to be stressed:

- Our last layer will use the softmax function: `model.add(Dense(10, activation = 'softmax'))`.
- The two parameters—15 (n_1) and 10 (n_2)—define the number of neurons in the different layers. Remember the second (output) layer must have ten neurons to be able to use the softmax function. But we will play with the value of n_1. Increasing n_1 will increase the complexity of the network.
- We set the categorical cross-entropy (`loss = 'categorical_crossentropy'`) as the loss function and the categorical accuracy (`metrics = ['categorical_accuracy']`) as the metrics. The reason for this choice is that we have hot-encoded the labels and therefore the categorical versions of these functions are needed.

Now let's try to perform the training as we did in last chapter for the single-neuron model. The code structure is always the same. Try to run the following code on your laptop:

```
model = build_model(tf.keras.optimizers.SGD(momentum = 0.0, learning_rate = 0.01))

EPOCHS = 1000

history = model.fit(
  data_train_norm, labels_train,
  epochs = EPOCHS, verbose = 0,
  batch_size = data_train_norm.shape[0]
)
```

We have set as the optimizer the standard version of the gradient descent. The biggest problem is that the model, as we coded it, will create a huge matrix for all observations (that are 60,000) and then will modify the weights and bias only after a complete sweep over all observations. This requires a lot of resources, memory, and CPU. If that was the only choice we have, we would have been doomed. Keep in mind that in the deep learning world, 60,000 examples of 784 features is not a big dataset at all. We need to find a way of letting this model learn faster.

Moreover, notice that, when training the model, we have set `batch_size = data_train_norm.shape[0]` inside the Keras `fit` method. Keras by default sets the batch size to 32 observations [5], but the batch gradient descent updates weights and biases after all training observations have been seen by the network. Therefore, we need to change this parameter to obtain the basic version of gradient descent.

In the same way, we need to set the `momentum = 0.0` inside the method `tf.keras.optimizers.SGD`. To summarize, since Keras does not include a function to perform the standard gradient descent, we used the stochastic gradient descent function, setting the momentum to zero and the batch size to the entire number of observations.

To do some basic error analysis, you also need the dev dataset, which we loaded and prepared in the previous paragraphs.

Do not get confused by the fact that the filename contains the word test. Sometimes the dev dataset is called the test dataset. When we discuss error analysis later in the book, we use three datasets: train, dev, and test. To remain consistent in the book, we use the name *dev* here as well, not to confuse you with different names in different chapters.

To calculate accuracy on the dev dataset, we use the `model.evaluate()` function and apply the built model on the dev dataset.

```
test_loss, test_accuracy = model.evaluate(data_test_norm, labels_test,
verbose = 0)
print('The accuracy on the test set is equal to: ', int(test_
accuracy*100), '%.')
```

which returns

```
The accuracy on the test set is equal to:  74 %.
```

To recap, we applied the model trained on the 60,000 observations to the dev test (made up of 10,000 observations) and then calculated the accuracy on both datasets.

A good exercise is to include this calculation in your model so that your `build_model()` function automatically returns the two values.

Gradient Descent Variations Performances

In Chapter 2, we looked at the different GD variations and discussed their advantages and disadvantages. Let's know see how they differ in a practical case.

Comparing the Variations

Let's summarize the findings for the three variations of gradient descent for 100 epochs (see Table 3-3).

Table 3-3. *Comparing the Performances of Three Variations of Gradient Descent*

Gradient Descent Variation	Running Time	Final Value of Cost Function	Accuracy
Batch gradient descent	0.35 min	1.86	43%
Stochastic gradient descent	60.23 min	0.26	91%
Mini-batch gradient descent (mini-batch size = 50)	1.70 min	0.26	90%

Now you can see that mini-batch gradient descent is definitely the best compromise in terms of execution time and classification performance. Now you should be convinced that it is currently the preferred method to be used as an optimizer in deep neural networks, among the different gradient descent types, since it can reach high performance, maintaining a good trade-off between performance and execution time.

Figure 3-14 shows how the cost function decreases with different mini-batch sizes. Note how, with respect to number of epochs, a smaller mini-batch size means a faster (not in time) decrease.

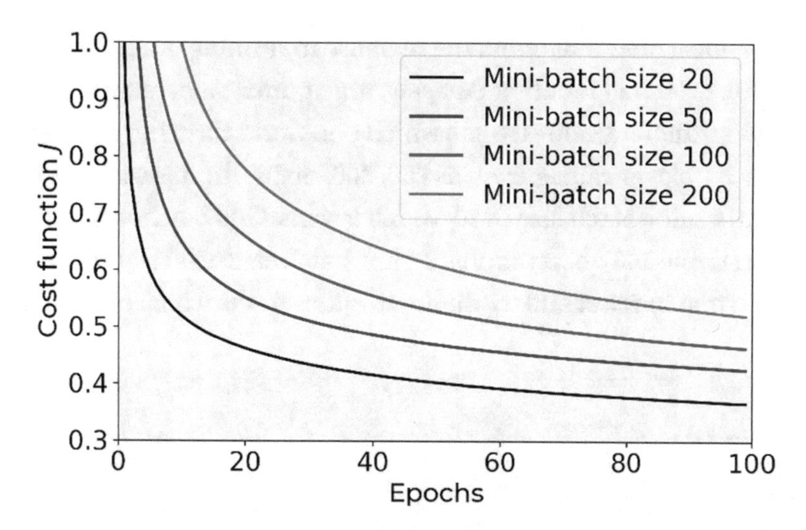

Figure 3-14. *A comparison of speed of convergence of the mini-batch gradient descent algorithm with different mini-batch sizes. The learning rate used for this figure was $\gamma = 0.0001$. Note that the time needed by each case is not the same. The smaller the mini-batch size, the more time the algorithm needs*

Tip The best compromise between running time and convergence speed (with respect to the number of epochs) is achieved using the mini-batch gradient descent. The optimal size of the mini-batches is dependent on your problem, but small numbers like 30 or 50 are a good place to start. You will find a compromise between running time and convergence speed.

To get an idea of how the running time depends on what value the cost function can reach after 100 epochs, see Figure 3-15 (in comparison to Chapter 2, the times are evaluated with a real dataset). Each point is labeled with the size of the mini-batch used in that run. You can see that decreasing the mini-batch size decreases the value of J after 100 epochs. This happens quickly, without increasing the running time significantly, until you arrive at a value for the mini-batch size that is around 20. At that point, the time starts to increase quickly and the value of J after 100 epochs does not decrease and flattens out.

The best compromise is to choose a value for the mini-batch size where the curve is closer to zero (a small running time and a small cost function value), and that is at a mini-batch size value of 20 in this specific case. This is the reason that those are the most common values. After that point, the increase in running time becomes very quick and is not helpful. Note that for other datasets, the optimal value may be very different. So it's worth trying different values to see which one works best. In very big datasets you may want to try bigger values, such as 200, 300, or 500. In this case, we have 60,000 observations and a mini-batch size of 50, which means 1200 batches. If you have much more data, for example 1e6 observations, a mini-batch size of 50 would give 20,000 batches. So, keep that in mind and try different values to see which one works best.

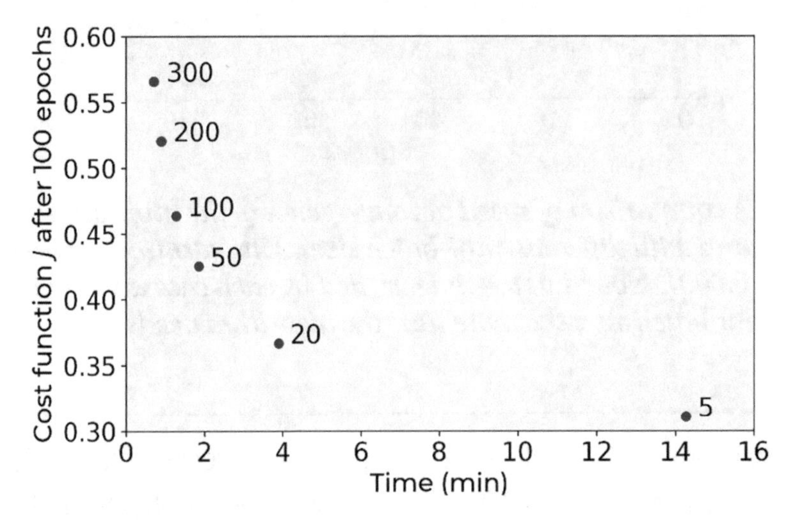

Figure 3-15. *The plot shows the value of the cost function after 100 epochs for the Zalando dataset vs. the running time needed to run through 100 epochs. Note that the points are single runs, and the plot is only indicative of the dependency. Running time and cost function have a small variance when evaluated over several runs. This variance is not shown in the plot*

As a tip, it is good programming practice to write a function that runs your evaluations. This way, you can tune your hyper-parameters (like the mini-batch size) without copy and pasting the same chunk of code over and over. The following function is one that you can use to train our model with different mini-batch sizes (of course, you can add more parameters to be optimized, such as the number of epochs, the learning rate, etc.):

```
def mini_batch_gradient_descent(mb_size):
  # build model
  model_mbgd = build_model(tf.keras.optimizers.SGD(momentum = 0.9,
  learning_rate = 0.0001))
```

```
# set number of epochs
EPOCHS = 100
# train model
history_mbgd = model_mbgd.fit(
  data_train_norm, labels_train,
  epochs = EPOCHS, verbose = 0,
  batch_size = mb_size,
  callbacks = [tfdocs.modeling.EpochDots()])
# save performances
hist_mbgd = pd.DataFrame(history_mbgd.history)
hist_mbgd['epoch'] = history_mbgd.epoch
return hist_mbgd
```

Tip Writing a function with the hyper-parameters as inputs is common practice. This allows you to test different models with different values for the hyper-parameters and check which ones are better.

Examples of Wrong Predictions

Running the model with batch gradient descent—with one hidden layer with 15 neurons for 1000 epochs and at a learning rate of 0.0001—will give us an accuracy on the training set of 86%. You can increase the accuracy by using more neurons in your hidden layer. For example, using 50 neurons, 1000 epochs, and a learning rate of 0.0001 will allow you to reach 87% on the training set and 85% on the test set. It is interesting to check a few examples of incorrectly classified images, to see if you can learn something from these errors. Figure 3-16 shows an example of incorrectly classified images for each class.

Figure 3-16. *One example of incorrectly classified images for each class. Over each image, the True class (labeled with "True") and the predicted (labeled with "Pred") class are reported. This model has one hidden layer with 15 neurons, has been run for 1000 epochs with a learning rate of 0.0001*

Some errors are understandable, such as where a coat was wrongly classified as a pullover. We could easily make the same mistake. The wrongly classified dress is, on the other hand, easy for a human to get right.

Weight Initialization

If you tried to run the code, you will have realized that the convergence of the algorithm is strongly variable, and it depends on the way you initialize your weights. In the previous sections, we focused on understanding how such a network works to not get distracted by additional information, but it is time to look at this problem a bit more closely, since it plays a fundamental role in many layers.

Basically, we want to avoid the gradient descent algorithm to explode and start returning nan. For example, in the first layer for the ith neuron, we need to calculate the ReLU activation function of the quantity (see the beginning of this chapter for an explanation if you forget why):

$$z_i = \sum_{j=1}^{n_x} \left(w_{ij}^{[1]} x_{i,j} + b_i^{[1]} \right)$$

Normally, in a deep network, the number of weights is quite big, so you can easily imagine that if the $w_{ij}^{[1]}$ are big, the quantity z_i can be quite big, and then the ReLU activation function can return a nan value since the argument is too big for Python to calculate it properly (remember that in a classification problems, you have a log() function and therefore values of the argument as zero for example are not acceptable). So, you want the z_i to be small enough to avoid an explosion of the output of the neurons, and big enough to avoid having the outputs die out and making the convergence a very slow process.

The problem has been researched extensively [6], and there are different initialization strategies depending on the activation function you are using. Let's outline a few of them in Table 3-4, where we assume that the weights will be initialized with a normal distribution with a mean of 0 and a standard deviation as given in the table (note that the standard deviation will depend on the activation function you want to use).

Table 3-4. *Some Examples of Weight Initialization for Deep Neural Networks*

Activation Function	Standard Deviation σ for a Given Layer	
Sigmoid	$\sigma = \sqrt{\frac{2}{n_{inputs} + n_{outputs}}}$	Usually called Xavier initialization
ReLU	$\sigma = \sqrt{\frac{4}{n_{inputs} + n_{outputs}}}$	Usually called He initialization

In a layer l the number of inputs will be the number of neurons of the preceding layer $l-1$ and the number of outputs will be the number of neurons in the layer coming next $l+1$. So, we will have

$$n_{inputs} = n_{l-1}$$

And

$$n_{outputs} = n_{l+1}$$

Very often, deep networks, as the one we discussed before, will have several layers all with the same number of neurons. Therefore you have for most of the layers $n_{l-1} = n_{l+1}$ and therefore you will have for Xavier initialization

$$\sigma_{Xavier} = \sqrt{1/n_{l+1}} \quad or \quad \sqrt{1/n_{l-1}}$$

And for ReLU activation functions the He initialization will be

$$\sigma_{He} = \sqrt{2/n_{l+1}} \quad or \quad \sqrt{2/n_{l-1}}$$

Let's consider the ReLU activation function (the one we used in this chapter). Every layer, as we have discussed, will have n_l neurons. A way of initializing the weights for our single hidden layer for example would then be

```
initializer = tf.keras.initializers.HeNormal()
layer = tf.keras.layers.Dense(15, kernel_initializer = initializer)
```

Typically, to more easily evaluate and construct the networks, the most typical initialization form used is for ReLU activation function

$$\sigma_{He} = \sqrt{2/n_{l-1}}$$

And

$$\sigma_{Xavier} = \sqrt{1/n_{l-1}}$$

For Sigmoid activation function.

Using this initialization can speed up training considerably and is the standard way that many libraries initialize weights (for example, the Caffe library).

In Keras, weight initialization is straightforward by means of the `tf.keras.initializers` function. Look at the Keras documentation to see which initialization strategies are available [7].

Adding Many Layers Efficiently

Typing all this code each time is a bit tedious and error prone. You can instead define a function that creates a layer. That can be done easily with this code

```
def model_nlayers(num_neurons, num_layers):
    # build model
    inputs = keras.Input(shape = 784) # input layer
    # first hidden layer
    dense = layers.Dense(num_neurons,
                         activation = 'relu')(inputs)
    # customized number of layers and neurons per layer
    for i in range(num_layers - 1):
        dense = layers.Dense(num_neurons,
                             activation = 'relu')(dense)
    # output layer
    outputs = layers.Dense(10, activation = 'softmax')(dense)
    model = keras.Model(inputs = inputs,
                        outputs = outputs,
                        name = 'model')
    # set optimizer and loss
    opt = tf.keras.optimizers.SGD(momentum = 0.9,
                                  learning_rate = 0.0001)
    model.compile(loss = 'categorical_crossentropy',
                  optimizer = opt,
                  metrics = ['categorical_accuracy'])

    # train model
    history = model.fit(
      data_train_norm, labels_train,
      epochs = 200, verbose = 0,
      batch_size = 20,
      callbacks = [tfdocs.modeling.EpochDots()])
```

```
    # save performances
    hist = pd.DataFrame(history.history)
    hist['epoch'] = history.epoch

    return hist
```

Let's go through the code:

- First we define the dimensions of the input layer.
- Then we define the first hidden layer (the number of neurons is given as the function's input).
- Then we add the other hidden layers one at a time. The number of layers is given as the function's input.
- We then add the output layer and we stack all the layers together inside the model.
- We then compile and train the model, returning its performance.

Notice that in the previous code, we used Keras functional API (see Appendix A if you are not sure how it works), a functionality that provides a more flexible way to create models with respect to the `tf.keras.Sequential` API. With this functionality, we easily created a model with a customizable number of layers and neurons per layer.

To create the networks, we can simply apply the function with a different number of neurons and layers as inputs:

```
res_10_1 = model_nlayers(10, 1)
res_10_2 = model_nlayers(10, 2)
res_10_3 = model_nlayers(10, 3)
res_10_4 = model_nlayers(10, 4)
res_100_4 = model_nlayers(100, 4)
```

The code is now much easier to understand, and you can use it to create networks as big as you want.

With the function defined, it's easy to run several models and compare them, as we have done in Figure 3-17, where we tested five different models:

- One layer and ten neurons each layer
- Two layers and ten neurons each layer

- Three layers and ten neurons each layer
- Four layers and ten neurons each layer
- Four layer and 100 neurons each layer

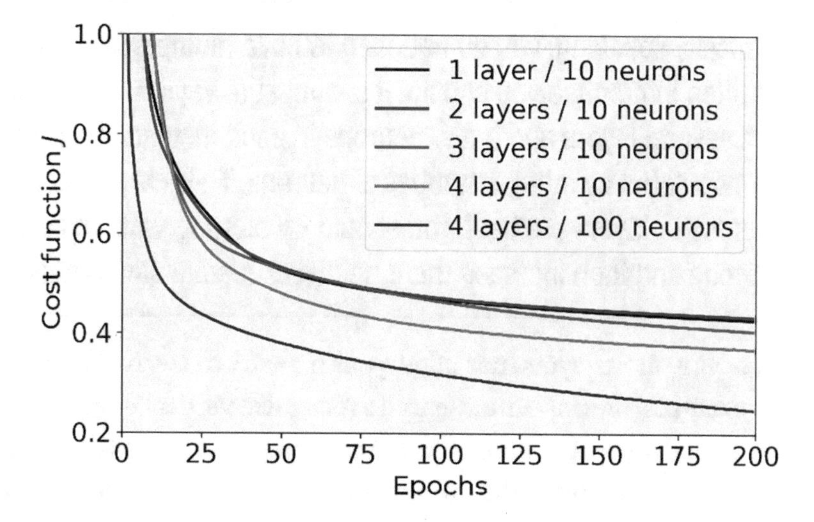

Figure 3-17. *The cost function vs. epochs for five models*

In case you are wondering, the model with four layers, each with 100 neurons, which seems much better than the others, is starting to go in the overfitting regime, with a train set accuracy of 91% and of 87% on the dev set (after only 200 epochs).

Advantages of Additional Hidden Layers

It is instructive to play with the models. Try varying the number of layers, the number of neurons, how you initialize the weights, and so on. If you invest some time you can reach an accuracy of over 90% in a few minutes of running time, but that requires some work. If you try several models you may realize that in this case using several layers does not seem to bring benefits versus a network with just one. This is often the case.

Theoretically speaking, a one-layer network can approximate every function you can imagine, but the number of neurons needed may be very large and therefore the model becomes less useful. Now the catch is that the ability of approximating a function does not mean that the network can do it, due for example to the sheer number of neurons involved or the time needed.

Empirically it has been shown that networks with more layers require a much smaller number of neurons to reach the same results and usually generalize better to unknown data.

Note Theoretically speaking, you do not need to have multiple layers in your networks, but often in practice you should. It is almost always a good idea to try a network with several layers and a few neurons in each instead of a network with one layer populated by a huge number of neurons. There is no set rule on how many neurons or layers are best. You should try starting with low numbers of layers and neurons and then increase them until your results stop getting better.

In addition, having more layers may allow your network to learn different aspects of your inputs. For example, one layer may learn to recognize vertical edges of an image, and another horizontal ones. Remember that in this chapter we discussed networks where each layer is identical (up to the number of neurons) to all the others. You will learn in Chapter 7 how to build networks whereby the layers perform very different tasks and are structured very differently from each other, making this kind of network much more powerful for certain tasks with respect to the models we discussed in this chapter.

As a simple example, imagine predicting the selling prices of houses. In this case a network with several layers may learn more information on how the features relate to the price. For example, the first layer may learn basic relationships, such as bigger houses mean higher prices. But the second layer may learn more complex features, such as a big house with a small number of bathrooms means a lower selling price.

Comparing Different Networks

Now you should know how to build neural networks with a huge number of layers or neurons. But it is relatively easy to lose yourself in a forest of possible models without knowing which ones are worth trying. Suppose you start with a network (as we have done in the previous sections) with one hidden layer with five neurons, one layer with ten neurons (for our softmax function), and our softmax neuron. Now suppose you have reached some accuracy and you would like to try different models. At first you should try increasing the number of neurons in your hidden layers to see what you can achieve. Figure 3-18 shows the cost function plotted as it decreases for different numbers of neurons. The calculations have been performed with a mini-batch gradient descent with a batch size of 50, one hidden layer with 1, 5, 15, and 30 neurons, and a learning rate of 0.0001. You can see how moving from one neuron to five immediately makes the convergence faster. But further increasing the number of neurons does not bring as much improvement. For example, increasing from 15 to 30 brings no improvement at all.

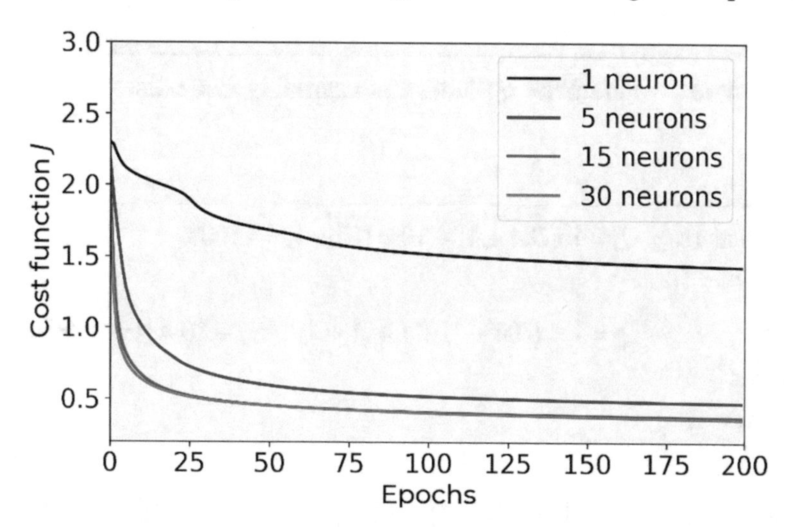

Figure 3-18. *The cost functionis decreasing vs. epochs for a neural network, with one hidden layer and 1, 5, 15, and 30 neurons. The calculations have been performed with mini-batch gradient descent with a batch size of 50 and a learning rate of 0.0001*

Let's first try to find a way of comparing those networks. Only comparing the number of neurons can be very misleading, as you will see shortly. Remember that your algorithm is trying to find the best combinations of weights and biases to minimize your cost function. But how many learnable parameters do we have in our model? We have the weights and the biases. You will remember from the theoretical discussion that we can associate a certain number of weights to each layer, and the number of learnable parameters in our layer l that we will indicate with $Q^{[l]}$ is given by the total number of elements in the matrix $W^{[l]}$, that is $n_l n_{l-1}$ (where we have $n_0 = n_x$ by definition) plus the number of biases we have (in each layer we will have n_l biases). The number $Q^{[l]}$ can then be written as follows

$$Q^{[l]} = n_l n_{l-1} + n_l = n_l(n_{l-1} + 1)$$

So that the total number of learnable parameters in our network (indicated here with Q) can be written as follows

$$Q = \sum_{j=1}^{L} n_l(n_{l-1} + 1)$$

Where by definition $n_0 = n_x$. Note that the Q parameter of our network is strongly architecture dependent. Let's calculate it with some practical examples (see Table 3-5).

Table 3-5. *Examples of Different Network Architectures and Their Corresponding Q Parameters*

Network Architecture	Parameter Q (Number of Learnable Parameters)	Number of Neurons
Network A: 784 features, 2 layers: $n_1 = 15$, $n_2 = 10$	$Q_A = 15(784 + 1) + 10 * (15 + 1) = 11935$	25
Network B: 784 features, 16 layers: $n_1 = n_2 = \ldots = n_{15} = 1$, $n_{16} = 10$	$Q_B = 1 * (784 + 1) + 1 * (1 + 1) + \ldots + 10 * (1 + 1) = 923$	25
Network C: 784 features, 3 layers: $n_1 = 10$, $n_2 = 10$, $n_3 = 10$	$Q_C = 10 * (784 + 1) + 10 * (10 + 1) + 10 * (10 + 1) = 8070$	30

Draw your attention to networks A and B. Both have 25 neurons, but the Q_A parameter is much bigger (more than a factor of ten) than Q_B. You can imagine how network A will be much more flexible in learning than network B, even if the number of neurons is the same.

Note Q in practice is not a measure of how complex or how good a network is. It may well happen that, of all the neurons, only a few will play a role, therefore calculating Q in this way does not tell the entire story. There is a vast amount of research on the so-called effective degrees of freedom of deep neural networks but that would go way beyond the scope of this book. But this parameter gives a good rule of thumb for deciding if the set of models you want to test are in a reasonable complexity progression.

Nonetheless, checking Q for the model you want to test may give you some hints about what you should neglect and what you should try. For example, let's consider the cases we have tested in Figure 3-18 and calculate the Q parameter for each network (see Table 3-6).

Table 3-6. *Network Architectures Tested in Figure 3-18 with their Corresponding Q Parameters*

Network Architecture	Parameter Q (Number of Learnable Parameters)	Number of Neurons
784 features, 1 layer with 1 neuron, 1 layer with ten neurons	$Q = 1 * (784 + 1) + 10 * (1 + 1) = 895$	11
784 features, 1 layer with 5 neuron, 1 layer with ten neurons	$Q = 5 * (784 + 1) + 10 * (5 + 1) = 3985$	15
784 features, 1 layer with 15 neuron, 1 layer with ten neurons	$Q = 15 * (784 + 1) + 10 * (15 + 1) = 11935$	25
784 features, 1 layer with 30 neuron, 1 layer with ten neurons	$Q = 30 * (784 + 1) + 10 * (30 + 1) = 23860$	40

From Figure 3-18, let's suppose we choose the model with 15 neurons as the candidate for the best model. Now let's suppose we want to try a model with three layers, all with the same number of neurons that should compete (and possibly be better) than our (for the moment) candidate model, with one layer and 15 neurons. What should we choose as a starting point for the number of neurons in the three layers? Let's indicate as model A the one with one layer and 15 neurons and model B as the model with three layers and an unknown number of neurons in each layer indicated with n_B. We can easily calculate the Q parameter for both networks

$$Q_A = 15*(784+1)+10*(15+1)=11935$$

And

$$Q_B = n_B*(784+1)+n_B*(n_B+1)+n_B*(n_B+1)+10*(n_B+1)=2n_B^2+797n_B+10$$

What value for n_B will give $Q_B \approx Q_A$? We can solve the equation

$$2n_B^2+797n_B+10=11935$$

You should be able to solve a quadratic equation, so we will only look at the solution here. This equation is solved for a value of $n_B = 14.4$, but since we cannot have 14.4 neurons, we will need to use the closest integer, $n_B = 14$. For $n_B = 14$, we will have $Q_B = 11560$, a value very close to 11935.

Note The fact that the two networks have the same number of learnable parameters does not mean that they can reach the same accuracy and does not even mean that if one learns very quickly the second will learn at all!

The model with three layers, each with 14 neurons, could be a good starting point for further testing.

Let's discuss another point that is important when dealing with a complex dataset. Consider the first layer. Let's suppose we consider the Zalando dataset and we create a network with two layers: the first with one neuron and the second with many. All the complex features that your dataset has may well be lost in your single first neuron, since it will combine all features into a single value and pass the same value to all the other neurons of the second layer.

Tips for Choosing the Right Network

You have discussed a lot of cases, you have seen a lot of formulas, but how can you decide how to design your network?

Unfortunately, there is no fixed set of rules. But you may consider the following tips:

- When considering a set of models (or network architectures) you want to test, a good rule of thumb is to start with a less complex one and move to more complex ones. A rule of thumb to estimate the relative complexity (to make sure that you are moving in the right direction) is the Q parameter .
- If you cannot reach good accuracy if any of your layers has a particular low number of neurons. This layer may kill the effective capacity of learning from a complex dataset of your network. Consider for example the case with one neuron in Figure 3-18. The model cannot reach low values for the cost function because the network is too simple to learn from a complex dataset like the Zalando one.

- Remember that a low or high number of neurons is always relative to the number of features you have. If you have only two features in your dataset, one neuron may well be enough, but if you have a few hundred (like in the Zalando dataset, where $n_x = 784$), you should not expect one neuron to be enough.
- Which architecture you need is also dependent on what you want to do. It's always worth checking the online literature to see what others have already discovered about specific problems. For example, it's well known that for image recognition, convolutional networks are very good, so they would be a good choice.

Final Tip When moving from a model with L layers to one with $L + 1$ layers, it's always a good idea to start with the new model and use a slightly smaller number of neurons in each layer and then increase it step by step. Remember that more layers have a chance of learning complex features more effectively, so if you are lucky fewer neurons may be enough. It is something worth trying. Always keep track of your optimizing metric for all your models. When you are not getting much improvement anymore, it may be worth trying completely different architectures (maybe convolutional neuronal networks, etc.).

Estimating the Memory Requirements of Models

For the calculation, let's consider a feed-forward neural network with N_L hidden layers, each having n neurons. Let's consider a concrete case to make things clearer. Suppose we are working with the MNIST dataset. In this case, the input data is composed of a vector of the 784 gray values of the images (each image is 28 × 28 pixels in gray values). The output layer of a network for classification will be composed of ten neurons (the ten classes in which the model tries to classify the images). The total number of weights N_W can be easily calculated and can be written as follows

$$N_W = \underbrace{784n + N_L n^2 + 10n}_{\text{Weights}} + \underbrace{N_L n}_{\text{Biases}} \underset{\text{large } n}{\sim} N_L n^2$$

Where the second part of the equation mean that, for large n, N_W will asymptotically grow quadratically in n, which is the number of neurons in each layer. In general, there are three components that need to be taken into account.

- **Parameters**: In memory, you need to keep the parameters, their gradients during backpropagation, and also additional information if the optimization is using momentum, Adagrad, Adam, or RMSProp optimizers. A good rule of thumb in order to account for all these factors[6] is to multiply the memory taken by the weights alone by roughly 3. With the notation we have used so far, the memory used from parameters (indicated with M_W) in Gb would be

$$M_W = \underbrace{3}_{\text{Correction Factor}} + \underbrace{64/8}_{\text{Conversion to byets for Floating point 64}} + \underbrace{\frac{N_W}{1024^3}}_{\text{Conversion to Gb}}$$

- **Activations**: Each neuron output must be stored, normally with their gradient for backpropagation. Conservatively only a mini-batch will need to be kept in memory. Calling S_{MB} the mini-batch size the memory needed for activations M_A in Gb can be written as

$$M_A = 2S_{MB}(2n+10)\frac{8}{1024^3}$$

- **Miscellaneous**: This part includes the data that must be loaded into memory and so on. For the purposes of a rough estimate, the memory taken here M_M will be estimated only with the dataset size. In the case of MNIST, that will be (in Gb) given by the following equation. Each pixel value, although originally an INT8, must be converted to floating-point 64-data type to perform the training

$$M_M = 6000 \times 784 \frac{8}{1024^3}$$

[6] Abadi, M.; Agarwal, A.; Barham, P.; Brevdo, E.; Chen, Z.; Citro, C.; Corrado, G.S.; Davis, A.; Dean, J.; Devin, M.; Ghemawat, S.;Goodfellow, I.; Harp, A.; Irving, G.; Isard, M.; Jia, Y.; Jozefowicz, R.; Kaiser, L.; Kudlur, M.; Levenberg, J.; Mané, D.; Monga, R.;Moore, S.; Murray, D.; Olah, C.; Schuster, M.; Shlens, J.; Steiner, B.; Sutskever, I.; Talwar, K.; Tucker, P.; Vanhoucke, V.; Vasudevan,V.; Viégas, F.; Vinyals, O.; Warden, P.; Wattenberg, M.; Wicke, M.; Yu, Y.; Zheng, X. "TensorFlow: Large-Scale Machine Learning on Heterogeneous Systems," 2015. Accessed 8th Sept. 2021

For example, to find out if a model will run on a limited memory device that will have M_D GB free, it is enough to check if the following equation has solutions in n or N_L

$$M_W + M_A + M_M = M_D$$

Note that this is just a rough indication and will not be precise since the amount of memory taken by a model may depend on software versions, the operating system, and many more factors. If we solve the last equation for n for example, we can get a very good estimate of the biggest FFNN network that could be run on such a device when applied to MNIST. For example, consider the case of a Raspberry Pi 4 with 2 GB of memory. Typically, such a system has roughly $M_D = 1.7$ GB free at any moment. So for a network with $N_L = 2$ and $S_{MB} = 128$, the last equation will give a solution of $n \approx 8200$. Indeed, trying to train a network with more than 8200 neurons on such a device will give a memory error on the device, since there is not enough memory to keep everything available in RAM. (If you test it, your results may vary, depending as mentioned on which version of the Raspberry Pi you have, which version of TensorFlow you are using, and so on.)

For practical purposes to get a rougher estimate, you can neglect the linear terms in n in the last equation and still get a usable guideline. For example, in the example discussed previously, neglecting the linear term would give an estimate of $n \approx 8700$. Higher than the actual one, but one that will give useful rough information about the maximum number of usable neurons. Finally, remember that only a practical test will guarantee that a specific model can run on a low-memory device.

General Formula for the Memory Footprint

In general, when working with a dataset of size n_x the formula for the number of parameters can be written as follows

$$N_W = \underbrace{n_x n + N_L n^2 + N_O n}_{\text{Weights}} + \underbrace{N_L n}_{\text{Biases}} \underset{\text{large } n}{\sim} N_L n^2$$

Where we indicate the number of neurons in the output layer as N_O. If you refer to the previous section, you can see that the formulas for M_W and M_A are unchanged, while M_M needs to be written as

$$M_M = m n_x \frac{8}{1024^3}$$

This accounts for the fact that the dataset size is not 60000 but m and that the input dimension is not 784 but n_x.

Exercises

EXERCISE 1 (LEVEL: EASY)

Try to build a multiclass classification model like the one you saw in this chapter, but with a different dataset, the MNIST database of handwritten digits (`http://yann.lecun.com/exdb/mnist/`). To download the dataset from TensorFlow, use the following code:

```
from tensorflow import keras
(x_train, y_train), (x_test, y_test) = keras.datasets.mnist.load_data()
```

EXERCISE 2 (LEVEL: MEDIUM)

Apply He weight initialization to the multiclass classification problem you saw in this chapter to see if you can speed up the learning phase.

EXERCISE 3 (LEVEL: DIFFICULT)

Try to optimize the feed-forward neural network built in this chapter to reach the best possible accuracy (without overfitting the training dataset!). Tune the number of epochs, the learning rate, the optimizer, the number of neurons, the layers, and the mini-batches. *Hint*: Write a function like the one we used to test different numbers of layers and neurons and give it as inputs all the tunable parameters.

EXERCISE 4 (LEVEL: DIFFICULT)

Consider the regression problem we solved with a model made by a single neuron (predicting radon activity in U.S. houses). Try to build a feed-forward neural network to solve the same regression task. See if you can get better prediction performance. *Hint*: You will need to change the loss function and the metrics to evaluate your results, and one-hot encoding will not be necessary anymore. As a starting point, you can find the entire code in the online version of the book at `https://adl.toelt.ai`.

References

[1] Burnham, K. P.; Anderson, D. R. (2002), *Model Selection and Multimodel Inference* (2nd ed.), Springer-Verlag.

[2] `https://en.wikipedia.org/wiki/Zalando`, last accessed 16.02.2021.

[3] `www.kaggle.com`, last accessed 16.02.2021.

[4] Xiao, Han, Kashif Rasul, and Roland Vollgraf. "Fashion-mnist: a novel image dataset for benchmarking machine learning algorithms." arXiv preprint arXiv:1708.07747 (2017).

[5] `https://keras.io/api/models/model_training_apis/#fit-method`, last accessed 14/03/2021.

[6] "Understanding the Difficulty of Training Deep Feedforward neural networks", X. Glorot, Y. Bengio (2010), `https://goo.gl/bHB5BM`.

[7] `https://www.tensorflow.org/api_docs/python/tf/keras/initializers`, last accessed 23.02.2021.

CHAPTER 4

Regularization

This chapter explains a very important technique often used when training deep networks: *regularization*. We look at techniques such as the ℓ_1 **and** ℓ_2 **methods**, dropout, and early stopping. You learn how these methods help prevent the problem of overfitting and help you achieve much better results from your models when applied correctly. We look at the mathematics behind the methods and at how to implement them correctly in Python and Keras.

Complex Networks and Overfitting

In the previous chapters, you learned how to build and train complex networks. One of the most common problems you will encounter when using complex networks is overfitting. In this chapter, we face an extreme case of overfitting and discuss a few strategies to avoid it. A perfect dataset for studying this problem is the Boston housing price dataset [1].

This dataset contains information collected by the U.S. Census Bureau concerning housing around the Boston area. Each record in the database describes a Boston suburb or town. The data was drawn from the Boston Standard Metropolitan Statistical Area (SMSA) in 1970. The attributes are defined as follows:

- CRIM: Per capita crime rate by town
- ZN: Proportion of residential land zoned for lots over 25,000 square feet
- INDUS: Proportion of non-retail business acres per town
- CHAS: Charles River dummy variable (1 if tract bounds river; 0 otherwise)
- NOX: Nitric oxides concentration (parts per 10 million)

U. Michelucci, *Applied Deep Learning with TensorFlow 2*, https://doi.org/10.1007/978-1-4842-8020-1_4

- RM: Average number of rooms per dwelling
- AGE: Proportion of owner-occupied units built prior to 1940
- DIS: Weighted distances to five Boston employment centers
- RAD: Index of accessibility to radial highways
- TAX: Full-value property-tax rate per $10,000
- PTRATIO: Pupil-teacher ratio by town
- $B - 1000(B_k - 0.63)^2 - B_k$: Proportion of African Americans by town
- LSTAT: % lower status of the population
- MEDV: Median value of owner-occupied homes in $1000s

Let's get the data

```
# sklearn libraries
from sklearn.datasets import load_boston
import sklearn.linear_model as sk
```

and then import the dataset

```
boston = load_boston()
features = np.array(boston.data)
target = np.array(boston.target)
```

The dataset has 13 features (contained in the `features` NumPy array) and the house price is contained in the `target` NumPy array. To normalize the features, we use this function

```
def normalize(dataset):
    mu = np.mean(dataset, axis = 0)
    sigma = np.std(dataset, axis = 0)
    return (dataset - mu)/sigma
```

To conclude the dataset preparation, we normalize it and then create a training and a dev dataset

```
features_norm = normalize(features)

np.random.seed(42)
rnd = np.random.rand(len(features_norm)) < 0.8

train_x = features_norm[rnd]
train_y = target[rnd]
dev_x = features_norm[~rnd]
dev_y = target[~rnd]

print(train_x.shape)
print(train_y.shape)
print(dev_x.shape)
print(dev_y.shape)
```

The `np.random.seed(42)` is there so that you will always get the same training and dev dataset (so the results are reproducible).

Then we build a complex neural network with four layers and 20 neurons for each layer. To build it, we define this function to build each layer, train the model, and validate it against the dev dataset:

```
def create_and_train_model_nlayers(data_train_norm, labels_train, data_dev_
norm, labels_dev, num_neurons, num_layers):
    # build model
    # input layer
    inputs = keras.Input(shape = data_train_norm.shape[1])
    # he initialization
    initializer = tf.keras.initializers.HeNormal()
    # first hidden layer
    dense = layers.Dense(
      num_neurons, activation = 'relu',
         kernel_initializer = initializer)(inputs)
    # customized number of layers and neurons per layer
    for i in range(num_layers - 1):
        dense = layers.Dense(
          num_neurons, activation = 'relu',
          kernel_initializer = initializer)(dense)
```

```
    # output layer
    outputs = layers.Dense(1)(dense)
    model = keras.Model(inputs = inputs, outputs = outputs,
                        name = 'model')
    # set optimizer and loss
    opt = keras.optimizers.Adam(learning_rate = 0.001)
    model.compile(loss = 'mse', optimizer = opt,
                  metrics = ['mse'])

    # train model
    history = model.fit(
      data_train_norm, labels_train,
      epochs = 10000, verbose = 0,
      batch_size = data_train_norm.shape[0],
      validation_data = (data_dev_norm, labels_dev))
    # save performances
    hist = pd.DataFrame(history.history)
    hist['epoch'] = history.epoch

    return hist, model
```

This function builds and trains a feed-forward neural network model and evaluates it on the training and dev sets. You learned about feed-forward neural networks implementation in the last chapter so you should now understand what it does.

We use the He initialization here, since we will use ReLU activation functions. In the output layer, we have one neuron with the identity activation function for regression (remember that, in Keras, when you do not specify an activation function, the default one will be the identity function). Additionally, we use the Adam optimizer.[1]

Now let's run the model with this code

```
hist, model = create_and_train_model_nlayers(train_x, train_y, dev_x,
dev_y, 20, 4)
```

[1] Don't worry if you don't understand how Adam works. Just think of it as a smarter gradient descent. Chapter 5 is dedicated to advanced optimizers.

As you may notice, to make things simpler, we avoided writing a function with many input parameters and simply hard-coded all the values in the body function (as for the learning rate, for example), since in this case we do not need to tune the parameters very much. Moreover, we are not using mini-batches here, since we only have a few hundred observations.

We are calculating the MSE for the training and dev datasets, as the typical cost function for linear regression problems. This way, we can check what is happening on both datasets at the same time. If you let the code run and plot the two MSEs—one for the training indicated with MSE_{train} and one for the dev dataset indicated with MSE_{dev}—you get Figure 4-1.

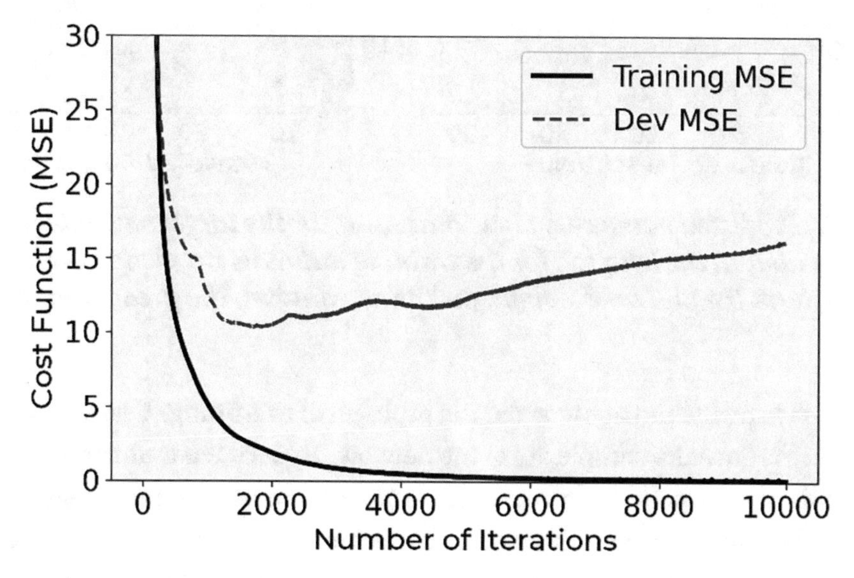

Figure 4-1. *The MSE for the training (continuous line) and the dev dataset (dashed line) for the neural network with four layers, each with 20 neurons*

Note how the training error goes to zero, while the dev error reaches a value of about 10 and then it starts increasing to a value of about 15, after dropping rapidly at the beginning. If you remember our basic error analysis introduction, you should know that this means that you are in a regime of *extreme overfitting* (when $MSE_{train} \ll MSE_{dev}$). The error on the training dataset is practically zero, while the one on the dev dataset is not. The model cannot generalize at all when applied to new data. Figure 4-2 shows the predicted value plotted versus the real value. Note how in the left plot, for the training data, the prediction is almost perfect, while on the plot on the right, for the dev dataset, it's not that good. Recall that a perfect model will give you predicted values exactly equal to the measured ones. So, when plotting one versus the other, they would all lie on the 45-degree line in the plot, as is happening in the left plot in Figure 4-2.

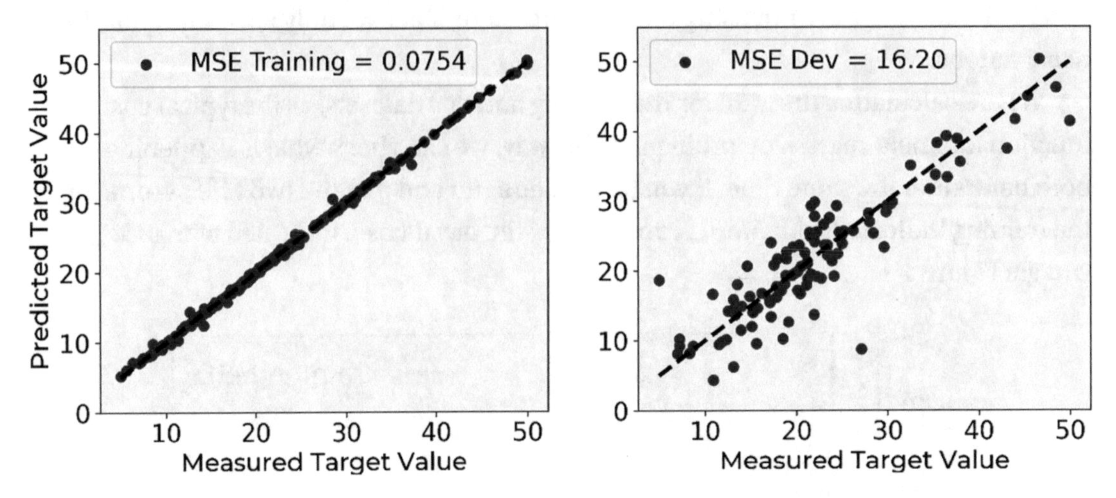

Figure 4-2. *Predicted value versus the real value for the target variable (the house price). Note how, in the left plot, for the training data, the prediction is almost perfect, while on the plot on the right, for the dev dataset, the predictions are more spread out*

What can we do in this case to avoid the problem of overfitting? One solution would be of course to reduce the complexity of the network. Reduce the number of layers and/or the number of neurons in each layer. But, as you can imagine, this strategy is very time consuming. You must try several network architectures to see how the training error and the dev error behave. In this case, this is still a viable solution, but if you are working on a problem where the training phase takes several days, this can be quite difficult and extremely time consuming. Several strategies have been developed to deal with this problem; the most common is called *regularization* and is the focus of this chapter.

What Is Regularization

Before going into the different methods, we must quickly discuss how the deep learning community interprets the term *regularization*. The term has deeply (pun intended) evolved over time. For example, in the traditional sense from the 90s, the term was reserved only to a penalty term in the loss function [2]. Lately, the term has gained a much more broader meaning. For example, Goodfellow [3] defines it as *"any modification we make to a learning algorithm that is intended to reduce its test error but not its training error."* Kukačka [4] generalizes the term even more and provides the definition: "Regularization is any supplementary technique that aims at making the model generalize better, i.e. produce better results on the test set." So be aware when using the term and always be precise with what you mean.

You may also have heard or read the claim that regularization has been developed to fight overfitting. This is also a way of understanding it. Remember, a model that is overfitting the training dataset is not generalizing well to new data. This definition is also in line with all the others.

This is just a matter of definitions but it's important to have heard them, so that you better understand what is meant when reading papers or books. This is a very active research area and, to give you an idea, Kukačka, in his review paper, lists 58 different regularization methods. It's important to understand that with their general definition, SGD (Stochastic Gradient Descent) is also considered a regularization method, something that not everyone agrees on. So be warned when reading research material about the variation with the term regularization.

In this chapter, we look at the three most common and known methods—ℓ_1, ℓ_2, and dropout, We also briefly talk about early stopping, although this method does not, technically speaking, fight overfitting. ℓ_1 and ℓ_2 achieve a so-called weight decay by adding a regularization term to the cost function, while dropout simply removes, in a random fashion, nodes from the network during the training phase. To understand the three methods properly, we need to study them in detail. Let's start with the most instructive one: ℓ_2 regularization.

At the end of the chapter, we look at a few other ideas on how to fight overfitting and get the model to generalize better. Instead of changing or modifying the model or the learning algorithm, we will consider strategies with the idea of modifying the training data to make learning more effective.

About Network Complexity

Let's spend a few moments briefly discussing the term we used very often: network complexity. You have read here, and can find almost everywhere, that with regularization you want to reduce network complexity. But what are we referring to really? It is very difficult to give a definition of network complexity, so much that nobody does it, actually. You can find several research papers on the problem of model complexity (note that this is not exactly network complexity), with roots in information theory. In this chapter, you see how the number of weights different from zero will change dramatically with the number of epochs, with the optimization algorithm and so on, therefore making this vague concept of complexity also dependent on how long you train your model. To make the story short, the term *network complexity* should be used only at a general level, since theoretically it's a very complex concept to define. A complete discussion of the subject is completely out of the scope of this book.

ℓ_p Norm

Before we start studying what ℓ_1 and ℓ_2 regularization is, we need to introduce the ℓ_p norm notation. We define the ℓ_p norm of a vector $\boldsymbol{x}$ with x_i components as

$$\|\boldsymbol{x}_p\| = \sqrt[p]{\sum_i |x_i|^p} \qquad p \in \mathbb{R}$$

where the sum is performed over all components of the vector $\boldsymbol{x}$.

Let's now start with the most instructive norm: the ℓ_2.

ℓ_2 Regularization

One of the most common regularization methods, ℓ_2 regularization, consists of adding a term to the cost function that has the goal of effectively reducing the capacity of the network to adapt to complex datasets. Let's first look at the mathematics behind the method.

Theory of ℓ_2 Regularization

When doing plain regression, the cost function is simply the MSE (Mean Squared Error)

$$L(\boldsymbol{w}) = \frac{1}{m}\sum_{i=1}^{m}(y_i - \hat{y}_i)^2$$

where y_i is our measured target variable, $\hat{y}_i$ is the predicted value, $\boldsymbol{w}$ is the vector of all the weights of our network including the bias, and m is the number of observations. Now let's define a new cost function $\tilde{L}(\boldsymbol{w},b)$

$$\tilde{L}(\boldsymbol{w}) = L(\boldsymbol{w}) + \frac{\lambda}{2m}\|\boldsymbol{w}\|^2{}_2$$

This additional term

$$\frac{\lambda}{2m}\|\boldsymbol{w}\|^2_2$$

is called the *regularization term* and is nothing more than the ℓ_2 norm squared and $\boldsymbol{w}$ multiplied by a constant factor $\lambda/2m$. λ is a called the *regularization parameter.*

Note The new regularization parameter λ is a new hyper-parameter that you need to tune to find its optimal value.

Now let's study the effect this term has on the GD (gradient descent) algorithm. Consider the update equation for the weight w_j

$$w_{j,[n+1]} = w_{j,[n]} - \gamma \frac{\partial \tilde{L}\left(\boldsymbol{w}_{[n]}\right)}{\partial w_j} = w_{j,[n]} - \gamma \frac{\partial L\left(\boldsymbol{w}_{[n]}\right)}{\partial w_j} - \frac{\gamma\lambda}{m} w_{j,[n]}$$

since

$$\frac{\partial}{\partial w_j}\|\boldsymbol{w}\|^2_2 = 2w_j$$

This gives us

$$w_{j,[n+1]} = w_{j,[n]}\left(1 - \frac{\gamma\lambda}{m}\right) - \gamma \frac{\partial L\left(\boldsymbol{w}_{[n]}\right)}{\partial w_j}$$

This is the equation that we need to use for the weights update. The difference with the one we already know from plain GD is that now the weight $w_{j,[n]}$ is multiplied by a constant $1 - \frac{\gamma\lambda}{m} < 1.$ This has the effect of shifting the weight values during the update toward zero, and therefore making the network less complex. This in turn helps to prevent overfitting. Let's see what is really happening to the weights by applying the method to the Boston housing dataset.

Keras Implementation

The implementation in Keras is extremely easy. The library performs all the computations for us, and we just have to decide which regularization we want to use, set the λ parameter, and apply it to each layer. The model construction remains the same.

Note In Keras, regularization is applied at the layer level, and not globally on the cost function. You will notice how it is added at each layer and not when defining the cost function. But the explanation here remains valid, and it works in the same way as we discussed. The reason for this is that it can be helpful to add regularization only on certain layers, for example the largest, and not on layers with only a few neurons.

We can do this with this function

```
def create_and_train_reg_model_L2(data_train_norm, labels_train, data_dev_
norm, labels_dev, num_neurons, num_layers, n_epochs, lambda_):
    # build model
    # input layer
    inputs = keras.Input(shape = data_train_norm.shape[1])
    # he initialization
    initializer = tf.keras.initializers.HeNormal()
    # regularization
    reg = tf.keras.regularizers.l2(l2 = lambda_)
    # first hidden layer
    dense = layers.Dense(
      num_neurons, activation = 'relu',
      kernel_initializer = initializer,
      kernel_regularizer = reg)(inputs)
    # customized number of layers and neurons per layer
    for i in range(num_layers - 1):
        dense = layers.Dense(
          num_neurons, activation = 'relu',
          kernel_initializer = initializer,
          kernel_regularizer = reg)(dense)
```

```
    # output layer
    outputs = layers.Dense(1)(dense)
    model = keras.Model(inputs = inputs, outputs = outputs,
                        name = 'model')
    # set optimizer and loss
    opt = keras.optimizers.Adam(learning_rate = 0.001)
    model.compile(loss = 'mse', optimizer = opt,
                  metrics = ['mse'])

    # train model
    history = model.fit(
      data_train_norm, labels_train,
      epochs = n_epochs, verbose = 0,
      batch_size = data_train_norm.shape[0],
      validation_data = (data_dev_norm, labels_dev))
    # save performances
    hist = pd.DataFrame(history.history)
    hist['epoch'] = history.epoch
    # print performances
    print('Cost function at epoch 0')
    print('Training MSE = ', hist['loss'].values[0])
    print('Dev MSE = ', hist['val_loss'].values[0])
    print('Cost function at epoch ' + str(n_epochs))
    print('Training MSE = ', hist['loss'].values[-1])
    print('Dev MSE = ', hist['val_loss'].values[-1])
    return hist, model
```

The main differences with respect to the previous function (the one that we used to build a network without regularization) are highlighted in bold.

With the `reg = tf.keras.regularizers.l2(l2 = lambda_)` line, we defined the ℓ_2 regularizer, setting the value for λ. Then we apply the regularizer to each layer, assigning it to the `kernel_regularizer`, which applies the penalty on the layer's kernel. The layers may also expose the keywork arguments `bias_regularizer` and `activity_regularizer`, which apply a penalty to the layer's bias and output, respectively, but are less often used. Here we employ only the `kernel_regularizer` argument.

Remember that in Python `lambda` is a reserved word, so we cannot use it. This is why we use `lambda_`.

Now let's train and evaluate our network to see what happens. This time we will print the MSE coming from the training (MSE_{train}) and from dev (MSE_{dev}) datasets to check what is going on. As mentioned, applying this method make many weights go to zero, effectively reducing the complexity of the network, and therefore fighting overfitting. Let's run the model for $\lambda = 0.0$, without regularization, and for $\lambda = 10.0$.

We can run this model with the following code

```
hist, model = create_and_train_reg_model_L2(train_x, train_y, dev_x, dev_y,
20, 4, 0.0)
```

and that gives us

```
Cost function at epoch 0
Training MSE =  653.5233764648438
Dev MSE =  623.965087890625
Cost function at epoch 5000
Training MSE =  0.2870051860809326
Dev MSE =  25.645526885986328
```

As expected, we are in an extreme overfitting regime ($MSE_{train} \ll MSE_{dev}$) after 5000 epochs. Now let's try this with $\lambda = 10.0$

```
hist, model = create_and_train_reg_model_L2(train_x, train_y, dev_x, dev_y,
20, 4, 10.0)
```

and that gives these results

```
Cost function at epoch 0
Training MSE =  2141.39599609375
Dev MSE =  2100.5986328125
Cost function at epoch 5000
Training MSE =  58.91643524169922
Dev MSE =  56.80580139160156
```

We aren't in an overfitting regime any more, since the two MSE values are of the same order of magnitude. The best way to see what is going on is to study the weights distribution for each layer. In Figure 4-3, the weights distribution for the four hidden layers are plotted. The light gray histogram is for the weights without regularization, and the darker (and much more concentrated around zero) histogram is for the weights with regularization.

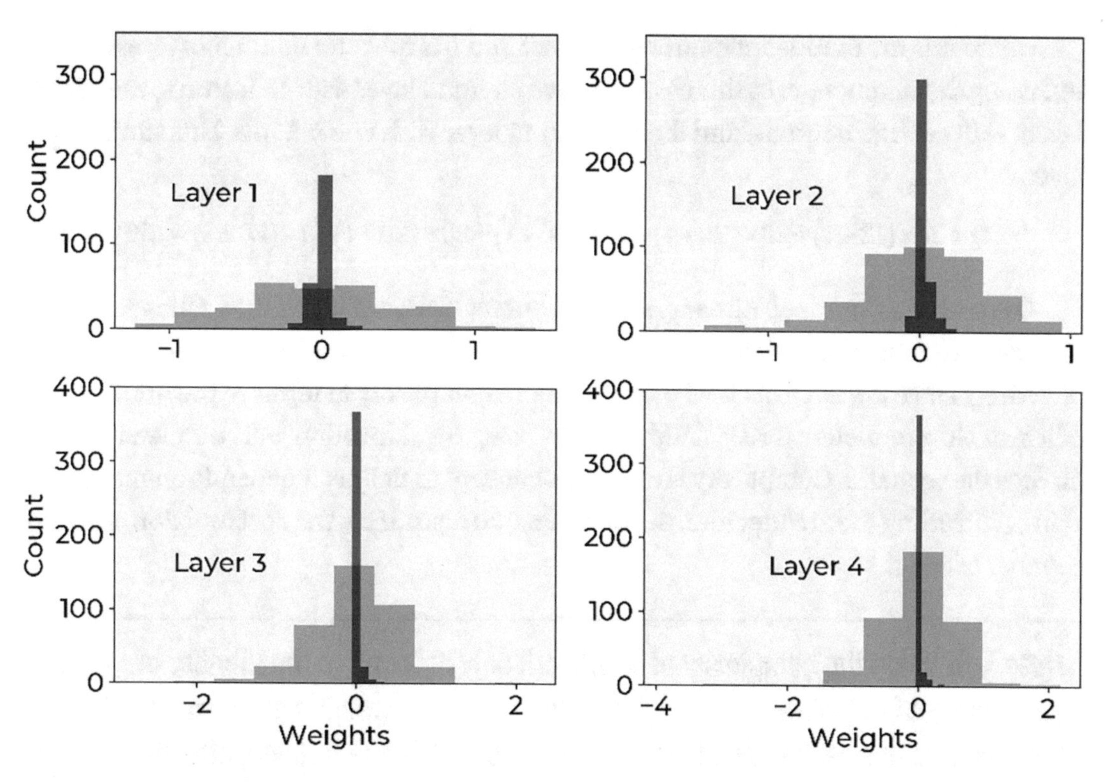

Figure 4-3. *Weights distribution for each layer. The light gray histogram is for the weights without regularization, and the darker (and much more concentrated around zero) histogram is for the weights with regularization*

You can clearly see how the weights, when we apply regularization, are much more concentrated around zero, meaning they are much smaller than without regularization. This makes the weight decay effect of regularization very evident. We now have the chance to have another brief digression about network complexity. We said that this method reduces the network complexity. We saw previously that you can consider the number of learnable parameters an indication of the complexity of a network, but you have also been warned that this can be very misleading. Now let's see why. Recall that the total number of learnable parameters we have in a network is given by the formula

$$Q = \sum_{j=1}^{L} n_l \left(n_{l-1} + 1 \right)$$

where n_l is the number of neurons in layer l and L is the total number of layers, including the output one. In this case, we have an input layer with 13 features, then four layers with each 20 neurons, and then an output layer with one neuron. Therefore Q is given by

$$Q = 20 \times (13+1) + 20 \times (20+1) + 20 \times (20+1) + 20 \times (20+1) + 1 \times (20+1) = 1561$$

Q is a very big number. Already, and without regularization, it is interesting to note that we have roughly 6% of the weights that after 1000 epochs are less than 10^{-3}, so effectively close to zero. This is why talking about complexity in terms of the number of learnable parameters is risky. Additionally, using regularization will completely change the scenario. Complexity is a difficult concept to define: it depends on many things, including the architecture, the optimization algorithm, the cost function, and the number of epochs trained.

Note Defining the complexity of a network only in terms of the number of weights is not completely correct. The total number of weights gives you an idea, but it can be misleading since many may be zero after the training, effectively disappearing from the network, and making it less complex. It is more correct to talk about *model complexity* instead of network complexity, since many more aspects are involved than simply how many neurons or layers the network has.

Incredibly enough, only half of the weights play a role in the predictions at the end. This is why defining the network complexity only with the Q parameter is misleading. Given your problem, your loss function, and optimizer, you may well end up with a network that when trained is much simpler than what it was during the construction phase. So be very careful when using the term complexity in the deep learning world. Be aware of the subtleties involved.

To give you an idea how effective regularization is in reducing the weights, Table 4-1 compares the percentage of weights less than 10^{-3} with and without regularization, after 1000 epochs in each hidden layer.

***Table 4-1.** Percentage of Weights Less Than 10^{-3} With and Without Regularization After 1000 Epochs*

Layer	Percent of Weight Less Than 10^{-3} for λ = 0.0	Percent of Weight Less Than 10^{-3} for λ = 3.0
1	0.77	1.54
2	0.0	28.25
3	1.0	40.0
4	0.25	45.75

But how should we choose λ? To get an idea (and remember that in the deep learning world there is no universal rule), it's useful to see what happens when you vary the λ parameter to your optimizing metric (in this case, the MSE). Figure 4-4 shows the behavior of MSE_{train} (continuous line) and of MSE_{dev} (dashed line) on the Boston dataset for this network varying λ after 1000 epochs.

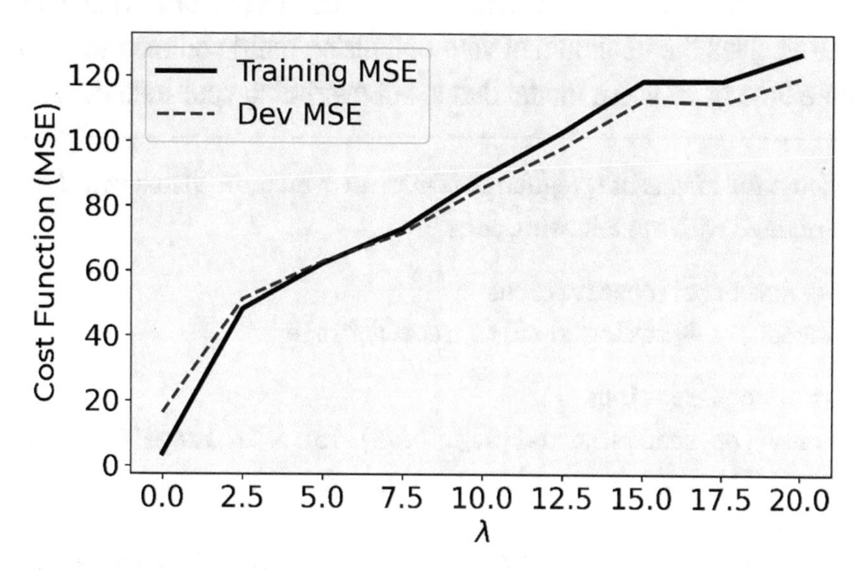

***Figure 4-4.** Behavior of the MSE for the training (continuous line) dataset and for the dev (dashed line) dataset for our network varying λ*

As you can see, with small values of λ (effectively without regularization), we are in an overfitting regime ($MSE_{train} \ll MSE_{dev}$), with the MSE_{train} and the MSE_{dev} increasing. Until $\lambda \approx 6$, the model overfits the training data, then the two values cross and the overfitting finishes. After that they grow together, at which point the model cannot capture the fine data structures anymore. After the crossing of the lines, the model is getting too simple to capture the features of the problem. Therefore, the errors grow together and the error on the training dataset gets bigger, since the model doesn't even fit the training data well. In this specific case, a good value to choose for λ is around 6, around the value when the two lines cross, since you are not in an overfitting region anymore (since $MSE_{train} \approx MSE_{dev}$). Remember that the main goal of having the regularization term is to get a model that generalizes in the best way possible when it's applied to new data. You can look at it in a different way: a value of $\lambda \approx 6$ gives you the minimum of MSE_{dev} outside the overfitting region (for $\lambda \lesssim 6$), so it's a good choice. Note that you may observe a very different behavior for your optimizing metric, so you have to decide what the best value is for λ on a case-by-case basis.

Note A good way for estimating the optimal value of the regularization parameter λ is to plot your optimizing metric (in this example, the MSE) for the training and the dev dataset and see how they behave for various values of λ. Then you choose the value that gives the minimum of your optimizing metric on the dev dataset and at the same time gives you a model that is not overfitting your training data.

Let's discuss the effects of ℓ_2 regularization in an even more visual way. Let's consider a dataset generated with the following code

```
nobs = 30 # number of observations
np.random.seed(42) # making results reproducible

# first set of observations
xx1 = np.array([np.random.normal(0.3, 0.15) for i in range (0, nobs)])
yy1 = np.array([np.random.normal(0.3, 0.15) for i in range (0, nobs)])
# second set of observations
xx2 = np.array([np.random.normal(0.1, 0.1) for i in range (0, nobs)])
yy2 = np.array([np.random.normal(0.3, 0.1) for i in range (0, nobs)])
# concatenating observations
c1_ = np.c_[xx1.ravel(), yy1.ravel()]
c2_ = np.c_[xx2.ravel(), yy2.ravel()]
c = np.concatenate([c1_, c2_])
```

```
# creating the labels
yy1_ = np.full(nobs, 0, dtype = int)
yy2_ = np.full(nobs, 1, dtype = int)
yyL = np.concatenate((yy1_, yy2_), axis = 0)
# defining training points and labels
train_x = c
train_y = yyL
```

Our dataset has two features: *x* and *y*. We generated two groups of points—xx1, yy1 and xx2, yy2—from a normal distribution. To the first, we assigned the label 0 (contained in the array yy1_) and to the second the label 1 (in the array yy2_). Now let's use a network like the one we described earlier (with four layers, each having 20 neurons) to do some binary classification on this dataset. We can take the same code given previously, modifying the output layer and the cost function. Recall that for binary classification, we need one neuron in the output layer with the sigmoid activation function

```
def create_and_train_regularized_model(data_train_norm, labels_train,
num_neurons, num_layers, n_epochs, lambda_):
    # build model
    # input layer
    inputs = keras.Input(shape = data_train_norm.shape[1])
    # he initialization
    initializer = tf.keras.initializers.HeNormal()
    # regularization
    reg = tf.keras.regularizers.l2(l2 = lambda_)
    # first hidden layer
    dense = layers.Dense(
      num_neurons, activation = 'relu',
      kernel_initializer = initializer,
      kernel_regularizer = reg)(inputs)
    # customized number of layers and neurons per layer
    for i in range(num_layers - 1):
        dense = layers.Dense(
          num_neurons, activation = 'relu',
          kernel_initializer = initializer,
          kernel_regularizer = reg)(dense)
```

```
    # output layer
    outputs = layers.Dense(1, activation = 'sigmoid')(dense)
    model = keras.Model(inputs = inputs, outputs = outputs,
                        name = 'model')
    # set optimizer and loss
    opt = keras.optimizers.Adam(learning_rate = 0.005)
    model.compile(loss = 'binary_crossentropy',
                  optimizer = opt, metrics = ['accuracy'])

    # train model
    history = model.fit(
      data_train_norm, labels_train,
      epochs = n_epochs, verbose = 0,
      batch_size = data_train_norm.shape[0])
    # save performances
    hist = pd.DataFrame(history.history)
    hist['epoch'] = history.epoch

    return hist, model
```

As you can see, the code is almost the same, except for the different cost function and the activation function of the output layer. Let's plot the decision boundary[2] for this problem. That means we will run our network on our dataset with the code

```
hist, model = create_and_train_regularized_model(train_x, train_y, 20, 4,
100, 0.0)
```

Figure 4-5 shows our dataset, whereby the white points are of one class and the black of the second. The gray area is the zone that the network classifies as being of one class and the white as the other class. You can see that the network can capture the complex structure of our data in a flexible way.

[2] In a statistical-classification problem with two classes, a decision boundary or decision surface is a surface that partitions the underlying space into two sets, one for each class (Source: Wikipedia at https://goo.gl/E5nELL).

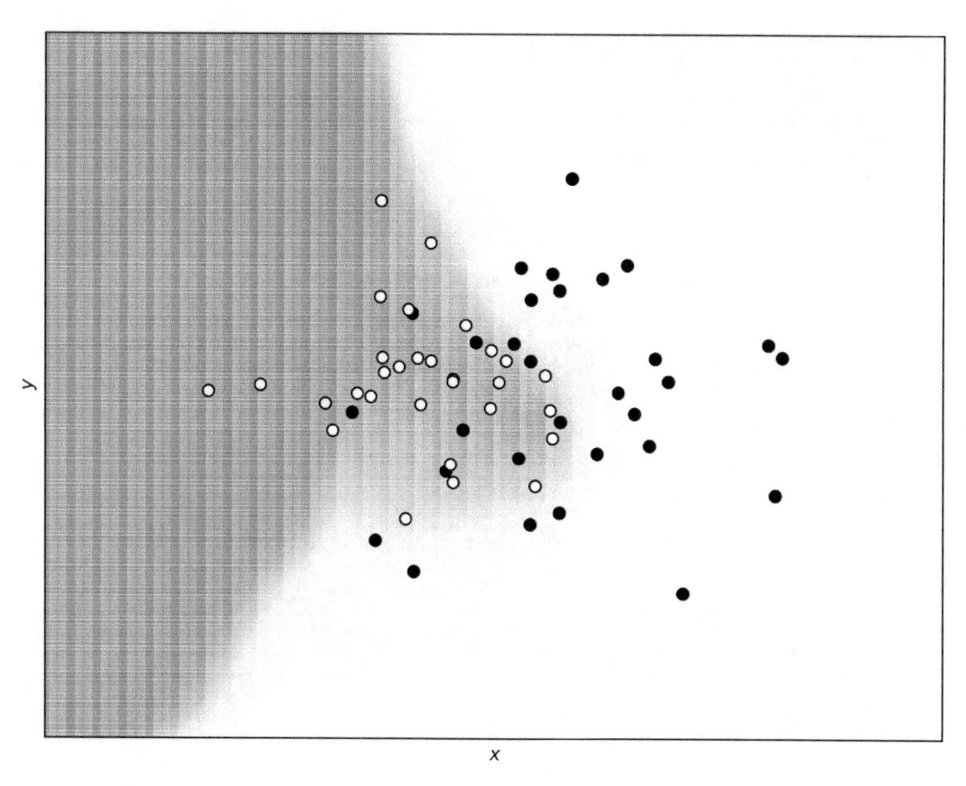

Figure 4-5. *Decision boundary without regularization. White points are of one class and black are the second. The gray area is the zone that the network classifies as being of one class and the white to the other. You can see that the network can capture the complex structure of this data*

Now let's apply regularization to the network, exactly as we did before, and see how the decision boundary is modified. We use a regularization parameter $\lambda = 0.04$.

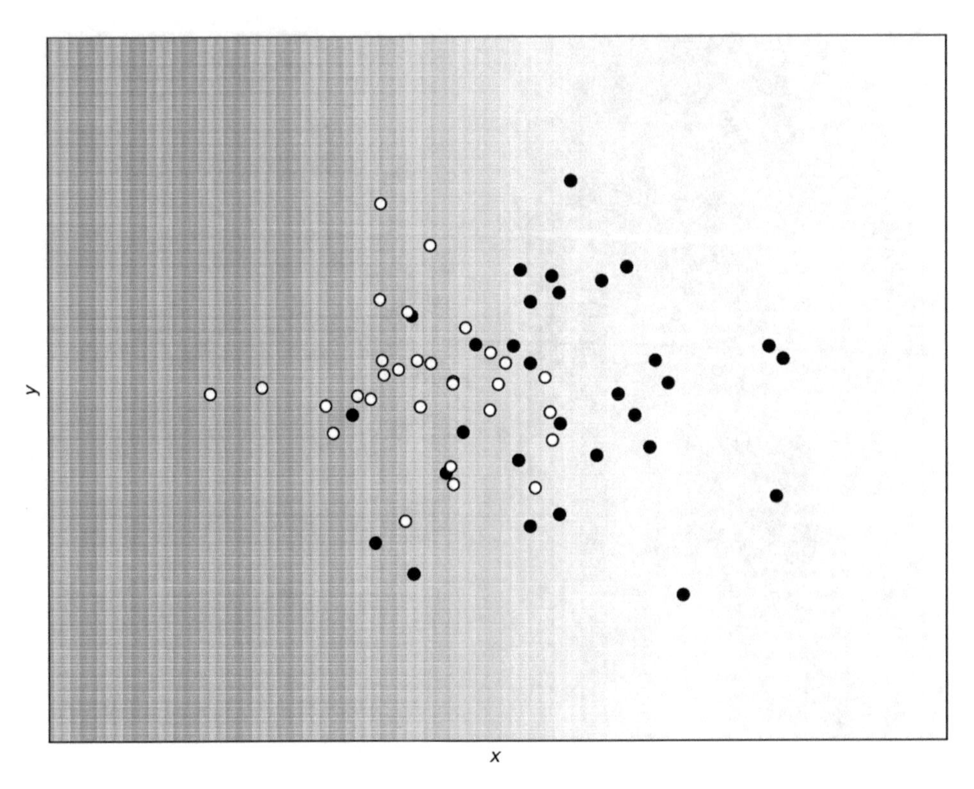

Figure 4-6. *The decision boundary as predicted by the network with ℓ_2 regularization and with a regularization parameter $\lambda = 0.04$*

You can clearly see in Figure 4-6 that the decision boundary is almost linear and is not able to capture the complex structure of the data anymore. Exactly what we expected: the regularization term makes the model simpler, and therefore less able to capture the fine structures. It is interesting to compare the decision boundary of our network with the result of logistic regression with just one neuron. We will not repeat the code here for space reasons (you can find the complete code version in the online version of the book), but if you compare the two decision boundaries in Figure 4-7 (the one coming from the network with one neuron that's linear), you can see that they are almost the same. The difference is that the regularized version presents a smoother decision boundary. To conclude, a regularization term of $\lambda = 0.04$ effectively gives the same results as a network with just one neuron.

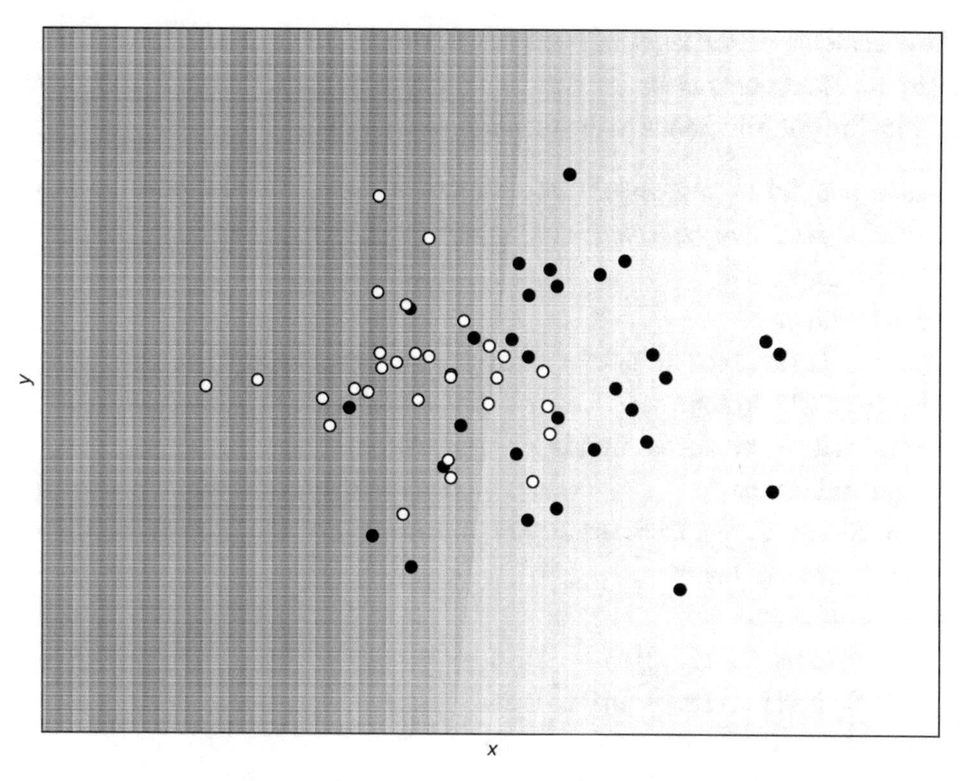

Figure 4-7. *The decision boundaries for the complex network with λ = 0.04 and for one with just one neuron. The two boundaries overlap nearly completely*

ℓ_1 Regularization

This section looks at a regularization technique that is very similar to ℓ_2 regularization. It is based on the same principle, adding a term to the cost function. This time, the mathematical form of the added term is different, but the method works very similarly to what you saw in the previous sections. Let's again look at the mathematics behind the algorithm.

Theory of ℓ_1 Regularization and Keras Implementation

ℓ_1 regularization also works by adding an additional term to the cost function

$$\tilde{L}(\boldsymbol{w}) = L(\boldsymbol{w}) + \frac{\lambda}{m}\|\boldsymbol{w}\|_1$$

The effect it has on the learning is effectively similar to the one described with ℓ_2 regularization. Keras has, as for ℓ_2, a function ready to be used. The code is the same as before, with the only difference in the regularized definition

```
def create_and_train_reg_model_L1(data_train_norm, labels_train, data_dev_
norm, labels_dev, num_neurons, num_layers, n_epochs, lambda_):
    # build model
    # input layer
    inputs = keras.Input(shape = data_train_norm.shape[1])
    # he initialization
    initializer = tf.keras.initializers.HeNormal()
    # regularization
    reg = tf.keras.regularizers.l1(l1 = lambda_)
    # first hidden layer
    dense = layers.Dense(
      num_neurons, activation = 'relu',
      kernel_initializer = initializer,
      kernel_regularizer = reg)(inputs)
    # customized number of layers and neurons per layer
    for i in range(num_layers - 1):
        dense = layers.Dense(num_neurons, activation = 'relu',
                             kernel_initializer = initializer,
                             kernel_regularizer = reg)(dense)
    # output layer
    outputs = layers.Dense(1)(dense)
    model = keras.Model(inputs = inputs,
                        outputs = outputs,
                        name = 'model')
    # set optimizer and loss
    opt = keras.optimizers.Adam(learning_rate = 0.001)
    model.compile(loss = 'mse',
                  optimizer = opt,
                  metrics = ['mse'])
```

```
    # train model
    history = model.fit(
      data_train_norm, labels_train,
      epochs = n_epochs, verbose = 0,
      batch_size = data_train_norm.shape[0],
      validation_data = (data_dev_norm, labels_dev))
    # save performances
    hist = pd.DataFrame(history.history)
    hist['epoch'] = history.epoch
    # print performances
    print('Cost function at epoch 0')
    print('Training MSE = ', hist['loss'].values[0])
    print('Dev MSE = ', hist['val_loss'].values[0])
    print('Cost function at epoch ' + str(n_epochs))
    print('Training MSE = ', hist['loss'].values[-1])
    print('Dev MSE = ', hist['val_loss'].values[-1])

    return hist, model
```

We can again compare the weights distribution between the model without the regularization term ($\lambda = 0.0$) and with regularization ($\lambda = 3.0$) in Figure 4-8. We used the Boston dataset for the calculation. We trained the model with the call

```
hist_reg, model_reg = create_and_train_reg_model_L1(train_x, train_y,
dev_x, dev_y, 20, 4, 1000, 3.0)
```

once with $\lambda = 0.0$ and once with $\lambda = 3.0$.

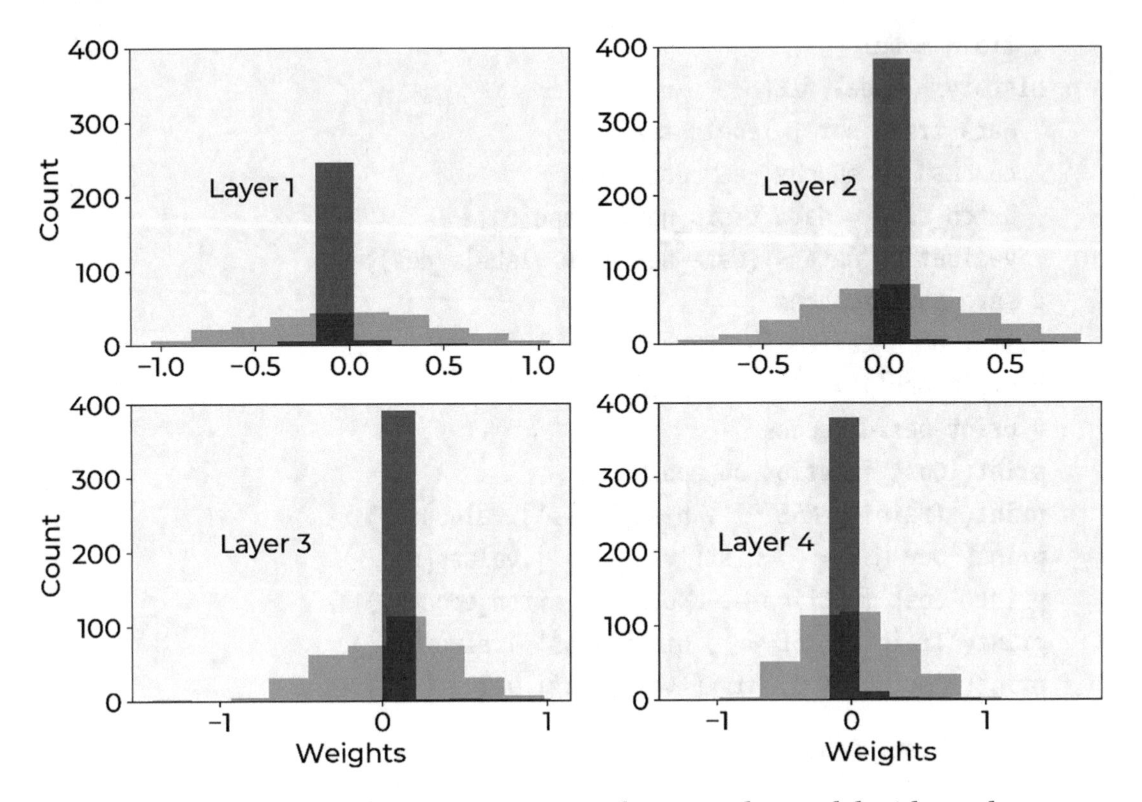

Figure 4-8. *Weight distribution comparison between the model without the ℓ_1 regularization term ($\lambda = 0.0$, light gray) and with ℓ_1 regularization ($\lambda = 3.0$, dark gray)*

As you can see, ℓ_1 regularization has the same effect as ℓ_2. It reduces the effective complexity of the network, reducing many weights to zero.

To give you an idea of how effective regularization is in reducing the weights, Table 4-2 compares the percentage of weights less than 10^{-3} with and without regularization after 1000 epochs.

***Table 4-2.** Percentage of Weights Less Than 10^{-3} With and Without Regularization After 1000 Epochs*

Layer	Percent of Weight Less Than 10^{-3} for $\lambda = 0.0$	Percent of Weight Less Than 10^{-3} for $\lambda = 3.0$
1	0.0	90.77
2	0.5	94.50
3	0.0	96.75
4	0.0	94.50

Are the Weights Really Going to Zero?

It is instructive to see why the weights are going to zero. For illustrative purposes in Figure 4-9, you can see weight $w_{12,5}^{[3]}$ (from layer 3) plotted versus the number of epochs for our artificial dataset with two features, ℓ_2 regularization, $\gamma = 10^{-3}$, $\lambda = 0.1$, after 1000 epochs. You can see how it quickly decreases to zero. The value after 1000 epochs is $2 \cdot 10^{-21}$, so for all purposes zero.

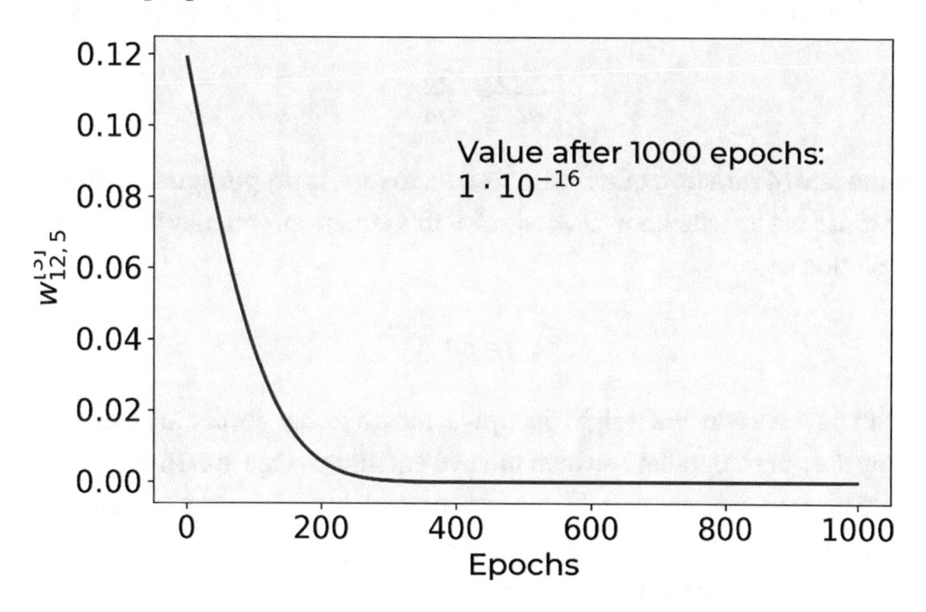

***Figure 4-9.** Weight $w_{12,5}^{[3]}$ plotted versus the epochs for our artificial dataset with two features, ℓ_2 regularization, $\gamma = 10^{-3}$, $\lambda = 0.1$, trained for 1000 epochs*

If you were wondering, the weight goes to zero almost exponentially. A way of understanding why is to consider the weight update equation for one weight

$$w_{j,[n+1]} = w_{j,[n]}\left(1 - \frac{\gamma\lambda}{m}\right) - \frac{\gamma \partial L\left(\boldsymbol{w}_{[n]}\right)}{\partial w_j}$$

Let's now suppose we find ourselves close to the minimum, in a region where the derivative of the cost function *J* is almost zero, so that we can neglect it. In other words, let's suppose

$$\frac{\partial L\left(\boldsymbol{w}_{[n]}\right)}{\partial w_j} \approx 0$$

We can rewrite the weight update equation as

$$w_{j,[n+1]} - w_{j,[n]} = -w_{j,[n]}\frac{\gamma\lambda}{m}$$

Now the rate of variation of the weight with respect to the iteration number is proportional to the weight itself. For those of you with knowledge of differential equations, you may realize that we can draw a parallel to the following equation

$$\frac{dx(t)}{dt} = -\frac{\gamma\lambda}{m}x(t)$$

Now the rate of variation of $x(t)$ with respect to time is proportional to the function itself. For those of you who know how to solve this equation, you may know that a generic solution is

$$x(t) = Ae^{-\frac{\gamma\lambda}{m}(t - t_0)}$$

You can now see why the weight decay will have a decay similar to an exponential function by drawing a parallel between the two equations. Figure 4-10 shows the weight decay we discussed together with a pure exponential decay. The two curves are not identical, as expected, since especially at the beginning the gradient of the cost function is surely not zero. But the similarity is remarkable and should give you an idea of how fast the weights can go to zero (really quickly).

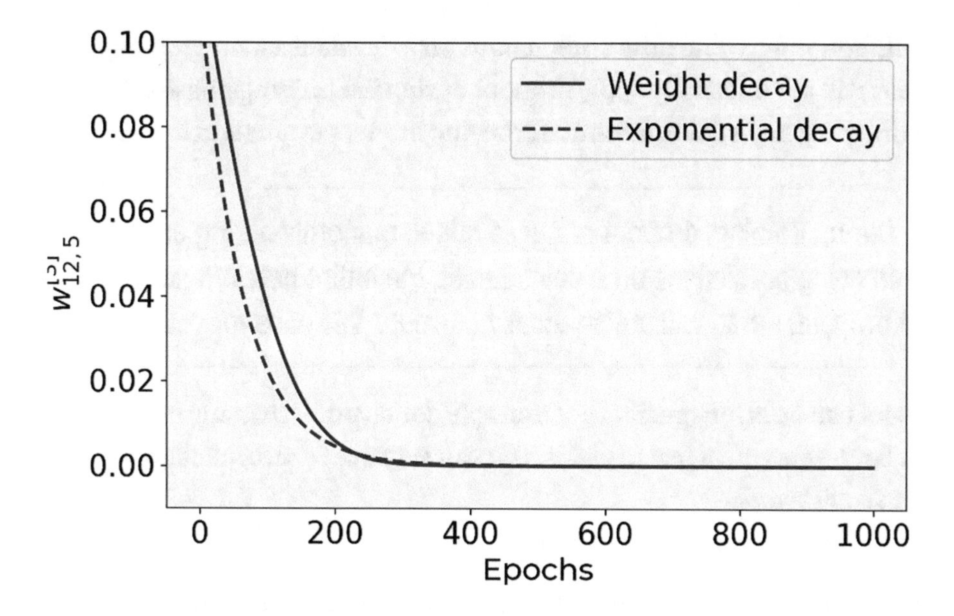

Figure 4-10. *Weight $w^{[3]}_{12,5}$ plotted versus the epochs for our artificial dataset with two features, ℓ_2 regularization, $\gamma = 10^{-3}$, $\lambda = 0.1$, trained for 1000 epochs (continous line), together with a pure exponential decay (dashed line) for illustrative purposes*

Note that when using regularization, you end up having tensors with a lot of zero elements, called sparse tensors. You can then profit from special routines that are extremely efficient with sparse tensors. This is something to keep in mind when you start moving toward more complex models, but a subject too advanced for this book.

Dropout

The basic idea of *dropout* is different: during the training phase you remove nodes from layer l randomly with a probability $p^{[l]}$. In each iteration you remove different nodes, effectively training at each iteration a different network (when using mini-batches, you train a different network for each batch, for example).

In Keras, you simply add how many dropout layers you want after the layer you need to drop, using the following function: `keras.layers.Dropout(rate)`. You must put as input the layer you want to drop, and you must set the `rate` parameter. This parameter can assume float values in the range [0, 1), since it represents the fraction of the input units to drop. Therefore, it is not possible to drop all the units (setting a rate equal to 1). Usually, the `rate` parameter is set the same for all networks (but technically speaking, can be layer specific).

Very importantly, when doing predictions on a dev dataset, no dropout should be used! Keras will automatically apply dropout during the training phase of the model, without dropping any additional unit during the model's evaluation on a different set.

Note During training, dropout removes nodes randomly during each iteration. But when doing predictions on a dev dataset, the entire network needs to be used without dropout. Keras will automatically consider this case for you.

Dropout can be layer-specific. For example, for layers with many neurons, `rate` can be small. For layers with a few neurons, you can set `rate = 0.0`, effectively keeping all neurons in such layers.

The implementation in Keras is easy.

```
def create_and_train_reg_model_dropout(data_train_norm, labels_train, data_
dev_norm, labels_dev, num_neurons, num_layers, n_epochs, rate):
    # build model
    # input layer
    inputs = keras.Input(shape = data_train_norm.shape[1])
    # he initialization
    initializer = tf.keras.initializers.HeNormal()
    # first hidden layer
    dense = layers.Dense(
      num_neurons, activation = 'relu',
      kernel_initializer = initializer)(inputs)
    # first dropout layer
    dense = keras.layers.Dropout(rate)(dense)
    # customized number of layers and neurons per layer
    for i in range(num_layers - 1):
        dense = layers.Dense(
          num_neurons, activation = 'relu',
          kernel_initializer = initializer)(dense)
        # customized number of dropout layers
        dense = keras.layers.Dropout(rate)(dense)
    # output layer
    outputs = layers.Dense(1)(dense)
```

```
    model = keras.Model(inputs = inputs,
                        outputs = outputs,
                        name = 'model')
    # set optimizer and loss
    opt = keras.optimizers.Adam(learning_rate = 0.001)
    model.compile(loss = 'mse', optimizer = opt,
                  metrics = ['mse'])

    # train model
    history = model.fit(
      data_train_norm, labels_train,
      epochs = n_epochs, verbose = 0,
      batch_size = data_train_norm.shape[0],
      validation_data = (data_dev_norm, labels_dev))
    # save performances
    hist = pd.DataFrame(history.history)
    hist['epoch'] = history.epoch
    # print performances
    print('Cost function at epoch 0')
    print('Training MSE = ', hist['loss'].values[0])
    print('Dev MSE = ', hist['val_loss'].values[0])
    print('Cost function at epoch ' + str(n_epochs))
    print('Training MSE = ', hist['loss'].values[-1])
    print('Dev MSE = ', hist['val_loss'].values[-1])

    return hist, model
```

As you can see, you must put a dropout layer (highlighted in bold) after the layer you want to modify, setting the `rate` parameter.

Now let's analyze what happens to the cost function when using dropout. Let's run our model applied to the Boston dataset for two values of the `rate` variable: 0.0 (without dropout) and 0.5. In Figure 4-11, you can see that when applying dropout, the cost function is very irregular. It oscillates wildly. The two models have been evaluated with the calls

```
hist_reg, model_reg = create_and_train_reg_model_dropout(train_x, train_y,
dev_x, dev_y, 20, 4, 8000, 0.50)
```

for rate = 0.0 and for 0.50.

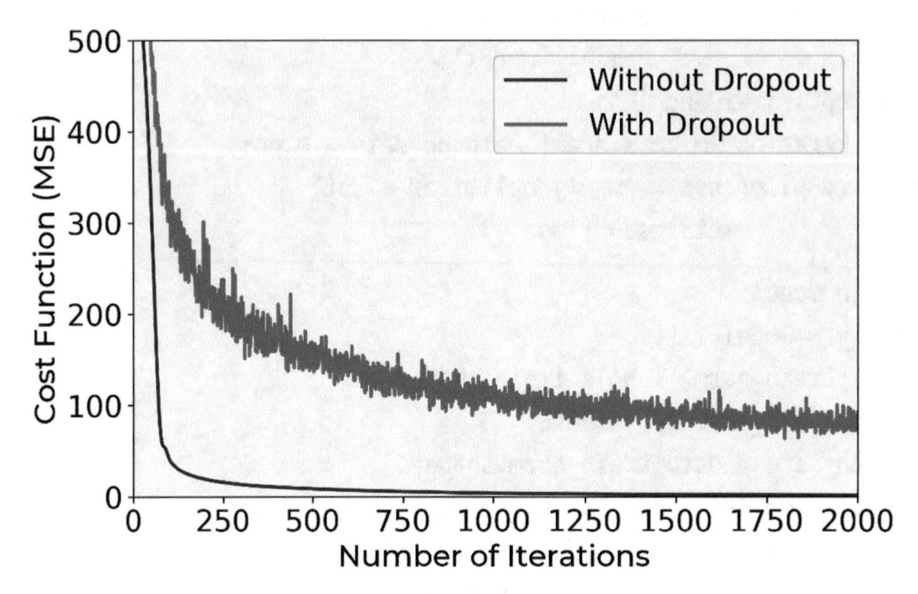

***Figure 4-11.** The cost function for the training dataset for our model for two values of the rate variable: 0.0 (no dropout) and 0.50. $\gamma = 0.001$. The models have been trained for 8000 epochs. No mini-batch has been used. The oscillating line is the one evaluated with regularization*

Figure 4-12 shows the evolution of the MSE for the training and dev datasets in case of dropout (`rate = 0.5`).

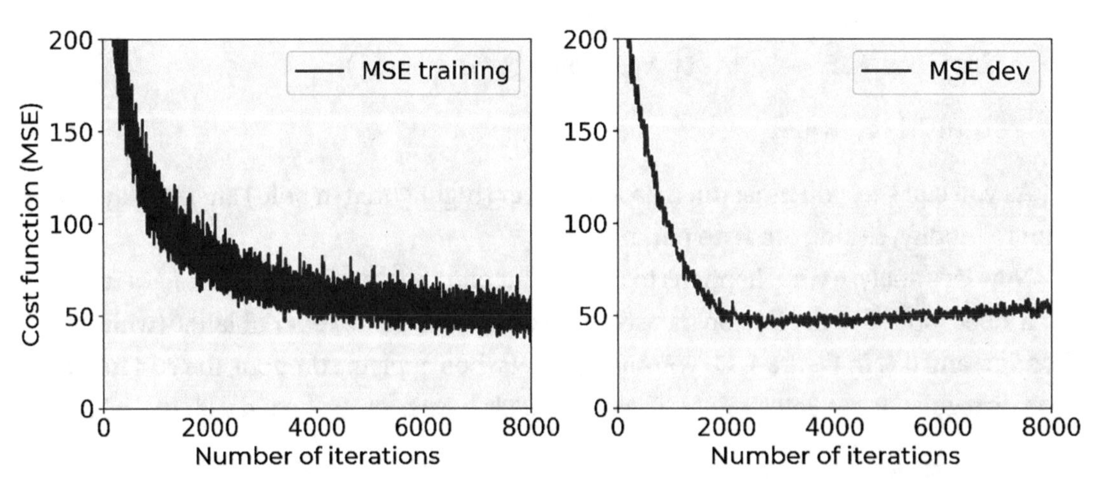

***Figure 4-12.** MSE for the training and dev datasets in case of dropout (rate = 0.50)*

Figure 4-13 shows the same plot but without dropout. The difference is quite striking. Very interesting is the fact that, without dropout, MSE_{dev} grows with epochs, while when using dropout it is rather stable.

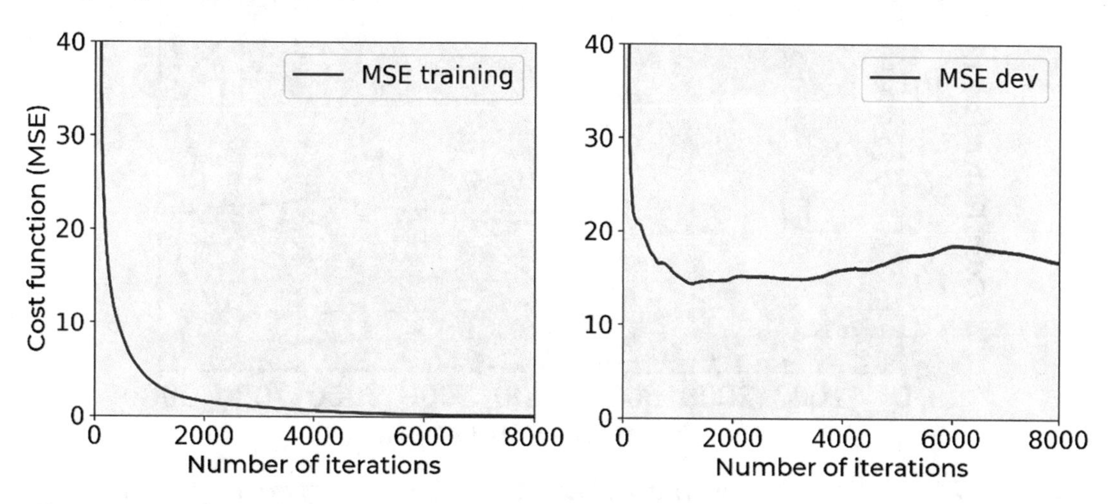

Figure 4-13. *MSE for the training and dev datasets in case of no dropout (rate = 0.0)*

Figure 4-13 shows that the MSE_{dev} grows after dropping at the beginning. The model is in a clear extreme overfitting regime ($MSE_{train} \ll MSE_{dev}$), and it generalizes worst when applied to new data. In Figure 4-12, you can see how the MSE_{train} and MSE_{dev} are of the same order of magnitude and the MSE_{dev} does not continue to grow, so we have a model that is a lot better at generalizing than the one shown in Figure 4-13.

Note When applying dropout, your metric (in this case, the MSE) will oscillate, so do not be surprised when trying to find the best hyper-parameters if you see your optimizing metric oscillating.

Early Stopping

Early stopping is another technique that is sometimes used to fight overfitting. Strictly speaking, this method does nothing to avoid overfitting; it simply stops the learning before the overfitting problem becomes too bad. Consider the example in the last section. Figure 4-14 shows the MSE_{train} and MSE_{dev} plotted on the same plot.

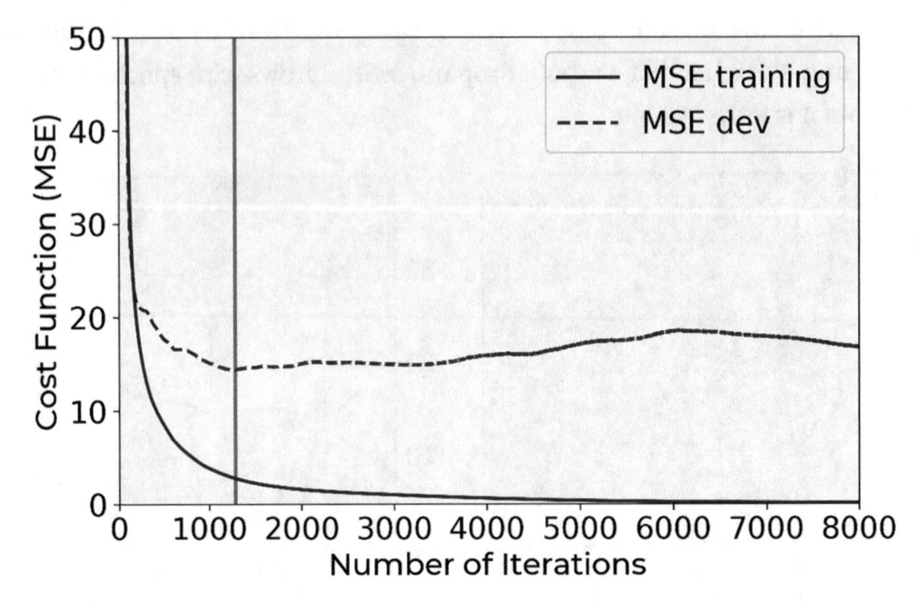

Figure 4-14. *MSE for the training and dev datasets in case of no dropout (rate = 0.0). Early stopping consists of stopping the learning phase at the iteration when the MSE_{dev} is at a minimum (indicated with a vertical line in the plot)*

Early stopping simply consists of stopping the training at the point when the MSE_{dev} has its minimum. (In Figure 4-14, the minimum is indicated by a vertical line.) Note that this is not an ideal way of solving the overfitting problem. Your model will still most probably generalize very badly to new data. It is usually preferable to use other techniques. Additionally, this is also time consuming and a manual process that is very error prone. You can get a good overview of the different contexts by checking the Wikipedia page at `https://goo.gl/xnKo2s`.

Additional Methods

All the methods we have discussed so far consist of, in some form or another, making the model less complex. You keep the data as it is and modify your model. But we can also try to do the opposite: leave the model as it is and work on the data. Here are two common strategies that help prevent overfitting (but are not easily applicable):

- **Get more data.** This is the simplest way of fighting overfitting. Unfortunately, very often in real life this is not possible. If you are classifying pictures of cats taken with a smartphone you may think of getting more data from the web. Although this may seem like a perfectly good idea, you may discover that the images have different quality, that possibly not all the images are really cats (what about cat toys?), you may only find images of white cats, and so on. Basically, your additional observations may come from a very different distribution than your original data and that will be a problem. So, when getting additional data, consider this problem well before proceeding.

- **Augment your data.** For example, if you are working with images you can generate additional images by rotating, stretching, shifting, and otherwise editing your original images. That is a very common technique that may help.

The problem of making the model generalize better on new data is one of machine learning's biggest goal. It is a complicated problem that requires experience and tests. Lots of tests. A lot of research is currently going on to solve such problems when working on very complex problems.

Exercises

EXERCISE 1 (LEVEL: EASY)

Try to determine which architecture (number of layers and number of neurons) does not overfit the Boston dataset. When does the network start overfitting? Which network would provide a good result? At a minimum, try the following combinations:

Number of Layers	Number of Neurons in Each Layer
1	3
1	5
2	3
2	5

EXERCISE 2 (LEVEL: MEDIUM)

Find the minimum value for λ (in the case of ℓ_2) for which the overfitting stops. Perform a set of tests using the function hist, `model = create_and_train_reg_model_L2(train_x, train_y, dev_x, dev_y, 20, 4, 0.0)`, varying the value of λ from 0 to 10.0 in regular increments (you can decide which values you want to test). Use at a minimum the values: 0, 0.5, 1.0, 2.0, 5.0, 7.0, 10.0, and 15.0. After that, plot the value of the cost function on the training dataset and on the dev dataset vs. λ.

EXERCISE 3 (LEVEL: MEDIUM)

In the ℓ_1 regularization example applied to the Boston dataset, plot the amount of weights close to zero in hidden layer 3 vs. λ. Considering only layer 3, plot the quantity `(np.sum(np.abs(weights3) < 1e-3)) / weights3.size * 100.0` we have evaluated before and calculate it for several values of λ. Consider at least these values: 0, 0.5, 1.0, 2.0, 5.0, 7.0, 10.0, and 15.0. Plot the value vs. λ. What shape does the curve have? Does it flatten out?

EXERCISE 4 (LEVEL: VERY DIFFICULT)

Implement ℓ_2 regularization from scratch.

References

[1] Delve (Data for Evaluating Learning in Valid Experiments), "The Boston Housing Dataset," `www.cs.toronto.edu/~delve/data/boston/bostonDetail.html`, 1996, last accessed 22.03.2021.

[2] Bishop, C.M, (1995) Neural Networks for Pattern Recognition, Oxford University Press.

[3] Goodfellow, I.J. et al., Deep Learning, MIT Press.

[4] Kukačka, J. et al., Regularization for *Deep Learning: A Taxonomy,* arXiv: 1710.10686v1, available at `https://goo.gl/wNkjXz`, last accessed 28.03.2021.

CHAPTER 5

Advanced Optimizers

In general, *optimizers* are algorithms that minimize a given function. Remember that training a neural network means simply minimizing the loss function. Chapter 1 looked at the gradient descent optimizer and its variations (mini-batch and stochastic). In this chapter, we look at more advanced and efficient optimizers. We look in particular at *Momentum, RMSProp,* and *Adam.* We cover the mathematics behind them and then explain how to implement and use them in Keras.

Available Optimizers in Keras in TensorFlow 2.5

TensorFlow has evolved a lot in the last few years. In the beginning, you could find gradient descent as a class, but it is not available anymore (not directly). Keras offers as of TensorFlow version 2.5 only advanced algorithms. In particular, you will find the following: Adadelta, Adagrad, Adam. Adamax, Ftrl, Nadam, RMSProp, and SGD (gradient descent with momentum). So plain gradient descent is not available anymore out of the box from Keras.

Note In general, you should determine which optimizers work best for your problem, but if you don't know where to start, Adam is always a good choice.

Advanced Optimizers

Up to now we have only discussed how gradient descent works. That is not the most efficient optimizer (although it's the easiest to understand), and there are some modifications to the algorithm that can make it faster and more efficient. This is a very active area of research, and you will find an incredible number of algorithms based on different ideas. This chapter looks at the most instructive and famous ones: Momentum, RMSProp, and Adam.

U. Michelucci, *Applied Deep Learning with TensorFlow 2,* https://doi.org/10.1007/978-1-4842-8020-1_5

Additional material that investigates the more exotic optimizers has been written by S. Ruder in a paper called "An overview of gradient descent optimization algorithms" (see `https://goo.gl/KgKVgG`). The paper is not for beginners and requires a strong mathematical background, but it provides an overview of the more exotic algorithms, including Adagrad, Adadelta, and Nadam. Additionally, it reviews weight-update schemes applicable in distributed environments like Hogwild!, Downpour SGD, and many more. Surely a read worth your time.

To understand the basic idea of *momentum* (and partially also of RMSProp and Adam), you first need to understand what exponentially weighted averages are. So, let's start with some mathematics.

Exponentially Weighted Averages

Let's suppose you are measuring a quantity θ (it could be the temperature where you live, for example) over time—once a day, for example. You will have a series of measurements that you can indicate with θ_i, where i goes from 1 to a certain number N. Now bear with me if this does not make much sense at first. Let's define recursively a quantity v_n, as follows

$$
\begin{aligned}
v_0 &= 1 \\
v_1 &= \beta v_0 + (1-\beta)\theta_1 \\
v_2 &= \beta v_1 + (1-\beta)\theta_2
\end{aligned}
$$

and so on with β a real number and $0 < \beta < 1$. Generally, we could write the nth term with a recursive formula, as follows

$$v_n = \beta v_{n-1} + (1-\beta)\theta_n$$

Now let's write all the terms v_1, v_2 and so on as a function of β and θ_i (so not recursively). For v_2 we have

$$v_2 = \beta\left(\beta v_0 + (1-\beta)\theta_1\right) + (1-\beta)\theta_2 = \beta^2 + (1-\beta)(\beta\theta_1 + \theta_2)$$

For v_3 we have

$$v_3 = \beta^3 + (1-\beta)\left[\beta^2\theta_1 + \beta\theta_2 + \theta_3\right]$$

Generalizing, we obtain

$$v_n = \beta^n + (1-\beta)\left[\beta^{n-1}\theta_1 + \beta^{n-2}\theta_2 + \ldots + \theta_n\right]$$

Or in a more elegant way (without the three dots), we get

$$v_n = \beta^n + (1-\beta)\sum_{i=1}^{n}\beta^{n-i}\theta_i$$

Now let's try to understand what this formula means. First of all, note that the term β^n disappears when you choose $v_0 = 0$. Let's do that (set $v_0 = 0$) and consider what remains

$$v_n = (1-\beta)\sum_{i=1}^{n}\beta^{n-i}\theta_i$$

Are you still with me? Now comes the interesting part. Let's define the *convolution* between two sequences.[1] Let's consider two sequences, x_n and h_n. The convolution between the two (which we indicate with the symbol $*$) is defined by

$$x_n * h_n = \sum_{k=-\infty}^{\infty} x_k h_{n-k}$$

Since we only have a finite number of measurements for our quantity θ_i, we can define

$$\theta_k = 0 \qquad k > n, k \leq 0$$

Therefore, we can write v_n as a convolution as follows

$$v_n = \theta_n * b_n$$

Where we have defined

$$b_n = (1-\beta)\beta^n$$

[1] Generally speaking, a sequence is an enumerated collection of objects.

To get an idea of what that means, let's plot θ_n, b_n, and v_n. To do that, let's assume θ_n has a Gaussian shape (the exact form is not relevant, is just for illustrative purposes), and let's use $\beta = 0.9$ (see Figure 5-1).

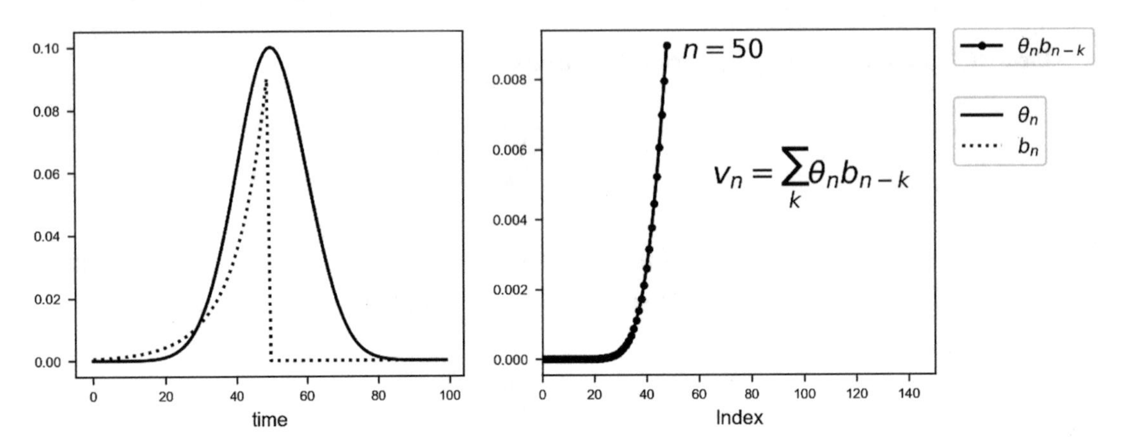

***Figure 5-1.** The left side is a plot showing θ_n (solid line) and b_n (dotted line) together. The right side is a plot showing the points that need to be summed to obtain v_n for $n = 50$*

Let's discuss briefly Figure 5-1. The Gaussian curve (θ_n) will be convoluted with b_n to obtain v_n. The result can be seen in the right plot. All those terms $(1-\beta)\beta^{n-i}\theta_i$ for $i = 1, \ldots, 50$ (plotted in the right plot) will be summed to obtain v_{50}. Note that v_n is the average of all θ_n for $n = 1, \ldots, 50$. Each term is multiplied by a term (b_n); that is, 1 for $n = 50$ and then this decreases rapidly for n decreasing toward 1. Basically, this is a weighted average, with an exponentially decreasing weight. The terms farther from $n = 50$ are less and less relevant, while the terms close to $n = 50$ get more weight. This is also a moving average. For each n, all the preceding terms are added and multiplied by a weight (b_n).

I would like now to show you why there is this factor $1 - \beta$ in b_n. Why not just choose β^n? The reason is very simple. The sum of b_n over all positive ns is equal to 1. Let's see why.

$$\sum_{k=1}^{\infty} b_k = (1-\beta)\sum_{k=1}^{\infty}\beta^n = (1-\beta)\lim_{N\to\infty}\frac{1-\beta^{N+1}}{1-\beta} = (1-\beta)\frac{1}{1-\beta} = 1$$

We used the fact that for $\beta < 1$ we have $\lim_{N\to\infty} \beta^{N+1} = 0$ and that for a geometric series we have

$$\sum_{k=1}^{n} ar^{k-1} = \frac{a\left(1-r^n\right)}{1-r}$$

The algorithm we described to calculate v_n is nothing else than the convolution of our quantity θ_i, with a series that has a sum equal to one and has the form $(1-\beta)\beta^i$.

Note The exponentially weighted average v_n of a series of a quantity θ_n is the convolution $v_n = \theta_n * b_n$ of our quantity θ_i with $b_n = (1-\beta)\beta^n$, where b_n has the property that its sum over the positive values of n is equal to 1. It is the moving average, where each term is multiplied by the weights given by the sequence b_n.

As you choose smaller and smaller β values, the number of points θ_n that have a weight that's significantly different than zero decreases, as you can see in Figure 5-2, where we have plotted the series b_n for different values of β.

$$n = 0.$$

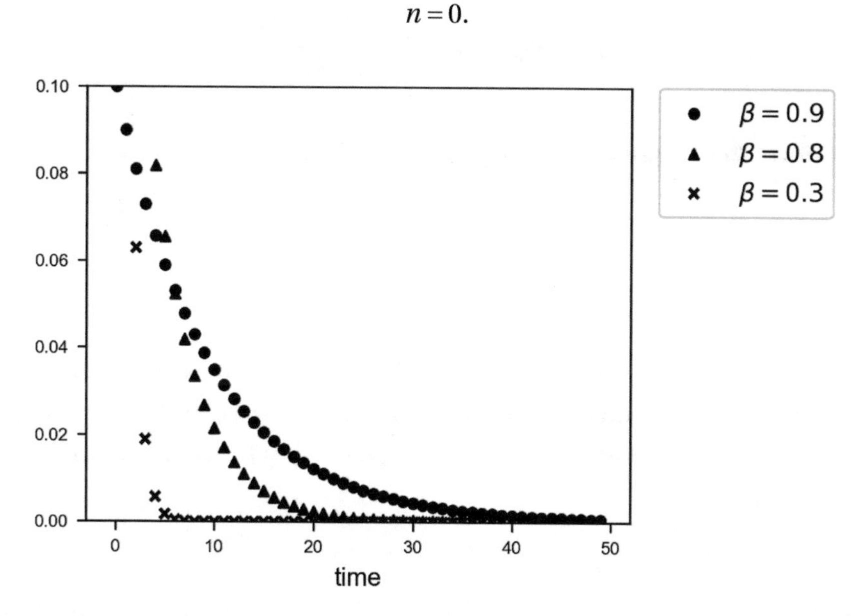

Figure 5-2. *The series b_n for three values of β: 0.9, 0.8, and 0.3. Note that as β gets smaller, the series is significantly different than zero for an increasingly smaller number of values*

This method is at the very core of the Momentum optimizer and more advanced learning algorithms, and you will see in the next sections how it works in practice.

Momentum

Recall that in plain gradient descent the weight updates are calculated with these equations

$$
\begin{cases}
\boldsymbol{w}_{[n+1]} = \boldsymbol{w}_{[n]} - \gamma \nabla_{\mathbf{w}} L\left(\boldsymbol{w}_{[n]}, b_{[n]}\right) \\
b_{[n+1]} = b_{[n]} - \gamma \dfrac{\partial L\left(\boldsymbol{w}_{[n]}, b_{[n]}\right)}{\partial b}
\end{cases}
$$

Where the subscript in square brackets indicates the iteration. The idea behind the Momentum optimizer is to use exponentially weighted averages of the corrections of the gradient for the weight updates. Mathematically, we formulate the previous statement as follows

$$
\begin{cases}
\boldsymbol{v}_{w,[n+1]} = \beta \boldsymbol{v}_{w,[n]} + (1-\beta) \nabla_{\mathbf{w}} L\left(\boldsymbol{w}_{[n]}, b_{[n]}\right) \\
v_{b,[n+1]} = \beta v_{b,[n]} + (1-\beta) \dfrac{\partial L\left(\boldsymbol{w}_{[n]}, b_{[n]}\right)}{\partial b}
\end{cases}
$$

Then perform the weight and bias updates with these equations

$$
\begin{cases}
\boldsymbol{w}_{[n+1]} = \boldsymbol{w}_{[n]} - \gamma \boldsymbol{v}_{w,[n]} \\
b_{[n+1]} = b_{[n]} - \gamma v_{b,[n]}
\end{cases}
$$

Where $\boldsymbol{v}_{w,[0]} = \mathbf{0}$ and $v_{b,[0]} = 0$ are usually chosen. Now that means, instead of using the derivatives of the cost functions with respect to the weights, we update the weights with a moving average of the derivatives. Usually, experience shows that a bias correction could theoretically be neglected.

Note The Momentum algorithm uses an exponential weighted average of the derivates of the cost function with respect to the weights for the weight updates. This way, in addition to the derivatives at a given iteration being used, *the past behavior is also considered.* It may happen that the algorithm oscillates around the minimum instead of converging directly. This algorithm can escape from plateaus much more efficiently than standard gradient descent.

Sometimes you find a slightly different formulation in books or blogs, which is (we report only the equation for the weights $\boldsymbol{w}$ for brevity)

$$\boldsymbol{v}_{w,[n+1]} = \gamma \boldsymbol{v}_{w,[n]} + \eta \nabla_{\mathbf{w}} L\left(\boldsymbol{w}_{[n]}, b_{[n]}\right)$$

The idea and meaning remain the same; it's simply a slightly different mathematical formulation. We find that the method described is easier to understand with the notion of sequence convolution and of weighted averages than this second formulation. Another formulation that you will find (the one that TensorFlow uses) is

$$v_{w,[n+1]} = \eta^t v_{w,[n]} + \nabla_{\mathbf{w}} L\left(\boldsymbol{w}_{[n]}, b_{[n]}\right)$$

Where η^t is called by TensorFlow momentum (the superscript t indicates that this variable is used by TensorFlow). In this formulation, the weight update assumes the form

$$\boldsymbol{w}_{[n+1]} = \boldsymbol{w}_{[n]} - \gamma^t v_{w,[n+1]} = \boldsymbol{w}_{[n]} - \gamma^t \left(\eta^t v_{w,[n]} + \nabla_{\mathbf{w}} L\left(\boldsymbol{w}_{[n]}, b_{[n]}\right)\right) = \boldsymbol{w}_{[n]} - \gamma^t \eta^t v_{w,[n]} - \gamma^t \nabla_{\mathbf{w}} L\left(\boldsymbol{w}_{[n]}, b_{[n]}\right)$$

Where again the superscript t indicates that the variable is the one used by TensorFlow. Although it seems different, this formulation is equivalent to the formulation shown earlier:

$$\boldsymbol{w}_{[n+1]} = \boldsymbol{w}_{[n]} - \gamma \beta v_{w,[n]} - \gamma (1 - \beta) \nabla_{\mathbf{w}} L\left(\boldsymbol{w}_{[n]}, b_{[n]}\right)$$

The TensorFlow formulation and the one discussed earlier are equivalent if we choose

$$\begin{cases} \eta = \dfrac{\beta}{1-\beta} \\ \gamma^t = \gamma(1-\beta) \end{cases}$$

This can be seen by simply comparing the two different equations for the weight updates. Values around $\eta = 0.9$ in TensorFlow implementations are used and they typically work well. The momentum almost always converges faster than with plain gradient descent.

Note Comparing the different parameters in the different optimizers is wrong. When talking about the learning rate, for example, it has a different meaning in the different algorithms. What you should compare is the best convergence speed you can achieve with several optimizers, regardless of the choice of parameters. Comparing the GD for a learning rate of 0.01 with Adam (you will see it later) for the same learning rate does not make much sense. You should compare the optimizers with the parameters to see which gives you the best and fastest convergence.

RMSProp

Let's move on to something a bit more complex, but usually more efficient. Let's look at the mathematical equations and then we will compare them to the others we have seen so far. At each iteration, we need to calculate

$$\begin{cases} \boldsymbol{S}_{\boldsymbol{w},[n+1]} = \beta_2 \boldsymbol{S}_{\boldsymbol{w},[n]} + (1-\beta_2)\nabla_{\boldsymbol{w}} J(\boldsymbol{w},b) \circ \nabla_{\boldsymbol{w}} L(\boldsymbol{w},b) \\ S_{b,[n+1]} = \beta_2 S_{b,[n]} + (1-\beta_2)\dfrac{\partial L(w,b)}{\partial b} \circ \dfrac{\partial L(w,b)}{\partial b} \end{cases}$$

where the symbol $\circ$ indicates an element-wise product. Then we will update our weights with the equations

$$\begin{cases} \boldsymbol{w}_{[n+1]} = \boldsymbol{w}_{[n]} - \dfrac{\gamma \nabla_{\boldsymbol{w}} L(\boldsymbol{w},b)}{\sqrt{\boldsymbol{S}_{\boldsymbol{w},[n+1]}} + \epsilon} \\ b_{[n+1]} = b_{[n]} - \gamma \dfrac{\partial L(\boldsymbol{w},b)}{\partial b} \dfrac{1}{\sqrt{S_{b,[n]}} + \epsilon} \end{cases}$$

So first you do an exponential weighted average of the quantities $\boldsymbol{S}_{\boldsymbol{w},[n+1]}$ and $S_{b,[n+1]}$ and then you use them to modify the derivates that you use to do your weight updates. The ϵ, which is usually $\epsilon = 10^{-8}$, is there to prevent the denominator from going to 0 when the quantities $\boldsymbol{S}_{\boldsymbol{w},[n+1]}$ and $S_{b,[n+1]}$ go to 0. The idea is that if the derivative is big, the S quantities are big. Therefore, the factors $1/\sqrt{\boldsymbol{S}_{\boldsymbol{w},[n+1]}} + \epsilon$ and $1/\sqrt{S_{b,[n]}} + \epsilon$ will be smaller and therefore the learning will slow down. The other way around is also true, so if the derivatives are small, the learning will be faster. This algorithm will make the learning faster for the parameters that are slowing it down. In TensorFlow is particular easy to use it simply with this code:

```
optimizer = tf.keras.optimizers.RMSprop(learning_rate=0.1)
```

Adam

The last algorithm we look at is called Adam (Adaptive Moment estimation). It combines the ideas of RMSProp and Momentum into one optimizer. Like Momentum, it uses an exponential weighted average of past derivatives and, like RMSProp, it uses the exponentially weighted averages of past squared derivatives.

You need to calculate the same quantities that you need for Momentum and for RMSProp and then calculate the following quantities

$$\boldsymbol{v}^{corrected}_{\boldsymbol{w},[n]} = \frac{\boldsymbol{v}_{\boldsymbol{w},[n]}}{1-\beta_1^n}$$

$$v^{corrected}_{b,[n]} = \frac{v_{b,[n]}}{1-\beta_1^n}$$

And in the same way

$$\boldsymbol{S}_{\boldsymbol{w},[n]}^{corrected} = \frac{\boldsymbol{S}_{\boldsymbol{w},[n]}}{1-\beta_2^n}$$

$$S_{b,[n]}^{corrected} = \frac{S_{b,[n]}}{1-\beta_2^n}$$

Where we use β_1 for the hyper-parameter, used in Momentum, and β_2 for the one we used in RMSProp. Then, similar to what we did in RMSProp, we update our weights with the equations

$$\begin{cases} \boldsymbol{w}_{[n+1]} = \boldsymbol{w}_{[n]} - \dfrac{\gamma \boldsymbol{v}_{\boldsymbol{w},[n]}^{corrected}}{\sqrt{\boldsymbol{S}_{\boldsymbol{w},[n+1]}^{corrected}} + \epsilon} \\ b_{[n+1]} = b_{[n]} - \gamma \dfrac{v_{b,[n]}^{corrected}}{\sqrt{S_{b,[n]}^{corrected}} + \epsilon} \end{cases}$$

TensorFlow does everything for you if you simply use the following line

```
optimizer = tf.keras.optimizers.Adam(learning_rate=0.001, beta_1=0.9,
beta_2=0.999, epsilon=1e-07)
```

where in this case the typical values for the parameters have been chosen: $\gamma = 0.001$, $\beta_1 = 0.9$, $\beta_2 = 0.999$, and $\epsilon = 10^{-7}$. Note how, since this algorithm adapts the learning rate to the situation, we can start with a bigger learning rate to speed up the convergence.

Comparison of the Optimizers' Performance

It is interesting to see how the optimizers behave to better understand why, for example, Adam is so efficient. To do that, let's create a toy problem. Let's consider a dataset of 30 tuples (x_i, y_i) with

$$x_i = -1 + \frac{2}{30} i$$

And

$$y_i = 2 + \frac{1}{2} x_i$$

Let's compare gradient descent, Adam, and RMSProp. The goal of the problem is, starting from the 30 data points, to determine the linear relationships between x_i and y_i. In other words, to determine the two parameters—2 and 1/2—in the last equation. For this example, we can write the linear relation as follows

$$y_i = w_0 + w_1 x_i$$

The optimizers will try to minimize the MSE with respect to w_0 and w_1. You will find the entire code to do this comparison in the online version of the book at `https://adl.toelt.ai`. In the online code, you will see an implementation of the gradient descent from scratch. In Figure 5-3, you can see the path that the GD algorithm takes in parameter space to reach the minimum of the MSE function in 200 iterations.

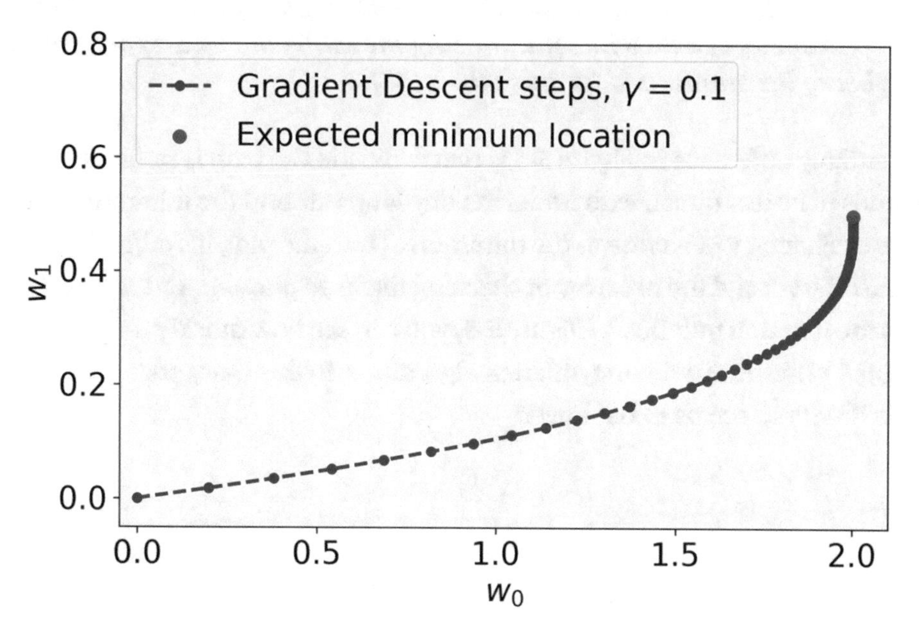

Figure 5-3. *The path that the GD algorithm follows in parameter spaces while minimizing the MSE when starting from (0,0). The number of iterations used for this plot is 200*

Figure 5-4 shows the different paths taken by the different optimizers when solving the same problem.

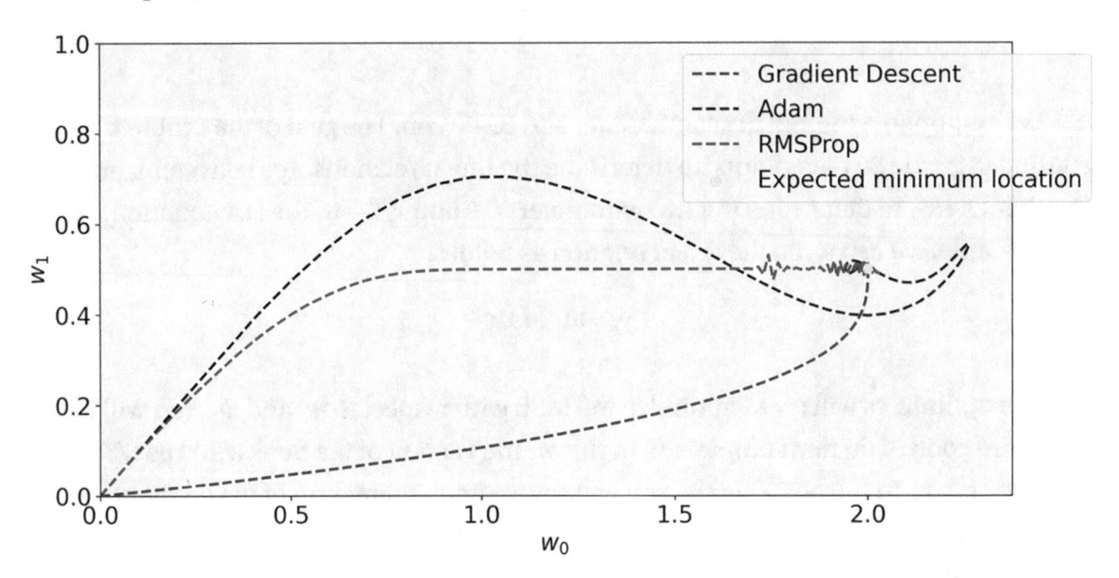

Figure 5-4. *The path that the GD algorithm, Adam, and RMSProp optimizers follow in parameter spaces while minimizing the MSE when starting from (0,0). The number of iterations used for this plot is 200*

One striking difference in Figure 5-4 is that while the GD path is rather direct, the others tends to be less direct, with Adam making loops around the minimum and RMSProp oscillating when close to the minimum. From the plot, it's difficult to see which one is faster, and the best way of checking that is to plot $w_0^{[i]}$ and $w_1^{[i]}$ vs. i, where i indicates the iteration number. In Figure 5-5, you can see how quickly $w_0^{[i]}$ converges to the expected value of 2.0 with the different algorithms. In this case, Adam and GD are on par, while RMSProp seems to be slower.

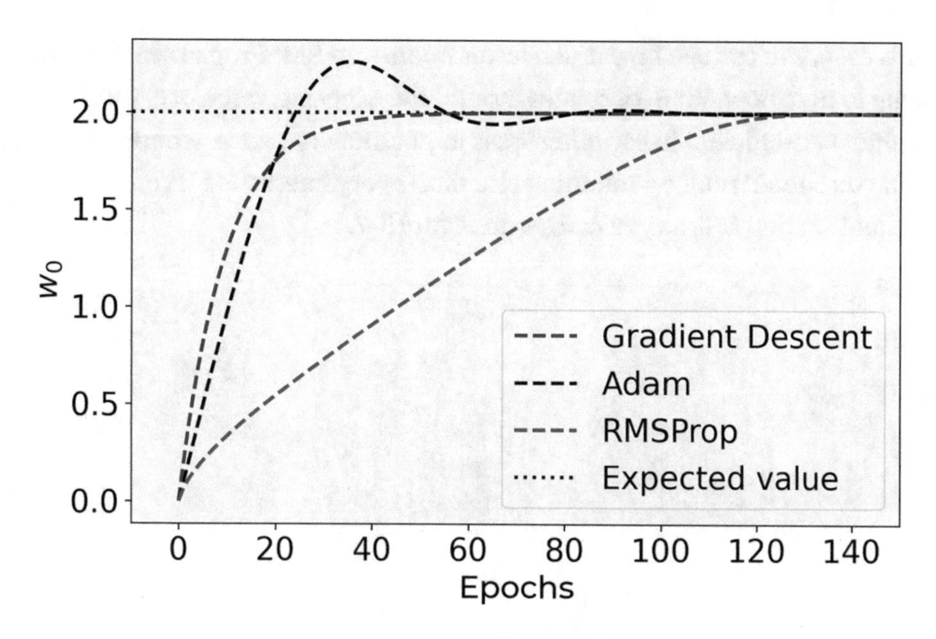

***Figure 5-5.** The plot of $w_0^{[i]}$ vs. the number of iterations with different optimizers*

A different picture emerges when considering $w_1^{[i]}$. Figure 5-6 shows how quickly the different algorithms converge.

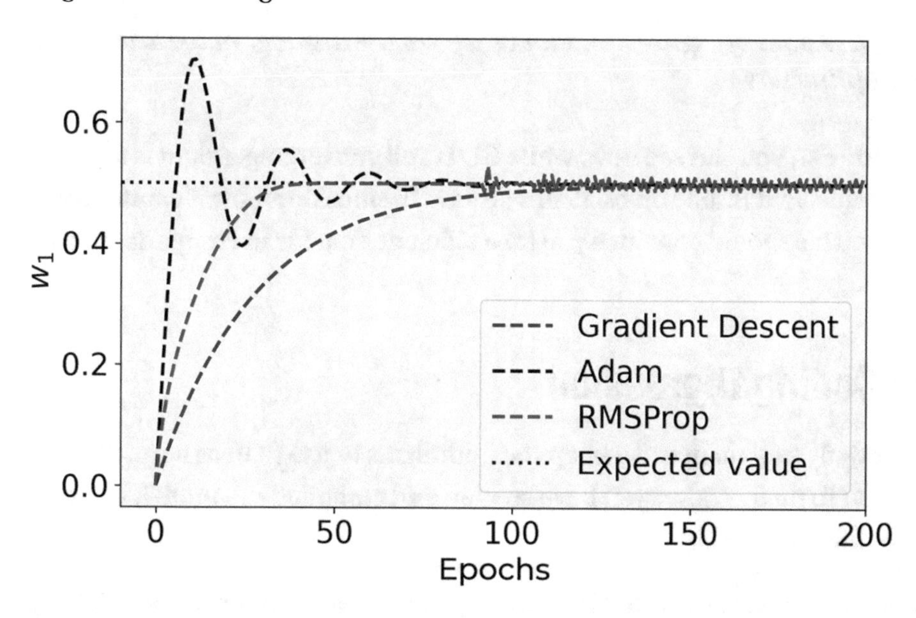

***Figure 5-6.** The plot of $w_1^{[i]}$ vs. the number of iterations with different optimizers*

In this case, you can see how much faster Adam and RMSProp are in converging. It is interesting to note how Adam oscillates around the expected value of 0.5. This property makes Adam very efficient in escaping areas in parameters' space, where training can get stuck. But you should notice something else that is very interesting if you zoom in on the picture after iteration 150, as you can see in Figure 5-7.

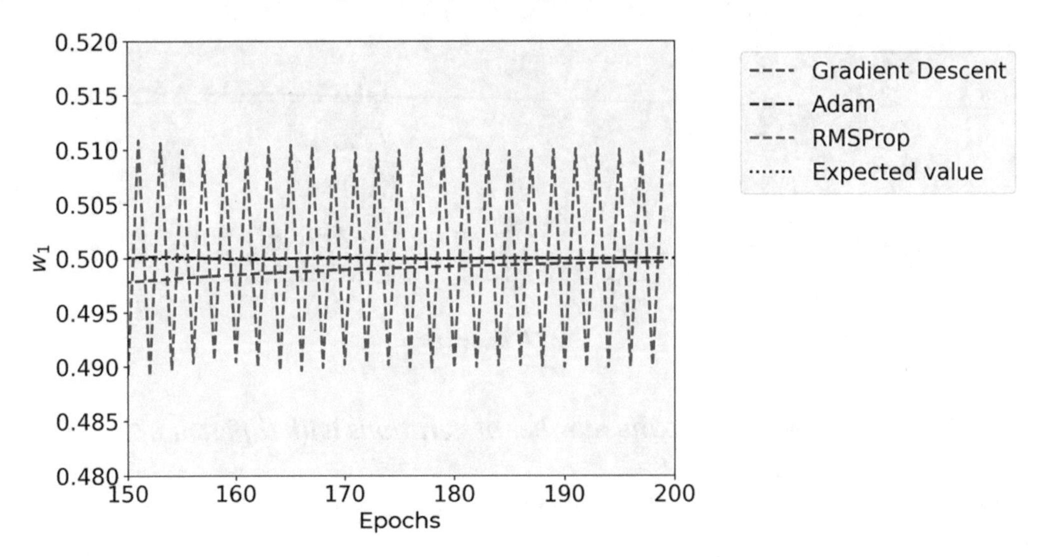

***Figure 5-7.** Zoom of Figure 6-5. Plot of $w_1^{[i]}$ vs. the number of iterations with different optimizers*

In Figure 5-7, you can see how, while GD is still converging, Adam is already at the expected value, and RMSProp oscillates around it, since probably it cannot converge completely. This should convince you how efficient Adam is in comparison with other optimizers.

Small Coding Digression

I want to briefly explain how to set up the optimizers to use (0,0) as a starting point for the updates. To do that, after you have created and compiled the model, you can set the weights of the model with

```
model.set_weights([np.array([w0_start]).reshape(1,1),np.array([w1_start]).
reshape(1,)])
```

where `w0_start = 2.0` and `w1_start = 0.5`.

After doing that, you need to save the value of the weights after each iteration. The easiest way to do that is to implement your update as a custom training loop by using `GradientTape()`. You will find the entire code online, but your training loops will look like this

```
for epoch in range(200):

    with tf.GradientTape() as tape:

        # Run the forward pass of the layer.
        ypred = model(x_, training=True)
        loss_value = loss_fn(y_, ypred)

    grads = tape.gradient(loss_value, model.trainable_weights)
    optimizer.apply_gradients(zip(grads, model.trainable_weights))

    w1_rmsprop_list.append(float(model.get_weights()[0][0]))
    w0_rmsprop_list.append(float(model.get_weights()[1][0]))
```

where `w1_rmsprop_list` and `w0_rmsprop_list` are lists that contain the values of the weights for each iteration. In this way, you can test any other optimizer you are interested in.

Which Optimizer Should You Use?

To make the story short, you should use Adam, as it is generally considered faster and better than the other methods. That does not mean that this is always the case. There are recent research papers that indicate how such optimizers could generalize poorly on new datasets (check out, for example, `https://goo.gl/Nzc8bQ`). And there are other papers that simply use GD with a dynamical learning rate decay. It mostly depends on your problem. But generally, Adam is a very good starting point.

Note If you are unsure which optimizer to start with, use Adam. It's generally considered faster and better than the other methods.

To give you an idea of how good it can be, let's apply it to the Zalando dataset. We will use a network with four hidden layers, each with 20 neurons. The model we use is the one discussed at the end of Chapter 3. Figure 5-8 shows that the cost function converges faster when using Adam optimization in comparison to GD. Additionally, in 100 epochs, GD reaches an accuracy of 86%, while Adam reaches 90%. Note that we have not changed anything in the model except the optimizer!

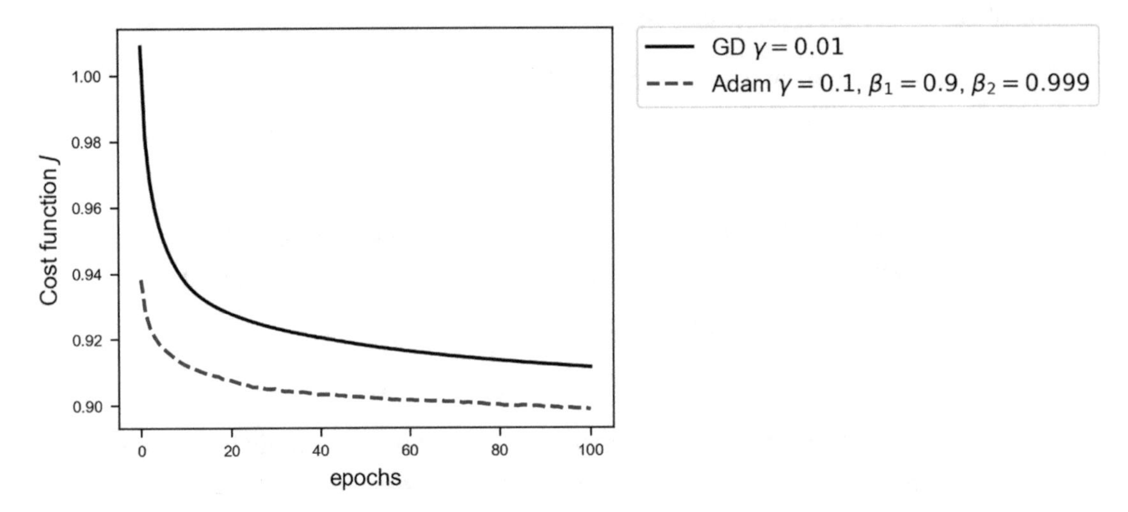

Figure 5-8. *The cost function for the Zalando dataset for a network with four hidden layers, each with 20 neurons. The continous line is plain GD with a learning rate of* $\gamma = 0.01$*, and the dashed line is Adam optimization with* $\gamma = 0.1$*,* $\beta_1 = 0.9$*,* $\beta_2 = 0.999$*, and* $\epsilon = 10^{-8}$

As suggested, when testing complex networks on big datasets, the Adam optimizer is a good place to start. But you should not limit your tests to this optimizer alone. A test of other methods is always worthwhile. Maybe for your problem other approaches will work better!

CHAPTER 6

Hyper-Parameter Tuning

Hyper-parameters can be loosely defined as parameters that do not change during training. For example, the number of layers in an FFNN, the number of neurons in each layer, activation functions, learning rate,[1] and so on.

This chapter looks at the problem of finding the best hyper-parameters to get the best results from your models. Doing this is called hyper-parameter tuning. We first describe what a black-box optimization problem is, and how those classes of problems relate to hyper-parameter tuning. We look at the three most common methods for tackling those kinds of problems: *grid search, random search,* and *Bayesian optimization.* You will learn, with examples, which one works in which conditions. At the end of the chapter, you will see how you can use those techniques to tune a deep model with a real-world dataset (the Zalando dataset). The goal of this chapter is not to give you an overview of all possible Python libraries that can help you with hyper-parameter tuning, but at the end of the chapter, you should know what hyper-parameter tuning is, why it's important, and how it works, at least in its most fundamental form.

Black-Box Optimization

The problem of hyper-parameter tuning is just a subclass of a much more general type of problems: *black-box optimizations*. A black-box function $f(x)$

$$f(x): \mathbb{R}^n \to \mathbb{R}$$

[1] Note that the learning rate can sometimes be changed during training, for example when applying decaying rate approaches. But more often, when using more advanced optimizers such as Adam, the learning rate is assigned at the beginning and does not change during training.

U. Michelucci, *Applied Deep Learning with TensorFlow 2*, https://doi.org/10.1007/978-1-4842-8020-1_6

is a function whose analytic form is unknown. A black-box function can be evaluated to obtain its value for all values of x on which it's defined, but no other information, such as its gradient, can be obtained. Generally, we talk about the global optimization of a black-box function (sometimes called a black-box problem) when we try to find the maximum or minimum of $f(x)$ given certain constraints. Here are some examples of these kinds of problems:

- Finding the hyper-parameter for a given machine learning model that maximizes the chosen optimizing metric
- Finding a maximum or minimum of a function, which can only be evaluated numerically or with code that we cannot look at. In some industry contexts, there is legacy code that is very complicated, and there are some functions that must be maximized based on its outcome.
- Finding the best place to drill for oil. Your function is how much oil you can find, and your x is your location.
- Finding the best combination of parameters for situations that are too complex to model, such as when launching a rocket in space. How do you optimize the amount of fuel, the diameter of each stage of the rocket, the precise trajectory, and so on.

This is a very fascinating class of problems that has produced smart solutions. We will look at three of them: grid search, random search, and Bayesian optimization.

Why is finding the best hyper-parameters for neural networks a black-box problem? Since we cannot calculate information like the gradients of our network output with respect to the hyper-parameters (for example, the number of layers in an FFNN), especially when using complex optimizers or custom functions, we need other approaches to find the best hyper-parameters that maximize the chosen optimizing metric. Note that if we could have the gradients, we could use an algorithm such as gradient descent to find the maximum or minimum.

Note Our black-box function *f* will be our neural network model (including things like the optimizer, cost function form, etc.) which gives as output our optimizing metric, given the hyper-parameters as input. In this case, *x* will be the array containing the hyper-parameters.

The problem may seem quite trivial, so why not try all the possibilities? Well, this may be possible in the examples we have looked at in the past chapters, but if you are working on a problem and training your model takes a week, this may present a challenge. Since typically you will have several hyper-parameters, trying all possibilities will not be feasible. Let's look at an example to understand it better. Let's suppose we are training a model of a neural network with several layers. We may decide to consider the following hyper-parameters to see which combination works best:

- Learning rate: Let's suppose we want to try the values $n \cdot 10^{-4}$ for $n = 1, \ldots, 10^2$ (100 values)
- Regularization parameter: 0, 0.1, 0.2, 0.3, 0.4, and 0.5 (six values)
- Choice of optimizer: GD, RMSProp, or Adam (three values)
- Number of hidden layers: 1, 2, 3, 5, and 10 (five values)
- Number of neurons in the hidden layers: 100, 200, and 300 (three values)

Consider that you will need to train your network

$$100 \times 6 \times 3 \times 5 \times 3 = 27000$$

times if you want to test all possible combinations. If your training takes five minutes, you will need 13.4 weeks of computing time. If the training takes hours or days, it becomes impossible. If the training takes one day for example, you will need 73.9 years to try all the possibilities. Most of the hyper-parameter choices will come from experience; for example, you can safely always use Adam, since it's the better optimizer out there (in almost all cases). But you will not be able to avoid trying to tune other parameters like the number of hidden layers or the learning rate. You can reduce the number of combinations you need with experience (as with the optimizer), or with some smart algorithms, as you see later in this chapter.

Notes on Black-Box Functions

Black-box functions are usually classified into two main classes:

- **Cheap functions**: Functions that can be evaluated thousands of times
- **Costly functions**: Functions that can only be evaluated a few times, usually fewer than 100 times

If the black-box function is cheap then the choice of the optimization method is not critical. For example, we can evaluate the gradient with respect to the x numerically, or simply search the maximum by evaluating the functions on a large number of points. If the function is costly, we need much smarter approaches. One of these is *Bayesian optimization,* which we discuss later in this chapter to give you an idea on how those methods work and how complex they are.

Note Neural networks are almost always, especially in the deep learning world, costly functions.

For costly functions we need to find methods that solve our problem with the smallest number of evaluation possible.

The Problem of Hyper-Parameter Tuning

Before looking at how we can find the best hyper-parameters, we need to quickly go back to neural networks and discuss what can we tune in deep models. Typically, when talking about hyper-parameters, beginners think only of numerical parameters, such as the learning rate or regularization parameter, for example. Remember that the following can also be varied to see if you can get better results:

- **Number of epochs:** Sometimes simply training your network longer will give you better results.
- **Choice of optimizer:** You can try choosing a different optimizer. If you are using plain gradient descent, you may try Adam and see if you get better results.
- **Varying the regularization method:** As discussed previously, there are several ways of applying regularization. Varying the method may well be worth trying.
- **Choice of activation function:** Although the activation function used in the past chapters for neurons in hidden layers was ReLU, others may work a lot better. Trying for example sigmoid or Swish may help you get better results.

- **Number of layers and number of neurons in each layer:** Try different configurations. Try layers with different numbers of neurons for example.
- **Learning rate decay methods:** Try different learning rate decay methods (if you are not using optimizers that do this already).
- **Mini-batch size:** Vary the size of the mini-batches. When you have a small amount of data you can use batch gradient descent. When you have a lot of data, mini-batches are more efficient.
- Weight initialization methods.

Let's classify the parameters we can tune in our models in the following three categories:

1. Parameters that are continuous real numbers, or in other words, that can assume any value. Examples: learning rate and regularization parameter.
2. Parameters that are discrete but can theoretically assume an infinite number of values. Examples: number of hidden layers, number of neurons in each layer, or number of epochs.
3. Parameters that are discrete and can only assume a finite number of possibilities. Examples: optimizer, activation function, and learning rate decay method.

For category 3 there is not much to do except try all the possibilities. They typically will change the model completely and it's impossible to model their effect, therefore making a test the only possibility. This is also the category where experience may help the most. It's widely known that the Adam optimizer is almost always the best choice, so you may concentrate your efforts somewhere else at the beginning. For categories 1 and 2, it's a bit more difficult, and you need to use some smart approaches to find the best values.

Sample Black-Box Problem

To try our hands at solving a black-box problem, let's create a "fake" black-box problem. Find the maximum of the function $f(x)$ given by the formula

$$g(x) = \cos\frac{x}{4} - \sin\frac{x}{4} - \frac{5}{2}\cos\frac{x}{2} + \frac{1}{2}\sin\frac{x}{2}$$

$$h(x) = -\cos\frac{x}{3} - \sin\frac{x}{3} - \frac{5}{2}\cos\frac{2}{3}x + \frac{1}{2}\sin\frac{2}{3}x$$

$$f(x) = 10 + g(x) + \frac{1}{2}h(x)$$

pretending not to know the formula itself. The formula will let you see the results, but let's pretend they are unknown. You may wonder why we use such a complicated formula. We wanted to have something with a few maxima and minima to show how the methods work on a non-trivial example. $f(x)$ can be implemented in Python with the code

```
def f(x):
    tmp1 = -np.cos(x/4.0)-np.sin(x/4.0)
           -2.5*np.cos(2.0*x/4.0)+0.5*np.sin(2.0*x/4.0)
    tmp2 = -np.cos(x/3.0)-np.sin(x/3.0)
           -2.5*np.cos(2.0*x/3.0)+0.5*np.sin(2.0*x/3.0)
    return 10.0+tmp1+0.5*tmp2
```

Figure 6-1 shows how $f(x)$ looks.

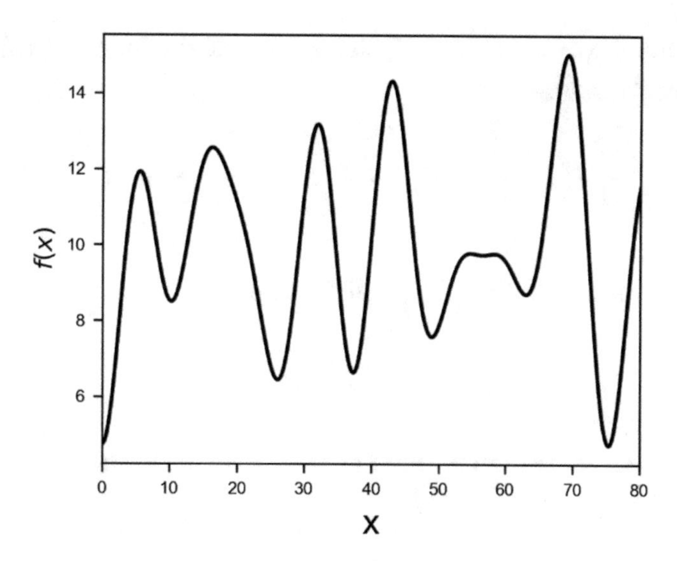

Figure 6-1. *A plot of the function f(x) described in the text*

The maximum is at an approximate value of $x = 69.18$ and has a value of 15.027. Our challenge is to find this maximum in the most efficient way possible, without knowing anything about $f(x)$ except its value at any point. When we say "efficient," we mean with the smallest number of evaluations possible.

Grid Search

The first method we look at, grid search, is also the less "intelligent". Grid search entails simply trying the function at regular intervals and seeing for which x the function $f(x)$ assumes the highest value. In this example, we want to find the maximum of the function $f(x)$ between two x values, x_{min} and x_{max}. What we do is simply take n points equally spaced between x_{min} and x_{max} and evaluate the function at those points. We define a vector of points

$$\boldsymbol{x} = \left(x_{min}, x_{min} + \frac{\Delta x}{n}, \ldots, x_{min} + (n-1)\frac{\Delta x}{n} \right)$$

where we defined $\Delta x = x_{max} - x_{min}$. Then we evaluate the function $f(x)$ at those points, obtaining a vector $\boldsymbol{f}$ of values

$$\boldsymbol{f} = \left(f(x_{min}), f\left(x_{min} + \frac{\Delta x}{n} \right), \ldots, f\left(x_{min} + (n-1)\frac{\Delta x}{n} \right) \right)$$

the estimate of the maximum $\left(\tilde{x}, \tilde{f}\right)$ will then be

$$\tilde{f} = \max_{0 \le i \le n-1} f_i$$

and supposing the maximum is found at $i = \tilde{i}$, we will also have

$$\tilde{x} = x_{min} + \frac{\tilde{i}\Delta x}{n}$$

Now, as you may imagine, the more points you use, the more accurate your maximum estimation will be. The problem is that, if the evaluation of $f(x)$ is costly, you will not be able to test as many points as you might like. You will need to find a balance between the number of points and the accuracy. Let's consider an example with the function $f(x)$ we described earlier. Let's consider $x_{max} = 80$ and $x_{min} = 0$ and let's take $n = 40$ points. We will have $\frac{\Delta x}{n} = 2$. We can create the vector $\boldsymbol{x}$ easily in Python with the code:

```
gridsearch = np.arange(0,80,2)
```

The array `gridsearch` will look like this

```
array([ 0, 2, 4, 6, 8, 10, 12, 14, 16, 18, 20, 22, 24, 26, 28, 30, 32, 34,
36, 38, 40, 42, 44, 46, 48, 50, 52, 54, 56, 58, 60, 62, 64, 66, 68, 70, 72,
74, 76, 78])
```

Figure 6-2 shows the function $f(x)$ as a continuous line, the crosses mark the points we sample in the grid search, and the black square marks the precise maximum of the function. The right plot shows a zoom around the maximum.

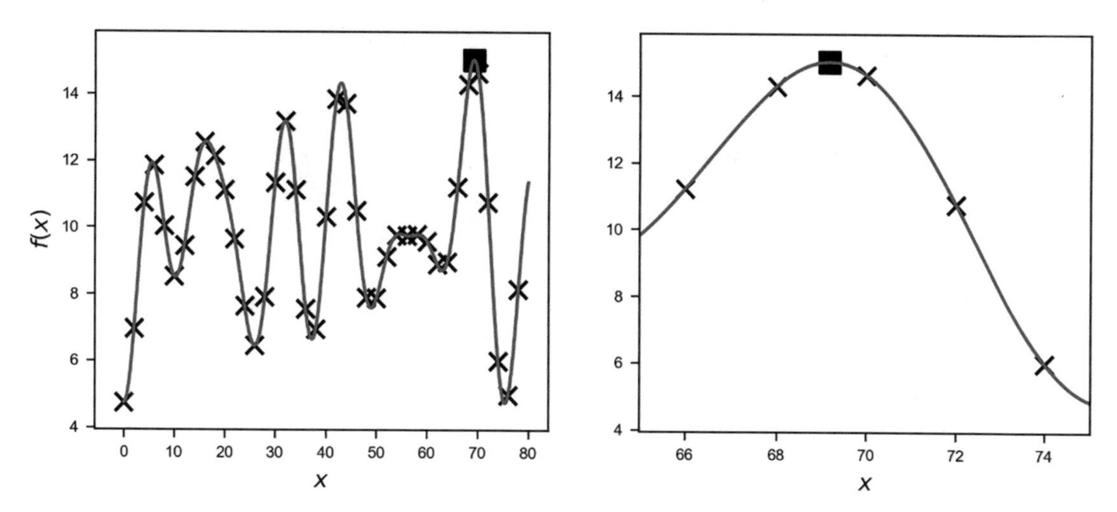

Figure 6-2. *The function f(x) on the range [0, 80]. The crosses mark the point we sample in the grid search, and the black square marks the maximum*

You can see how the points we sample in Figure 6-2 get close to the maximum, but don't get it exactly. Of course, sampling more points would get us closer to the maximum, but would cost us more evaluations of $f(x)$. We can find the maximum $\left(\tilde{x},\tilde{f}\right)$ easily with the trivial code

```
x = 0
m = 0.0
for i, val in np.ndenumerate(f(gridsearch)):
    if (val > m):
        m = val
        x = gridsearch[i]

print(x)
print(m)
```

That gives us

```
70
14.6335957578
```

Close to the actual maximum (69.18, 15.027), but not quite right. Let's try the previous example and vary the number of points we sample. We will vary the number of points sampled n from 4 to 160. For each case, we will find the maximum and its location as described earlier. We can do it with this code

```
xlistg = []
flistg = []

for step in np.arange(1,20,0.5):
    gridsearch = np.arange(0,80,step)

    x = 0
    m = 0.0
    for i, val in np.ndenumerate(maxim(gridsearch)):
        if (val > m):
            m = val
            x = gridsearch[i]

    xlistg.append(x)
    flistg.append(m)
```

in the lists `xlistg` and `flistg`, we will find the position of the maximum found and the value of the maximum for the various values of *n*.

Figure 6-3 shows the a plot of the distributions of the results. The black vertical line is the correct value of the maximum.

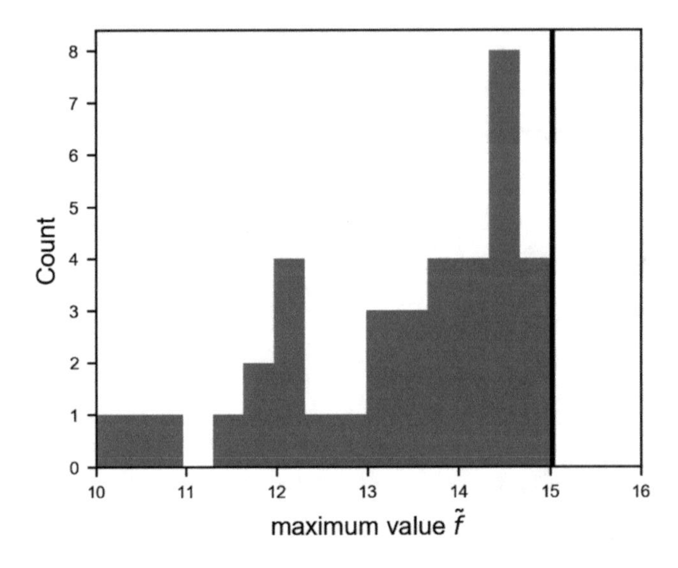

Figure 6-3. *The distribution of the results for $\tilde{f}$ obtained by varying the number of points n sampled in the grid search. The black vertical line indicates the real maximum of f(x)*

As you can see, the results vary quite a lot and can be very far from the correct value, as far as 10. This tells us that using the wrong number of points can return very wrong results. As you can imagine, the best results are the ones with the smallest step Δx, since it's more likely to be closer to the maximum. Figure 6-4 shows the value of the found maximum varies with the step Δx.

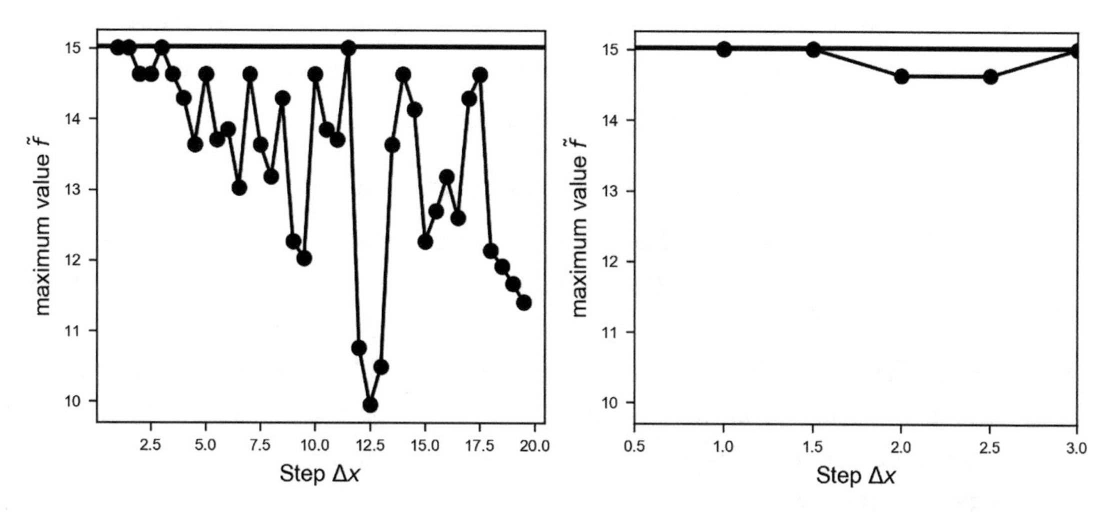

Figure 6-4. *The behavior of the found value of the maximum vs. the x step* Δx

In the zoom in the right plot of Figure 6-4, it's evident how smaller values of Δx get you better values of $\tilde{f}$. Note that a step of 1.0 means sampling 80 values of $f(x)$. If, for example, the evaluation takes one day, you will need to wait 80 days to get all the measurements you need.

Tip Grid search is efficient only when the black-box function is cheap. To get good results, a large number of sampling points is usually needed.

To make sure you are really getting the maximum, you should decrease the step Δx, or increase the number of sampling points, until the maximum value you find does not change appreciably anymore. In the previous example, as we see from the right plot in Figure 6-4, we are sure we are close the maximum when our step Δx gets smaller than roughly 2.0, or in other words when the number of sampled points is greater or roughly equal to 40. Remember that 40 may seems like a small number at first sight, but if $f(x)$ evaluates the metric of your deep learning model, and the training takes for example two hours, you are looking at 3.3 days of computer time.

Normally in the deep learning world, two hours is not much for training a model, so make a quick calculation before starting a long grid search. Additionally, keep in mind that when doing hyper-parameter tuning, you are moving in a multi-dimensional space (you are not optimizing only one parameter, but many), so the number of evaluations you need gets big very quickly.

Let's look at a quick example. Let's suppose you decide you can afford 50 evaluations of your black-box function. If you decide you want to try the following hyper-parameters:

- Optimizer (RMSProp, Adam or plain GD) (three values)
- Number of epochs (1000, 5000 or 10000) (three values)

You are already looking at nine evaluations. How many values of the learning rate can you then afford to try? Only five! And with five values, it's not probable that you'll get close to the optimal value. This shows how grid search is viable only for cheap black-box functions. Remember that, often, time is not the only problem. If you are using the Google Cloud platform to train your network, for example, you are paying the hardware you use by the second. Maybe you have lots of time at your disposal, but costs may go over budget very quickly.

Random Search

A strategy that is as "dumb" as grid search, but that works amazingly better, is a random search. Instead of sampling x points regularly in the range (x_{min}, x_{max}), you sample the points randomly. We can do it with this code

```
import numpy as np
randomsearch = np.random.random([40])*80.0
```

The `randomsearch` array will look like this

```
array([ 0.84639256, 66.45122608, 74.12903502, 36.68827838, 61.71538757,
69.29592273, 48.76918387, 69.81017465, 1.91224209, 21.72761762,
22.17756662, 9.65059426, 72.85707634, 2.43514133, 53.80488236, 5.70717498,
28.8624395 , 33.44796341, 14.51234312, 41.68112826, 42.79934087,
25.36351055, 58.96704476, 12.81619285, 15.40065752, 28.36088144,
30.27009067, 16.50286852, 73.49673641, 66.24748556, 8.55013954,
29.55887325, 18.61368765, 36.08628824, 22.1053749 , 40.14455129,
73.80825225, 30.60089111, 52.01026629, 47.64968904])
```

Depending on the seed you used, the actual numbers you get may be different. As we have done for grid search, Figure 6-5 shows the plot of $f(x)$, where the crosses mark the sampled points, and the black square is the maximum. On the right plot, you see a zoom around the maximum.

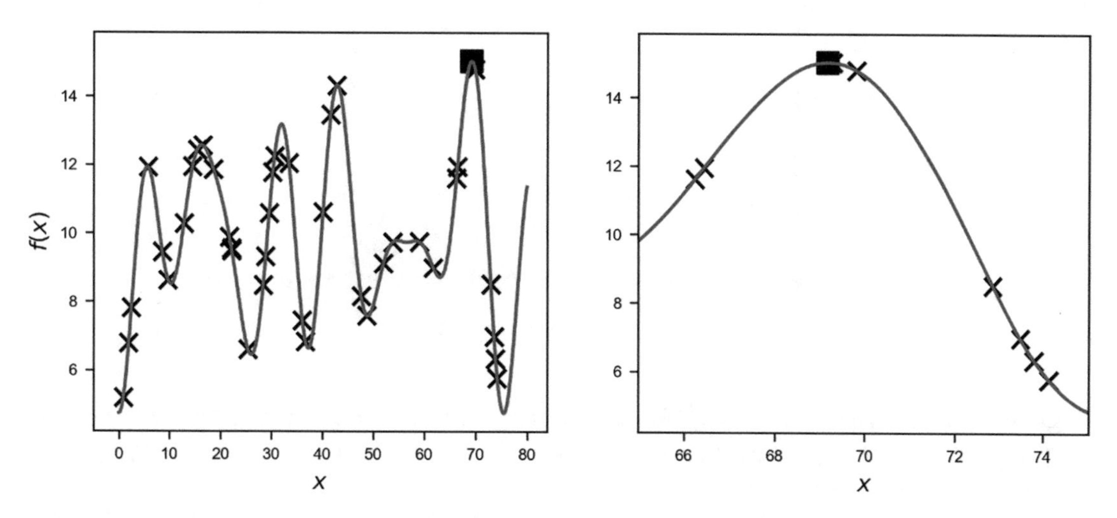

Figure 6-5. *The function f(x) on the range [0, 80]. The crosses mark the point we sampled with random search, and the black square marks the maximum*

The risk with this method is that, if you are very unlucky, your random chosen points are nowhere close the real maximum. But that probability is quite low. Note that if you take a constant probability distribution for your random points, you have the same probability of getting the points everywhere. Now it is interesting to see how this method performs. Let's consider 200 different random sets of 40 points, obtained by varying the random seed used in the code. The distributions of the maximum found $\tilde{f}$ are plotted in Figure 6-6.

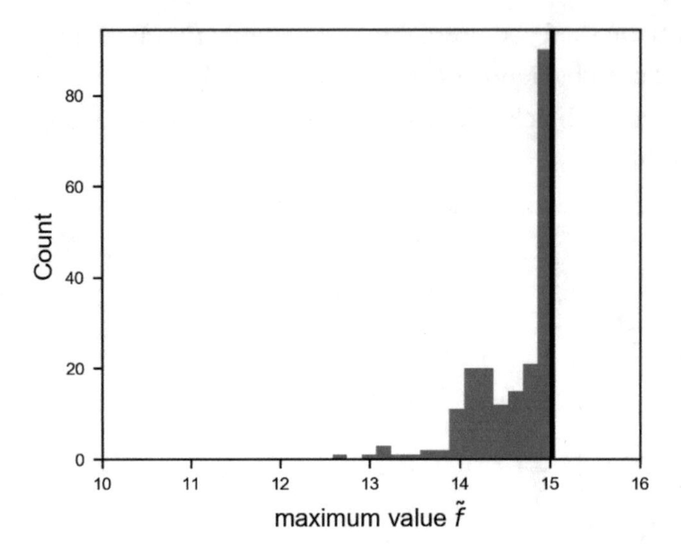

Figure 6-6. *The distribution of the results for* $\tilde{f}$ *obtained by 200 different random sets of 40 points sampled in the random search. The black vertical line indicates the real maximum of f(x)*

As you can see, regardless of the random sets used, you get very close to the real maximum. Figure 6-7 shows the distributions of the maximum found with random search and by varying the number of points sampled from 10 to 80.

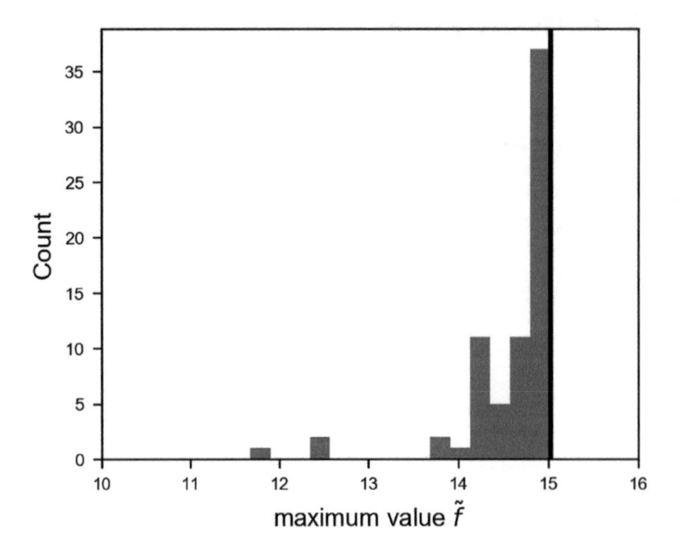

Figure 6-7. *The distribution of the results for* $\tilde{f}$ *obtained by varying the number of points n sampled in the random search, from 10 to 80. The black vertical line indicates the real maximum of f(x)*

If you compare this with grid search, you can see that random search is better at consistently getting results closer to the real maximum. Figure 6-8 compares the distribution you get for your maximum $\tilde{f}$ when using a different number of sampling points n with random and with grid search. In both cases, the plots were generated with 38 different sets, so the total count is the same.

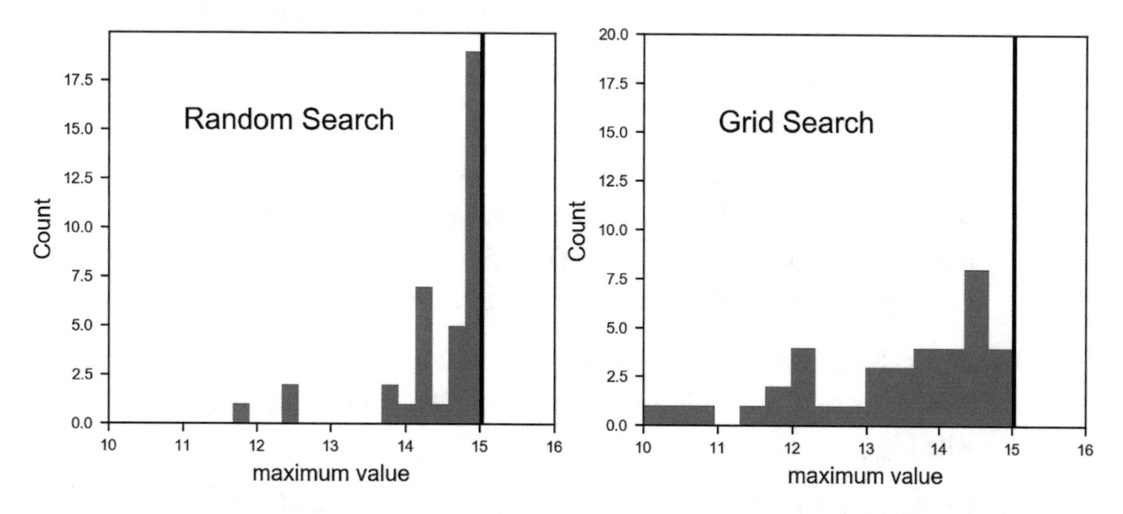

***Figure 6-8.** A comparison of the distribution of $\tilde{f}$ between grid (right) and random (left) search while varying the number of sampling points n. Both plots counts sum to 38, the number of different numbers of sampling points used. The correct value of the maximum is marked with the vertical black line in both plots*

It's easy to see how, on average, random search is better than grid search. The values you get are consistently closer to the right maximum.

Tip Random search is consistently better than grid search and you should use it every time it's possible. The difference between random search and grid search becomes even more marked when dealing with a multi-dimensional space for your variable *x*. Hyper-parameter tuning is nearly always a multi-dimensional optimization problem.

If you are interested in a very good paper on the random search scale at high-dimensional problems, you can read the paper by J. Bergstra and Y. Bengio, "Random Search for Hyper-Parameter Optimization" [1].

Coarse to Fine Optimization

There is still an optimization trick that helps with grid or random search. It is called "coarse to fine optimization." Let's suppose we want to find the maximum of $f(x)$ between x_{min} and x_{max}. We will look at the idea for random search, but it works in the same way as with grid search. The following steps give you the algorithm you need to follow.

1. Do a random search in the region $R_1 = (x_{min}, x_{max})$. Indicate the maximum found with (x_1, f_1).
2. Consider a smaller region around x_1, $R_2 = (x_1 - \delta x_1, x_1 + \delta x_1)$ for some δx_1 (which we discuss later) and do a random search in this region. The hypothesis is of course that the real maximum lies in this region. Indicate the maximum you find here with (x_2, f_2).
3. Repeat Step 2 around x_2, in the region indicated with R_3 with a δx_2 smaller than δx_1 and indicate the maximum you find in this step with (x_3, f_3).
4. Repeat step 2 again around x_3, in the region we indicate with R_4 with a δx_3 smaller than δx_2.
5. Continue the same way as many times as you need until the maximum (x_i, f_i) in the region R_{i+1} does not change any more.

Just one or two iterations are typically needed, but theoretically you could go on for a large number of iterations. The problem with this method is that you cannot be sure that your real maximum actually lies in your regions R_i. But this optimization has a big advantage if it does.

Let's consider the case where we do a standard random search. If we want to have on average a distance between the sampled points of 1% of $x_{max} - x_{min}$ we would need around 100 points if we decided to perform only one random search, and consequently we had to perform 100 evaluations. Now let's consider the algorithm we just described. We could start with just ten points in region $R_1 = (x_{min}, x_{max})$, where we indicate the maximum we find with (x_1, f_1). Then let's take $2\delta x = \frac{x_{\max} - x_{\min}}{10}$ and let's use ten points in region $R_2 = (x_1 - \delta x, x_1 + \delta x)$. In the interval $(x_1 - \delta x, x_1 + \delta x)$, we will have on average a distance between the points of 1% of $x_{max} - x_{min}$, but we just sampled our functions only

20 times instead of 100, so a factor of five fewer! For example, let's sample the function we used previously between $x_{min} = 0$ and $x_{max} = 80$ with ten points with this code

```
np.random.seed(5)
randomsearch = np.random.random([10])*80

x = 0
m = 0.0
for i, val in np.ndenumerate(f(randomsearch)):
    if (val > m):
        m = val
        x = randomsearch[i]
```

This gives us the maximum location and value of $x_1 = 69.65$ and $f_1 = 14.89$, so not bad, but not as precise as the real ones—69.18 and 15.027. Now let's sample ten points around the maximum we found in the regions $R_2 = (x_1 - 4, x_1 + 4)$

```
randomsearch = x + (np.random.random([10])-0.5)*8

x = 0
m = 0.0
for i, val in np.ndenumerate(maxim(randomsearch)):
    if (val > m):
        m = val
        x = randomsearch[i]
```

This gives us the result 69.189 and 15.027. Quite a precise result with only 20 evaluations of the function. If we do a plain random search with 500 (25 times more than what we just did) sampling points, we get $x_1 = 69.08$ and $f_1 = 15.022$. This result shows how this trick can be really helpful. But remember the risk: If your maximum is not in your regions R_i you will never be able to find it, since you are still dealing with random numbers. So, it is always a good idea to choose the regions $(x_i - \delta x_i, x_i + \delta x_i)$ that are relatively big to make sure that they contain your maximum.

How big they need to be depends, as almost everything in the deep learning world, on your dataset and the problem at hand and may be impossible to know in advance. Testing is unfortunately required. Figure 6-9 shows the sampled points on the function $f(x)$. In the plot on the left you see the first ten points, and on the right the region R_2 with the additional ten points. The small rectangle on the left plot marks the x region R_2.

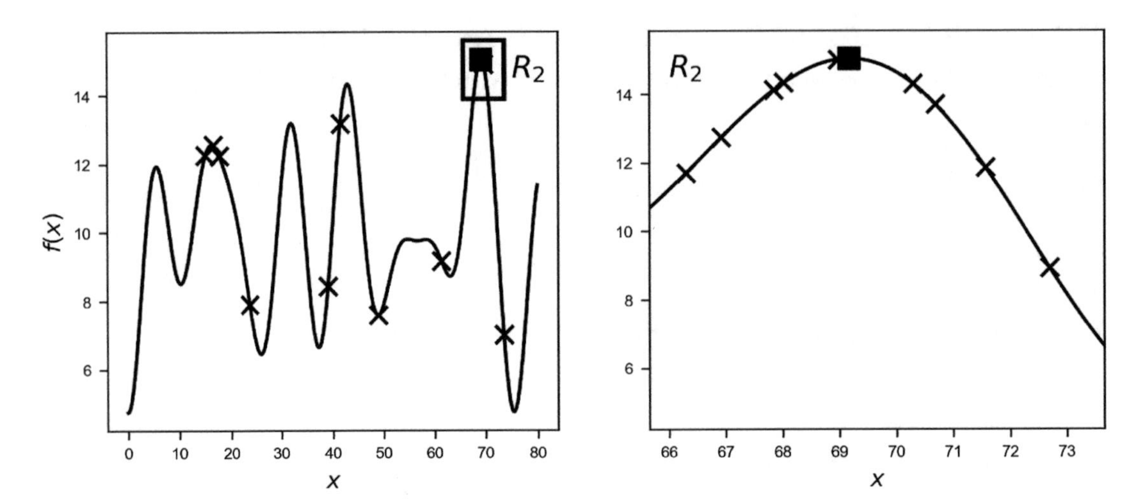

Figure 6-9. *The function f(x). The crosses mark the sampled points. On the left the ten points sampled in the region R_1 (entire range), on the right the ten points sampled in R_2. The black square marks the real maximum.The plot on the right is the zoom of the region R_2*

The decision of how many points you should sample at the beginning is crucial. We had luck here. Let's consider a different seed when choosing our initial ten random points and see what can happen. As you can see in Figure 6-10, choosing the wrong initial points leads to disaster!

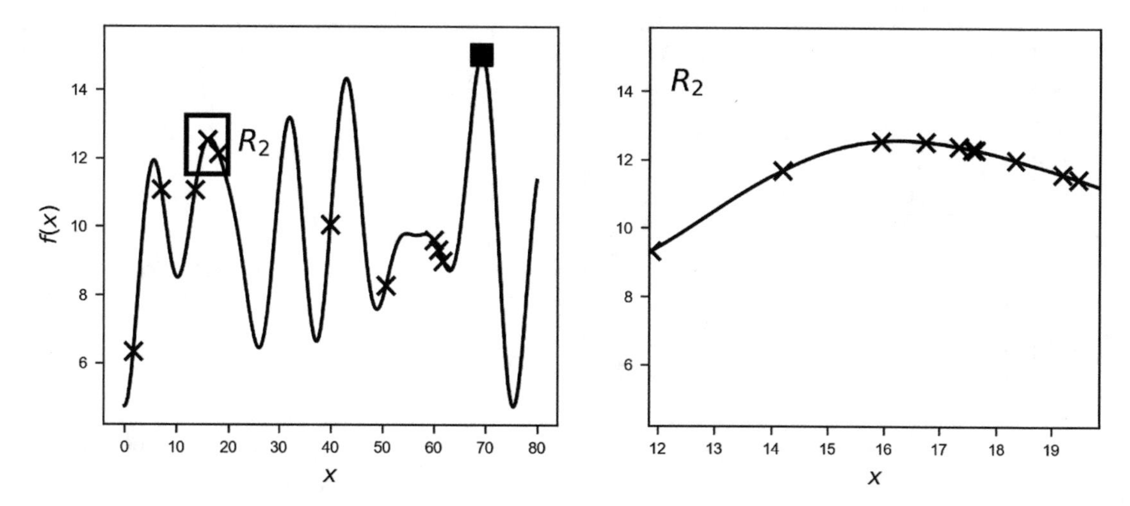

***Figure 6-10.** The function f(x). The crosses mark the sampled points. On the left the ten points sampled in the region R_1 (entire range), on the right the ten points sampled in R_2. The black square marks the real maximum. The algorithm finds a maximum very well around 16, but it's not the absolute maximum. The small rectangle on the left marks the region R_2. The plot on the right is the zoom on the region R_2*

In Figure 6-10, note how the algorithm finds the maximum at around 16, since in the initial sampled points the maximum value is around $x = 16$, as you can see on the plot on the left in Figure 6-10. No points are close to the real maximum around $x = 69$. The algorithm finds a maximum really well, simply not the absolute maximum. That is the danger you face when using this trick. It can even go worse than that. Consider Figure 6-11, where we just sampled one single point at the beginning. You can see on the plot on the left of Figure 6-11, how the algorithm completely misses any maximum. It simply gives as a result the highest value of the points marked by crosses on the points on the right plot: (58.4, 9.78).

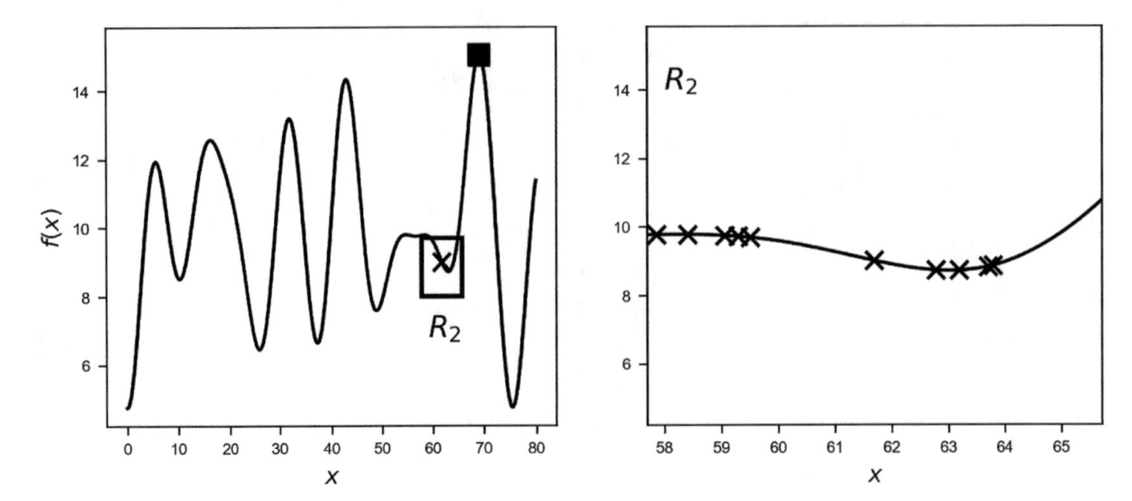

Figure 6-11. *The function f(x). The crosses mark the sampled points. On the left the one point sampled in the regions R_1 (entire range), on the right the ten points sampled in R_2. The black square marks the real maximum. The algorithm does not find a maximum since none is present in the region R_2. The plot on the right is the zoom on the region R_2*

If you decide to use this method, keep in mind that you will still need a good number of points at the beginning to get close to the maximum before refining your search. After you are relatively sure you have points around the maximum, you can use this technique to refine your search.

Bayesian Optimization

Nadaraya-Watson Regression

To understand Bayesian optimization, we have to look at some new mathematical concepts. Let's start with the Nadaraya-Watson regression, an idea from 1964[2]. The basic idea is quite simple. Given an unknown function $y(x)$ and given N points $x_i = 1, \ldots, N$, we indicate with $y_i = f(x_i)$ with $i = 1, \ldots, N$ the value of the function calculated at the different x_i. The idea of the Nadaraya-Watson regression is that we can evaluate the unknown function f at an unknown point x using the formula

$$y(x) = \sum_{i=1}^{N} w_i(x) y_i$$

where $w_i(x)$ are weights that are calculated according to the formula

$$w_i(x) = \frac{K(x, x_i)}{\sum_{j=1}^{N} K(x, x_j)}$$

where $K(x, x_i)$ is called a ***kernel***. Note that given how the weights are defined we have

$$\sum_{i=1}^{N} w_i(x) = 1$$

In the literature, you can find several kernels but the one we are interested in is the Gaussian one, often called the *Radial Basis Function* (RBF)

$$K(x, x_i) = \sigma^2 e^{-\frac{1}{2l^2}\|x - x_i\|^2}$$

The parameter l makes the Gaussian shape wider or narrower. The σ is typically the variance of your data, but in this case, it plays no role since the weights are normalized to 1. This is at the base of Bayesian optimization, as you will see later.

Gaussian Process

Before talking about Gaussian processes, we must define what a random process is. We talk of a *random process* when for any point $x \in \mathbb{R}^d$ we assign a random variable $f(x) \in \mathbb{R}$. A random process is Gaussian if for any finite number of points, their joint distribution is normal. That means that $\forall n \in \mathbb{N}$, and for $\forall x_1, \ldots x_n \in \mathbb{R}^d$ then the vector

$$\boldsymbol{f} \equiv \begin{pmatrix} f(x_1) \\ \vdots \\ f(x_n) \end{pmatrix} \sim \mathcal{N}$$

where with the notation we intend that the vector components follow a normal distribution, indicated with $\mathcal{N}$. Remember that a random variable with a Gaussian distribution is said to be normally distributed. From here comes the name Gaussian process. The probability distribution of the normal distribution is given by the function

$$g(x \mid \mu, \sigma^2) = \frac{1}{\sqrt{2\pi\sigma^2}} e^{\frac{(x-\mu)^2}{2\sigma^2}}$$

where μ is the mean or expectation of the distribution and σ is the standard deviation. We will use the following notation from here on

$$\textit{Mean value of } f: \ m$$

and the covariance of the random values will be indicated by K

$$\text{cov}\left[f(x_1), f(x_2)\right] = K(x_1, x_2)$$

the choice of the letter K is purposeful. We will assume in what follows that the covariance will have a Gaussian shape, and we will use for K the RBF function we defined earlier.

Stationary Process

For simplicity we will consider only stationary processes. A random process is stationary if its joint probability distribution does not change with time. That means that the mean and variance also will not change when shifted in time. We will also consider a process for which its distribution depends only on the relative position of the points. This leads to these conditions

$$\begin{aligned} K(x_1, x_2) &= \tilde{K}(x_1 - x_2) \\ Var\left[f(x)\right] &= \tilde{K}(0) \end{aligned}$$

Note that to apply the method we are describing, you must convert your data to be stationary, eliminating seasonality or trends in time for example.

Prediction with Gaussian Processes

Now we have reached the interesting part: given the vector $\boldsymbol{f}$, how can we estimate $f(x)$ at an arbitrary point x? Since we are dealing with random processes, we will estimate the probability that the unknown function assumes a given value $f(x)$. Mathematically we will predict the following quantity

$$p(f(x) | f)$$

Or, in other words, the probability of getting the value $f(x)$ given the vector $\boldsymbol{f}$, composed by all the points $f(x_1), \ldots, f(x_n)$.

Assuming that $f(x)$ is a stationary Gaussian process with $m = 0$, the prediction can be shown to be given by

$$p(f(x)|f) = \frac{p(f(x), f(x_1), \ldots, f(x_n))}{p(f(x_1), \ldots, f(x_n))} = \frac{\mathcal{N}(f(x), f(x_1), \ldots, f(x_n)|0, \tilde{C})}{\mathcal{N}(f(x_1), \ldots, f(x_n)|0, C)}$$

where with $\mathcal{N}(f(x), f(x_1), \ldots, f(x_n)|0, \tilde{C})$ we have indicated the normal distribution calculated on the points with an average of 0 and a covariance matrix $\tilde{C}$ of dimensions $n + 1 \times n + 1$, since we have $n + 1$ points in the numerator. The derivation is somewhat involved and is based on several theorems, such as Bayes' theorem. For more details you can refer to the (advanced) explanation by C.B. Do [3]. To understand what Bayesian optimization is, we can simply use the formula without derivation. C will have dimensions $n \times n$ since we have only n points in the denominator.

We have

$$C = \begin{pmatrix} K(0) & K(x_1 - x_2) & \cdots \\ K(x_2 - x_1) & \ddots & \vdots \\ \vdots & \cdots & K(0) \end{pmatrix}$$

And

$$\tilde{C} = \begin{pmatrix} K(0) & \boldsymbol{k}^T \\ \boldsymbol{k} & C \end{pmatrix}$$

where we have defined

$$\boldsymbol{k} = \begin{pmatrix} K(x - x_1) \\ \vdots \\ K(x - x_n) \end{pmatrix}$$

It can be shown[2] that the ratio of two normal distributions is again a normal distribution, which means we can write

$$p(f(x)|\boldsymbol{f}) = \mathcal{N}(f(x)|\mu,\sigma^2)$$

With

$$\mu = \boldsymbol{k}^T C^{-1} \boldsymbol{f}$$
$$\sigma^2 = K(0) - \boldsymbol{k}^T C^{-1} \boldsymbol{k}$$

The derivation of the exact form for μ and σ is quite long and would go beyond the scope of the book. Basically, now we know that, on average, our unknown function will assume the value μ in x with a variance σ. Let's now see how this method works in practice in Python. Let's define our kernel $K(x)$

```
def K(x, l, sigm = 1):
    return sigm**2*np.exp(-1.0/2.0/l**2*(x**2))
```

Let's simulate our unknown function with an easy one

$$f(x) = x^2 - x^3 + 3 + 10x + 0.07x^4$$

that can be implemented as follows

```
def f(x):
    return x**2-x**3+3+10*x+0.07*x**4
```

Let's consider the function in the range (0, 12). Figure 6-12 shows how the function looks.

[2] Remember that a normal distribution has an exponential form, and the ratio of two exponentials is still an exponential.

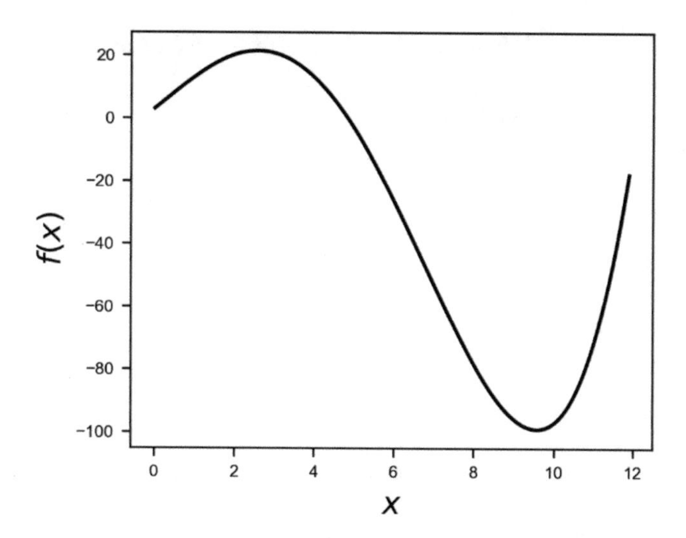

***Figure 6-12.** A plot of the unknown test function as described in the text*

Let's build first our $\boldsymbol{f}$ vector with five points

```
randompoints = np.random.random([5])*12.0
f_ = f(randompoints)
```

where we have used the seed 42 for the random numbers: `np.random.seed(42)`. Figure 6-13 shows the random points marked with a cross on the plot.

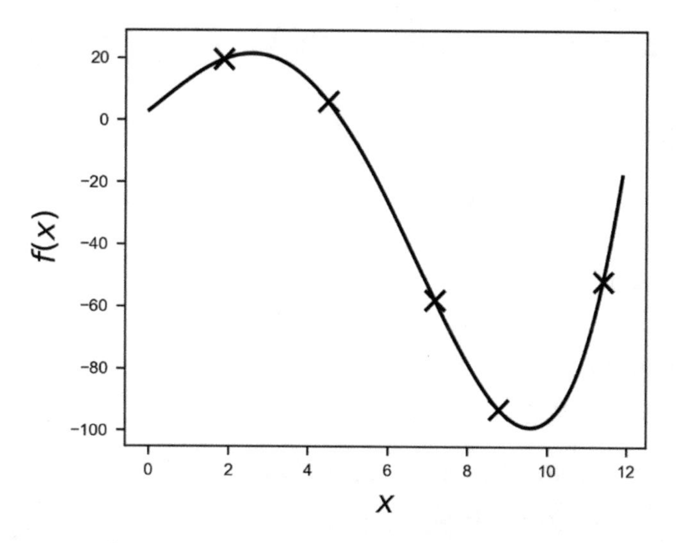

***Figure 6-13.** A plot of the unknown function. The crosses mark the random points chosen in the text*

We can apply the method described earlier with the following code

```
xsampling = np.arange(0,14,0.2)

ybayes_ = []
sigmabayes_ = []
for x in xsampling:

    f1 = f(randompoints)
    sigm_ = np.std(f1)**2
    f_ = (f1-np.average(f1))

    k = K(x-randompoints, 2 , sigm_)

    C = np.zeros([randompoints.shape[0],
                  randompoints.shape[0]])
    Ctilde = np.zeros([randompoints.shape[0]+1,
                       randompoints.shape[0]+1])

    for i1,x1_ in np.ndenumerate(randompoints):
        for i2,x2_ in np.ndenumerate(randompoints):
            C[i1,i2] = K(x1_-x2_, 2.0, sigm_)

    Ctilde[0,0] = K(0, 2.0, sigm_)
    Ctilde[0,1:randompoints.shape[0]+1] = k.T
    Ctilde[1:,1:] = C
    Ctilde[1:randompoints.shape[0]+1,0] = k

    mu = np.dot(np.dot(np.transpose(k), np.linalg.inv(C)), f_)
    sigma2 = K(0, 2.0,sigm_)-
             np.dot(np.dot(np.transpose(k),
                           np.linalg.inv(C)), k)
    ybayes.append(mu)
    sigmabayes_.append(np.abs(sigma2))

ybayes = np.asarray(ybayes_)+np.average(f1)
sigmabayes = np.asarray(sigmabayes_)
```

Take some time to study this so you understand it. In the `ybayes` list, we will find the values of $\mu(x)$ evaluated on the values contained in the `xsampling` array. Here are some hints that will help you understanding the code:

- We loop over a range of x points where we want to evaluate our function, with the code `for x in xsampling`.
- We build our $\boldsymbol{k}$ and $\boldsymbol{f}$ vectors with the code for each element of the vectors: `k = K(x-randompoints, 2 , sigm_)` and `f1 = f(randompoints)`. For the kernel, we chose a value for the parameter `l` as defined in the function of 2. We subtracted in the vector $\boldsymbol{f}$ the average to obtain $m(x) = 0$ to be able to apply the formulas as derived.
- We build the matrices C and $\tilde{C}$.
- We calculate μ with `mu = np.dot(np.dot(np.transpose(k), np.linalg.inv(C)), f_)` and the standard deviation.
- We reapply all the transformation that we did to make our process stationary in the opposite order, simply adding the average of $f(x)$ back to the evaluated surrogate function `ybayes = np.asarray(ybayes_)+np.average(f1)`.

Figure 6-14 shows how this method works. The dashed line is the predicted function obtained by plotting $\mu(x)$ as calculated in the code, when we have five points ($n = 5$). The gray area is the region between the estimated function and +/- σ.

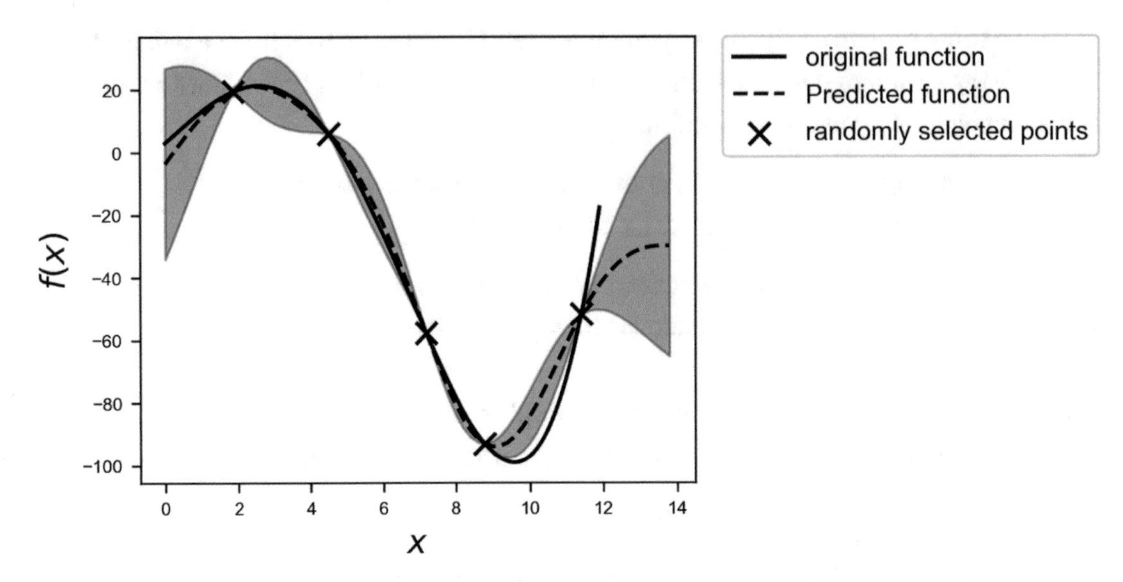

***Figure 6-14.** The predicted function, dashed line, evaluating $\mu(x)$. The gray area is the region between the estimated function and +/- σ*

Given the few points we have, this is actually not a bad result. Keep in mind that you still need a few points to be able to get a reasonable approximation. Figure 6-15 shows the result when we have only two points. The result is not as good. The gray area is the region around the estimated function and +/- σ. You can see that the farther you are from the points, the higher the uncertainty, or the variance, of the predicted function.

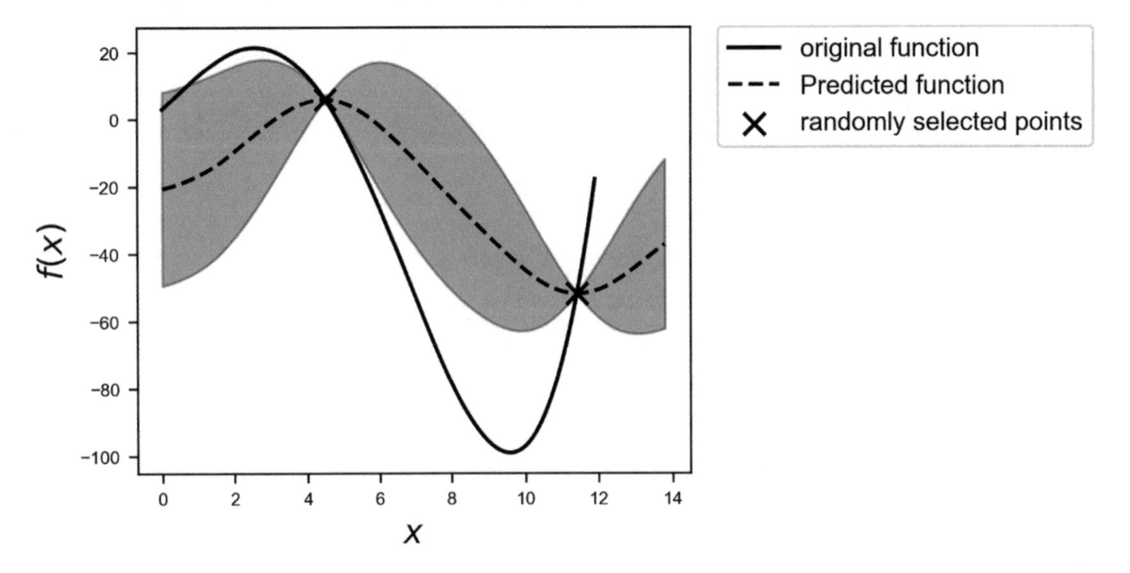

***Figure 6-15.** The predicted function, the dashed line, when we have only two points. The gray area is the region between the estimate function +/- σ*

Let's stop for a second and think about why we do all this. The idea behind it is to find a *surrogate function* $\tilde{f}$ that approximates our function f and that has the property

$$\max \tilde{f} \approx \max f$$

This surrogate function must have another very important property: it must be cheap to evaluate. That way, we can easily find the maximum of $\tilde{f}$ *and, by using the previous property, we will have the maximum of* f*, that by hypothesis is costly.*

As you have seen, it may be very difficult to know if we have enough points to find the right value of the maximum. After all, by definition, we do not have any idea how our black-box function looks. So how do we solve this problem?

The main idea behind the method can be described by the following algorithm:

1. Start with a small number of randomly chosen sample points (how many points will depend on your problem).
2. Use this set of points to get a surrogate function, as described previously.
3. Add a point to the set, with a specific method that we discuss later, and reevaluate the surrogate function.
4. If the maximum found with the surrogate function continues to change, continue adding points as in Step 3 until the maximum does not change or you run out of time or budget and cannot perform any more evaluations.

If the method we hinted at in Step 3 is smart enough, you will be able to find the maximum relatively quickly and accurately.

Acquisition Function

How do you choose the additional points mentioned in Step 3 in the previous section? The idea is to use an acquisition function. The algorithm works in this way:

1. Choose a function (and we will see a few possibilities in a moment) called an "acquisition function."
2. Then choose as additional point x, the one at which the acquisition function has a maximum.

There are several acquisition functions we can use. We will describe only one that we will use to see how this method works, but there are several that you may want to check out, including entropy search, probability of improvement, expected improvement, and upper confidence bound.

Upper Confidence Bound (UCB)

In the literature you find two variations of this acquisition function. We can write the function as

$$a_{UCB}(x)=\mathbb{E}\tilde{f}(x)+\eta\sigma(x)$$

where we indicate with $\mathbb{E}\tilde{f}(x)$ the "expected" value of the surrogate function on the x-range we have in our problem. The expected value is nothing more than the average of the function over the given x range. $\sigma(x)$ is the variance of the surrogate function that we calculate with our method at the point x. The new point we select is the one where $a_{UCB}(x)$ is maximum. $\eta > 0$ is a tradeoff parameter. *This acquisition function basically selects the points where the variance is biggest.* Look at Figure 6-15 again. The method would select the points where the variance is greater, so points as far as possible from the points we have already. In this way the approximation tends to get better and better. Another variation of the UCB acquisition function is the following

$$\tilde{a}_{UCB}(x)=\tilde{f}(\mathrm{x})+\eta\sigma(x)$$

This time, the acquisition function will make a tradeoff between choosing points around the surrogate function maximum and points where its variance is the biggest. This second method works best to quickly find good approximation of the maximum of f, while the first tends to give good approximation of f over the entire x range. The next section explains how these methods work.

Example

Let's start with the complex trigonometric function, as described at the beginning of the chapter and consider the x range $[0, 40]$. Our goal is to find its maximum and approximate the function. To facilitate the code, let's define two functions: one to evaluate the surrogate function and one to evaluate the new point. To evaluate the surrogate function, we can use the following function

```
def get_surrogate(randompoints):
    ybayes_ = []
    sigmabayes_ = []
    for x in xsampling:

        f1 = f(randompoints)
        sigm_ = np.std(f_)**2
        f_ = (f1-np.average(f1))
        k = K(x-randompoints, 2.0, sigm_ )

        C = np.zeros([randompoints.shape[0],
                       randompoints.shape[0]])
        Ctilde = np.zeros([randompoints.shape[0]+1,
                            randompoints.shape[0]+1])
        for i1,x1_ in np.ndenumerate(randompoints):
            for i2,x2_ in np.ndenumerate(randompoints):
                C[i1,i2] = K(x1_-x2_, 2.0, sigm_)

        Ctilde[0,0] = K(0, 2.0)
        Ctilde[0,1:randompoints.shape[0]+1] = k.T
        Ctilde[1:,1:] = C
        Ctilde[1:randompoints.shape[0]+1,0] = k

        mu = np.dot(np.dot(np.transpose(k),
                            np.linalg.inv(C)), f_)
        sigma2 = K(0, 2.0, sigm_)
                 - np.dot(np.dot(np.transpose(k),
                                  np.linalg.inv(C)), k)
        ybayes_.append(mu)
        sigmabayes_.append(np.abs(sigma2))

    ybayes = np.asarray(ybayes_)+np.average(f1)
    sigmabayes = np.asarray(sigmabayes_)

    return ybayes, sigmabayes
```

This function has the same code as we discussed in the example in the previous sections, but it's packed in a function that returns the surrogate function, contained in the array ybayes, and the σ^2, contained in the array `sigmabayes`. Additionally, we need a function that evaluates the new points using the acquisition function $a_{UCB}(x)$. We can get it with this function

```
def get_new_point(ybayes, sigmabayes, eta):

    idxmax = np.argmax(np.average(ybayes)
             +eta*np.sqrt(sigmabayes))
    newpoint = xsampling[idxmax]

    return newpoint
```

To make things simpler, we define the array that contains all the x values we want at the beginning outside the functions. Let's start with six randomly selected points. To see how this method is doing, let's start with some definitions

```
xmax = 40.0
numpoints = 6
xsampling = np.arange(0,xmax,0.2)
eta = 1.0

np.random.seed(8)
randompoints1 = np.random.random([numpoints])*xmax
```

In the `randompoints1` array, we have our first six selected random points. We can easily get the surrogate function with our function with

```
ybayes1, sigmabayes1 = get_surrogate(randompoints1)
```

Figure 6-16 shows the result. The dotted line is the acquisition function $a_{UCB}(x)$ normalized to fit in the plot.

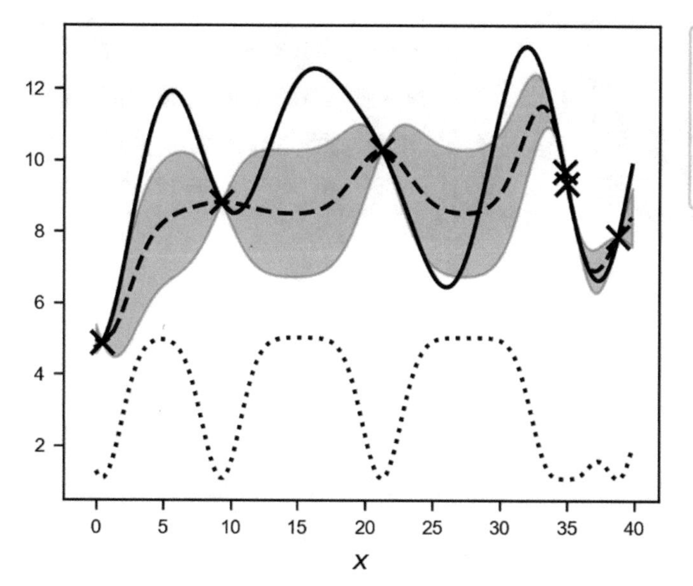

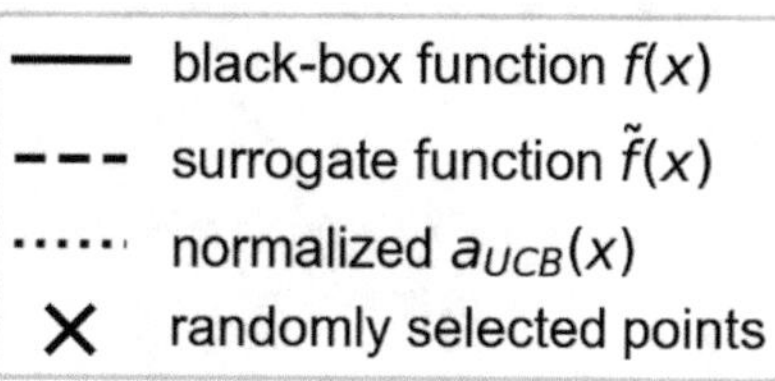

Figure 6-16. *An overview of the black-box function f(x) (the solid line), the randomly selected points (marked by crosses), the surrogate function (the dashed line), and the acquisition function $a_{UCB}(x)$ (the dotted line, shifted to fit in the plot). The gray area is the region contained between the lines $\tilde{f}(x)+\sigma(x)$ and $\tilde{f}(x)-\sigma(x)$*

The surrogate function is not very good yet, since we don't have enough points, and the big variance (the gray area) makes this evident. The only region that is well approximated is the region $x \gtrsim 35$. You can see how the acquisition function is big when the surrogate function is not approximating the black-box function well, and small when it does, as for example with $x \gtrsim 35$. So choosing a new point where $a_{UCB}(x)$ is at its maximum is equivalent to choosing the points where the function is less well approximated, or in more mathematical terms, where the variance is bigger. For a comparison, Figure 6-17 shows the same plot as in Figure 6-16, but with the acquisition function $\tilde{a}_{UCB}(x)$ and with $\eta = 3.0$.

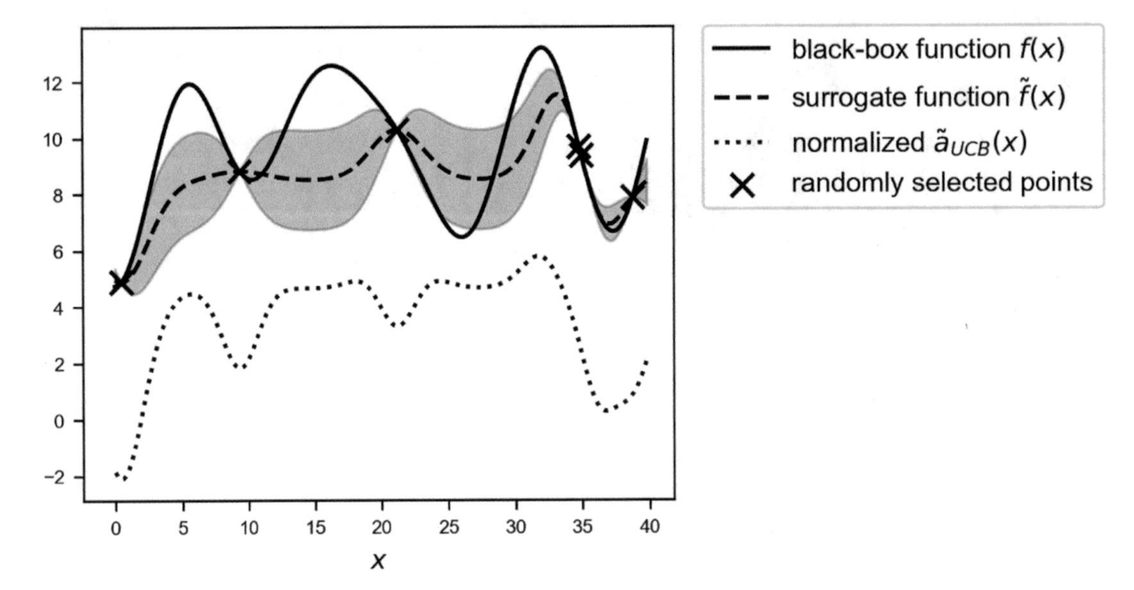

***Figure 6-17.** An overview of the black-box function f(x) (the solid line), the randomly selected points (marked by crosses), the surrogate function (the dashed line), and the acquisition function $\tilde{a}_{UCB}(x)$ (the dotted line, shifted to fit in the plot). The gray area is the region contained between the lines $\tilde{f}(x)+\sigma(x)$ and $\tilde{f}(x)-\sigma(x)$*

As you can see, $\tilde{a}_{UCB}(x)$ tends to have a maximum around the maximum of the surrogate function. Keep in mind that if η is big, then the maximum of the acquisition function will shift toward regions with high variance. But this acquisition function tends to find "a" maximum slightly faster. I said "a" since it depends on where the maximum of the surrogate function is, and not where the maximum of the black-box function is.

Let's see now what happens while using $a_{UCB}(x)$ with $\eta = 1.0$. For the first additional point, we need to run the following three lines of code

```
newpoint = get_new_point(ybayes1, sigmabayes1, eta)
randompoints2 = np.append(randompoints1, newpoint)
ybayes2, sigmabayes2 = get_surrogate(randompoints2)
```

For the sake of simplicity, we name each array for each step differently, instead of creating a list. But typically, you should make these iterations automatic. Figure 6-18 shows the result with the additional point, marked with a black circle.

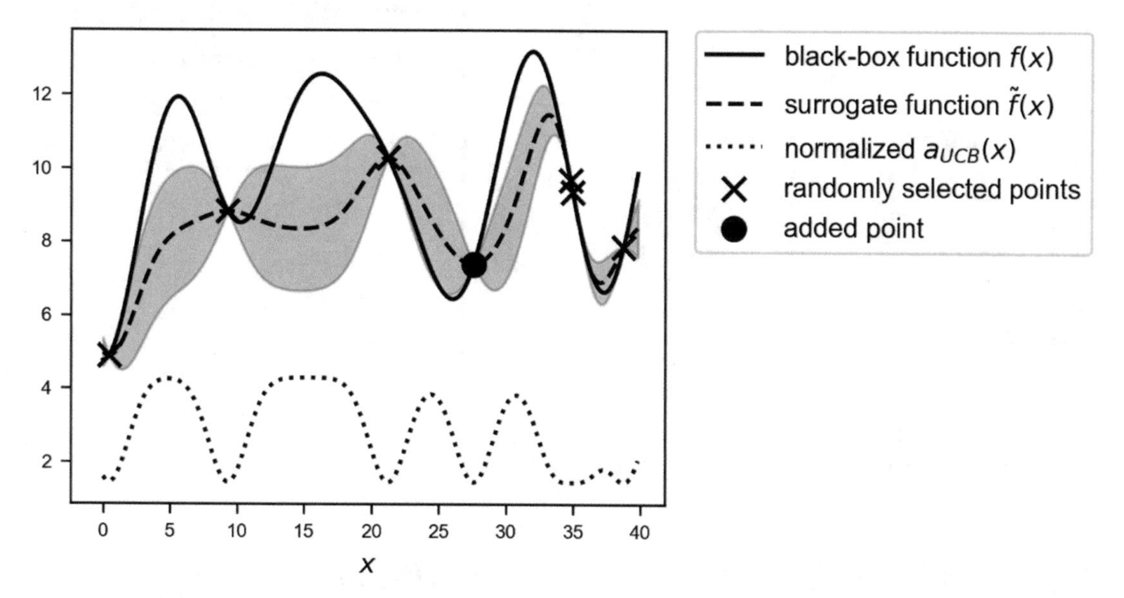

Figure 6-18. *An overview of the black-box function f(x) (the solid line), the randomly selected points (marked with crosses), and with the new selected point around x ≈ 27 (marked by a circle), the surrogate function (the dashed line), and the acquisition function $a_{UCB}(x)$ (the dotted line, shifted to fit in the plot). The gray area is the region contained between the lines $\tilde{f}(x)+\sigma(x)$ and $\tilde{f}(x)-\sigma(x)$*

The new point is around $x \approx 27$. Let's continue to add points. Figure 6-19 shows the results after adding five points.

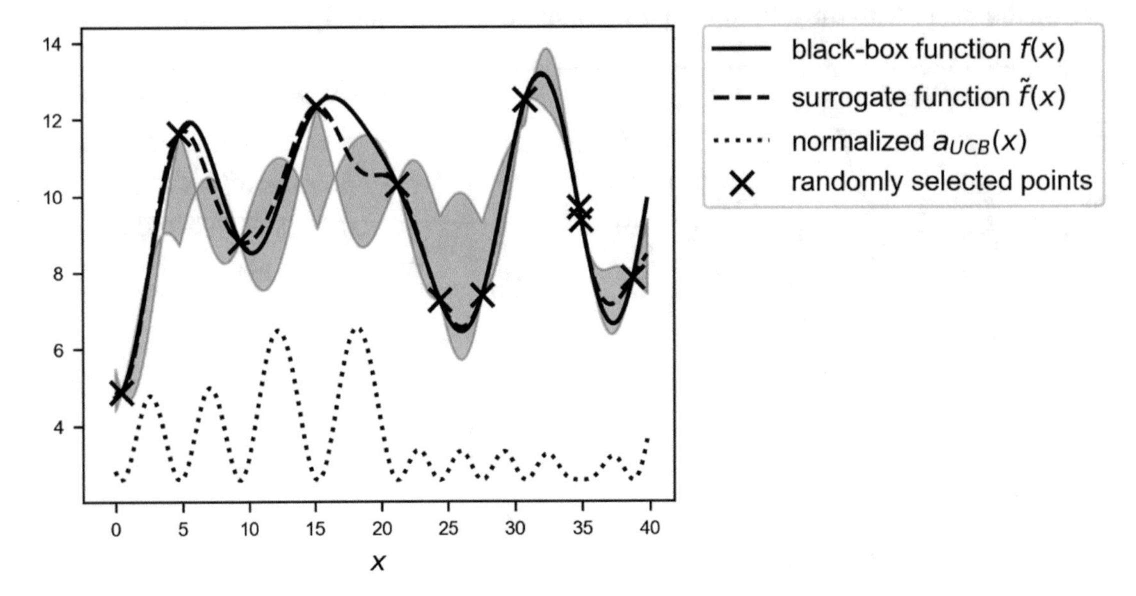

***Figure 6-19.** An overview of the black-box function f(x), the randomly selected points with the new six points, the surrogate function, and the acquisition function $\tilde{a}_{UCB}(x)$. The gray area is the region contained between the lines $\tilde{f}(x)+\sigma(x)$ and $\tilde{f}(x)-\sigma(x)$*

Look at the dashed line. The surrogate function now approximates the black-box function quite well, especially around the real maximum. Using this surrogate function, we can find a very good approximation of our original function with just 11 evaluations in total! Keep in mind we don't have any additional information about *f*, except the 11 evaluations.

Now let's see what happens with the acquisition function $\tilde{a}_{UCB}(x)$, and let's check how fast we can find the maximum. In this case, let's use $\eta = 3.0$, to get a better balance between the surrogate function's maximum and its variance. Figure 6-20 shows the result after adding one additional point, marked by a black circle. We already have a good approximation of the real maximum!

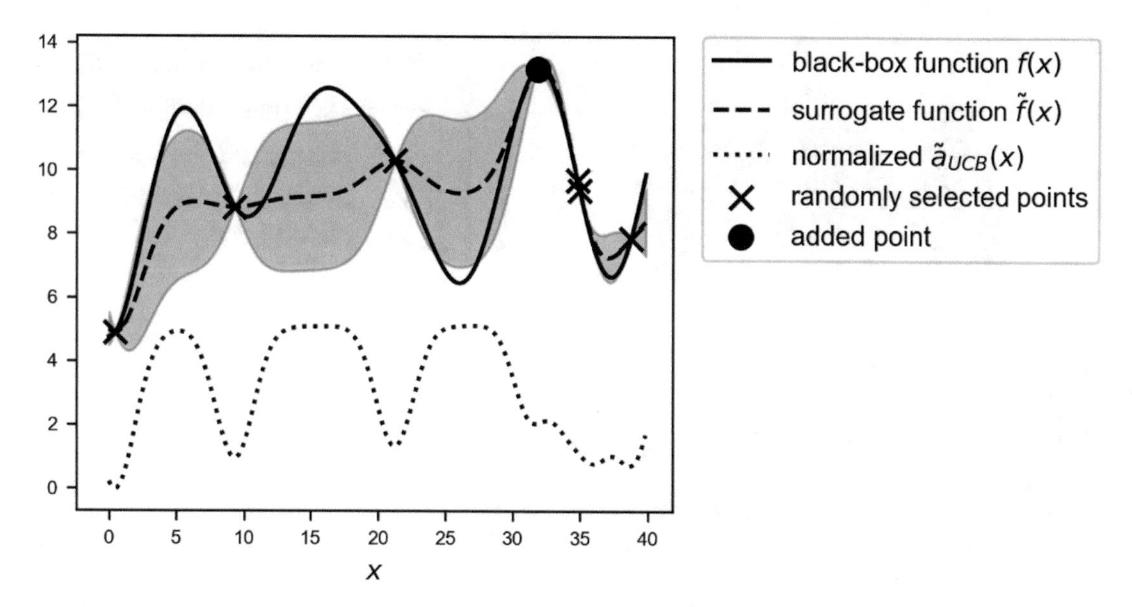

Figure 6-20. *An overview of the black-box function f(x), the randomly selected points with the additional selected points, the surrogate function, and the acquisition function* $\tilde{a}_{UCB}(x)$*. The gray area is the region contained between the lines* $\tilde{f}(x)+\sigma(x)$ *and* $\tilde{f}(x)-\sigma(x)$

Now let's add another point. Figure 6-21 shows that the additional point is still close to the maximum but has shifted in the direction of the area with a high variance around 30.

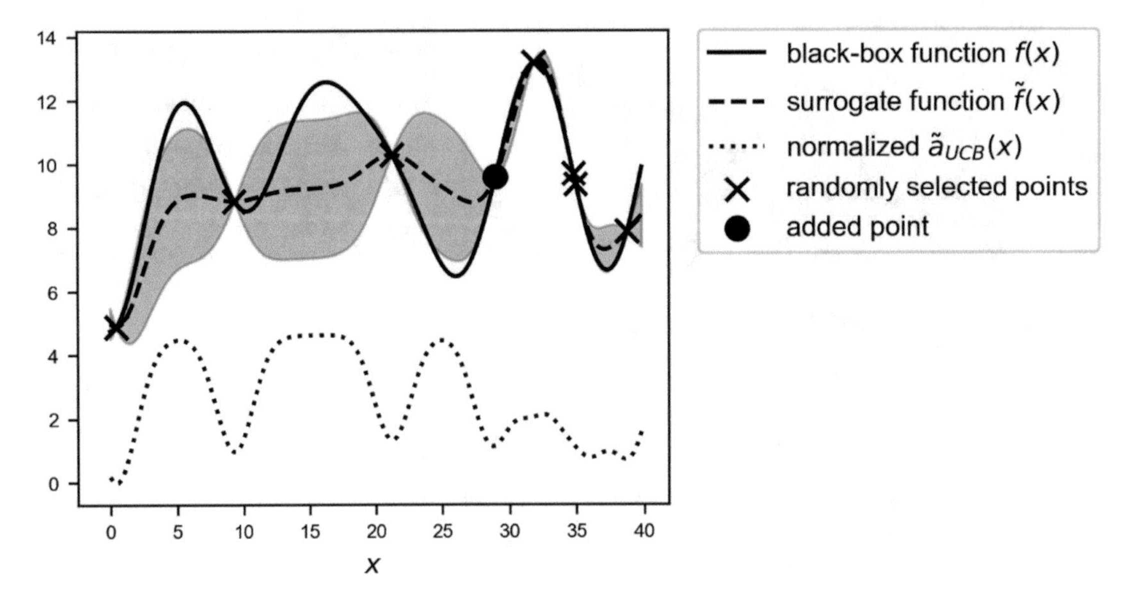

Figure 6-21. *An overview of the black-box function f(x), the randomly selected points with the additional selected points, the surrogate function, and the acquisition function $\tilde{a}_{UCB}(x)$. The gray area is the region contained between the lines $\tilde{f}(x)+\sigma(x)$ and $\tilde{f}(x)-\sigma(x)$*

If we chose a smaller η, the point would be closer to the maximum, and if we chose a bigger one, the point would be closer to the point with the highest variance between 25 and roughly 32. Let's add another point and see what happens. Figure 6-22 shows how the method now chooses a point close to another region with high variance, between 10 and roughly 22, again marked by a black circle.

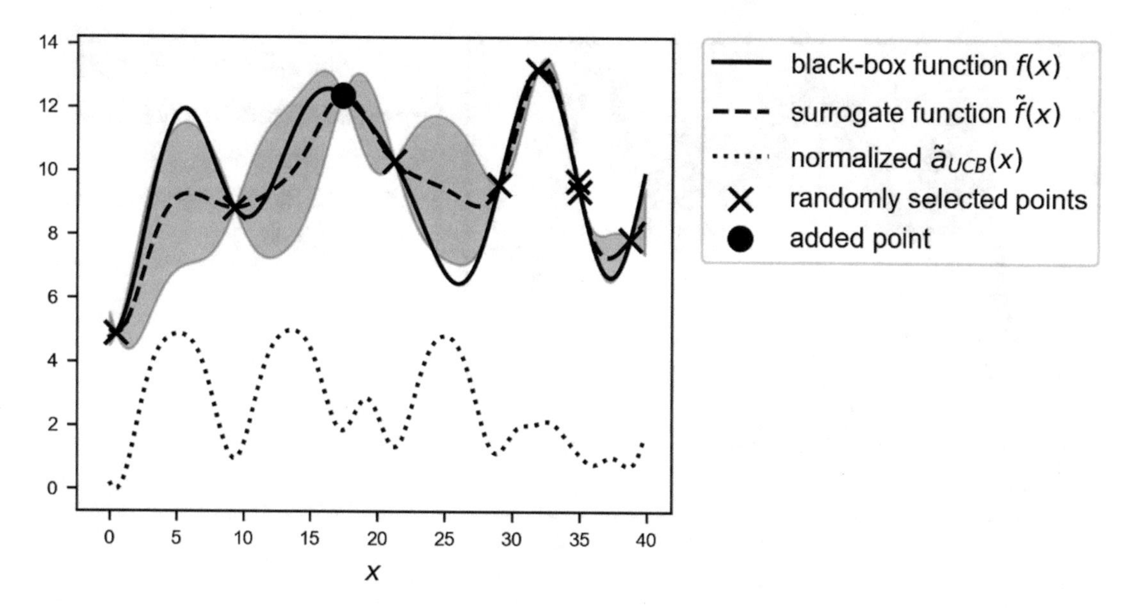

***Figure 6-22.** An overview of the black-box function f(x), the randomly selected points, the surrogate function, and the acquisition function $\tilde{a}_{UCB}(x)$. The additional selected point is marked by the black circle. The gray area is the region contained between the lines $\tilde{f}(x)+\sigma(x)$ and $\tilde{f}(x)-\sigma(x)$*

And finally, the method refines the maximum area around 15, as you can see in Figure 6-23, adding a point around 14, marked by the black circle.

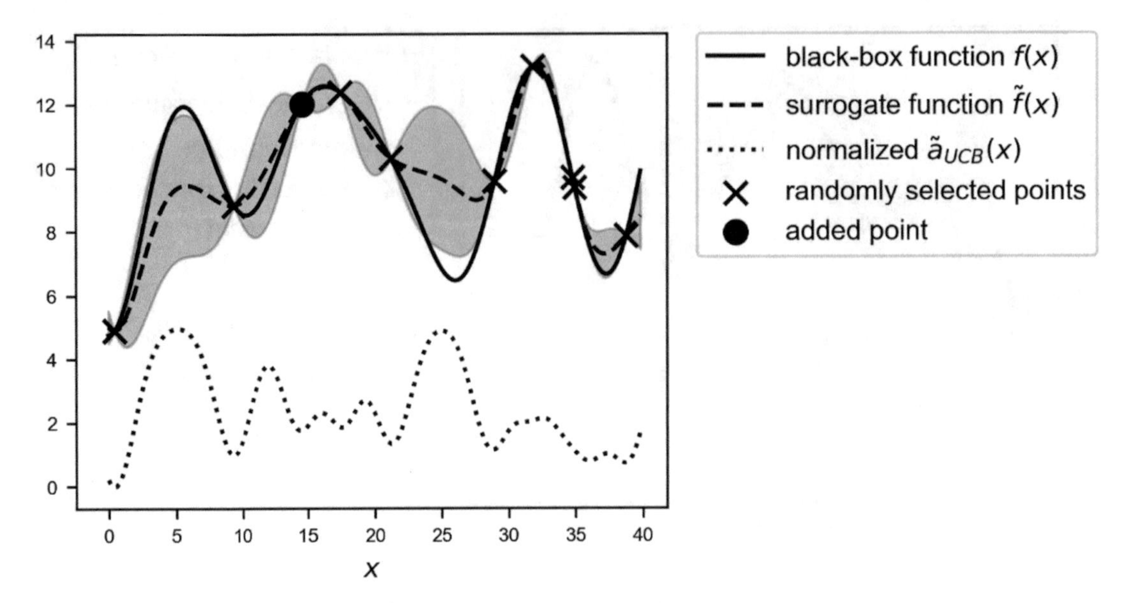

Figure 6-23. *An overview of the black-box function f(x), the randomly selected points with the additional selected points, the surrogate function, and the acquisition function* $\tilde{a}_{UCB}(x)$*. The gray area is the region contained between the lines* $\tilde{f}(x)+\sigma(x)$ *and* $\tilde{f}(x)-\sigma(x)$

As this discussion and comparison of the behavior of the two types of acquisition functions should have made clear by now, choosing the right acquisition function is dependent on which strategy you want to apply to approximate your black-box function.

Note Different types of acquisition functions will give different strategies in approximating the black-box function. For example, $a_{UCB}(x)$ will add points in regions with the highest variance, while $\tilde{a}_{UCB}(x)$ will add points finding a balance, regulated by η, between the maximum of the surrogate function and areas with high variance.

An analysis of all the different types of acquisition function would go far beyond the scope of this book. A good deal of research and reading published papers is required to get enough experience and to understand how different acquisition functions work and behave.

If you want to use Bayesian optimization with your TensorFlow model, you do not have to develop the method completely from scratch. You can try the library called `GPflowOpt` from N. Knudde et al. [4].

Sampling on a Logarithmic Scale

There is a last small subtlety that we need to discuss. Sometimes you will find yourself in a situation where you want to try a big range of possible values for a parameter, but you know from experience that the best value is probably in a specific range. Let's suppose you want to find the best value for the learning rate for your model, and you decide to test values from 10^{-4} to 1, but you know, or at least expect, that your best value lies between 10^{-3} and 10^{-4}. Now let's suppose you are working with grid search and you sample 1,000 points. You may think you have enough points, but you will get

- 0 points between 10^{-4} and 10^{-3}
- 8 points between 10^{-3} and 10^{-2}
- 89 points between 10^{-1} and 10^{-2}
- 899 points between 1 and 10^{-1}

You get a lot more points in the less interesting ranges, and none where you want them. Figure 6-24 shows the distribution of the points. Note that the x-axis uses a logarithm scale. You can clearly see how you get many more points for bigger values of the learning rate.

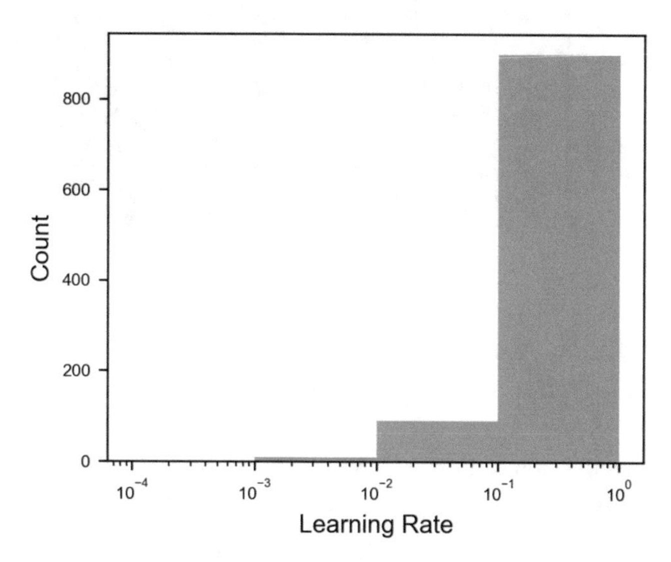

Figure 6-24. *The distribution of 1,000 points selected with grid search on a logarithmic x-scale*

You probably want to sample in a much finer way for smaller values of the learning rate than for bigger ones. What you should do is sample your points on a logarithmic scale. The basic idea is that you want to sample the same number of points between 10^{-4} and 10^{-3}, 10^{-3} and 10^{-2}, 10^{-1} and 10^{-2}, and 10^{-1} and 1. To do that, you can use the following Python code. First select a random number between 0 and a negative of the absolute value of the highest number of the power of 10 you have; in this case it's -4.

```
r = - np.arange(0,1,0.001)*4.0
```

Then your array with the selected points can be created with the following

```
points2 = 10**r
```

Figure 6-25 shows how the distributions of the points contained in the array `points2` is now completely flat, as we wanted.

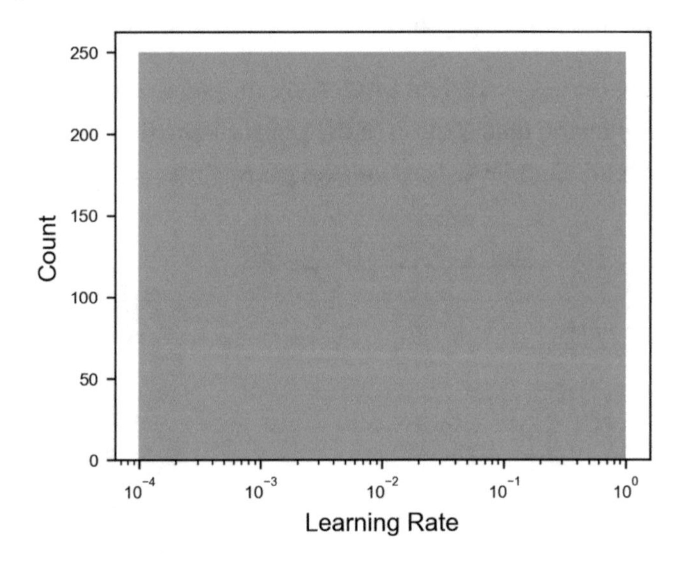

Figure 6-25. *The distribution of 1,000 points selected with grid search on a logarithmic x-scale with the modified selection method*

You get 250 points in each region, and you can easily check this code for the range 10^{-3} and 10^{-4}. For the other ranges simply change the numbers in the code.

```
print(np.sum((alpha <= 1e-3) & (alpha > 1e-4)))
```

Now you can see how you have the same number of points between the different powers of ten. With this simple trick you can ensure that you get enough points in region of your chosen range, where otherwise you would get almost no points. Remember that in this example, with 1,000 points, we get zero points between 10^{-3} and 10^{-4} using the standard method . This range is the most interesting for the learning rate, so you want to have enough points in this range to optimize your model. Note that the same applies to random search. It works in the exact same way.

Hyper-Parameter Tuning with the Zalando Dataset

To give you a concrete example of how hyper-parameter tuning works, let's apply what you have learned in a simple case. Let's start with the data, as usual. Let's use the Zalando dataset from Chapter 3. For a complete discussion, refer to that chapter. Let's quickly load and prepare the data and then discuss tuning.

First, as usual, we need the necessary libraries

```
# general libraries
import pandas as pd
import numpy as np
import matplotlib
import matplotlib.pyplot as plt
import matplotlib.font_manager as fm
from random import *
import time

# tensorflow libraries
from tensorflow.keras.datasets import fashion_mnist
import tensorflow as tf
from tensorflow import keras
from tensorflow.keras import layers
import tensorflow_docs as tfdocs
import tensorflow_docs.modeling
```

Then, to retrieve the dataset, we can simply run the following command

```
((trainX, trainY), (testX, testY)) = fashion_mnist.load_data()
```

Remember we have 60,000 observations in the train dataset and 10,000 in the dev dataset. After executing the command, we have two NumPy matrices (`trainX` and `testX`) containing all the pixel values that describe each of the training and test images and two NumPy arrays (`trainY` and `testY`) containing the associated labels.

Let's print the training dataset dimensions, as an example

```
print('Dimensions of the training dataset: ', trainX.shape)
print('Dimensions of the training labels: ', trainY.shape)
```

That will return as output

```
Dimensions of the training dataset:  (60000, 28, 28)
Dimensions of the training labels:  (60000,)
```

Remember that 784 represents the gray values of the image pixels (that have a size of 28×28 pixels).

Now we need to modify the data to obtain a "flattened" version of each image, meaning an array of 754 pixels, instead of a matrix of 28x28 pixels.

The following lines reshape the matrixes dimensions

```
data_train = trainX.reshape(60000, 784)
data_test = testX.reshape(10000, 784)
```

Finally let's normalize the input so that instead of having values from 0 to 255 (the grayscale values), it has only values between 0 and 1. This is very easy to do with this code

```
data_train_norm = np.array(data_train/255.0)
data_test_norm = np.array(data_test/255.0)
```

Before moving to developing our network, we need to solve a final problem. Labels must be provided in a different form, when performing a multiclass classification task. The Python code to do this is very simple:

```
labels_train = np.zeros((60000, 10))
labels_train[np.arange(60000), trainY] = 1

labels_test = np.zeros((10000, 10))
labels_test[np.arange(10000), testY] = 1
```

Now we have prepared the data as we need. For a complete discussion, refer to Chapter 3, where we spend a lot of time preparing the data for this dataset. Now let's move on to the model. Let's start with something easy. As a metric, let's use for this example the accuracy, since the dataset is balanced. Let's consider a network with just one layer and see what number of neurons provides the best accuracy.

Our hyper-parameter in this example will be the number of neurons in the hidden layer. Basically, we need to build a new network for each value of the hyper-parameter (the number of neurons in the hidden layer) and train it. We need two functions: one to build the network and one to train it. To build the model, we can define the following function

```
def build_model(number_neurons, learning_rate):
  # create model
  model = keras.Sequential()
  # add first hidden layer and set input dimensions
  model.add(layers.Dense(number_neurons, input_dim = 784,
                         activation = 'relu'))
  # add output layer
  model.add(layers.Dense(10, activation = 'softmax'))
  # compile model
  model.compile(loss = 'categorical_crossentropy',
               optimizer = tf.keras.optimizers.SGD(
                        learning_rate = learning_rate),
                  metrics = ['categorical_accuracy'])
    return model
```

You should understand this function, since we used the code several times in the book already. This function has an input parameter, number_neurons, that will contain, as the name indicates, the number of neurons in the hidden layer. The function to train the model will look like this

```
def train_model(number_neurons, learning_rate, number_epochs,
               mb_size):
  # build model
  model = build_model(number_neurons, learning_rate)
  # train model
  history = model.fit(
    data_train_norm, labels_train,
    epochs = number_epochs, verbose = 0,
    batch_size = mb_size,
    callbacks = [tfdocs.modeling.EpochDots()])
```

```
    # test model
    train_loss, train_accuracy = model.evaluate(
                                 data_train_norm, labels_train,
                                 verbose = 0)
    test_loss, test_accuracy = model.evaluate(
                                 data_test_norm, labels_test,
                                 verbose = 0)
    return train_accuracy, test_accuracy
```

You have already seen a very similar function several times already. The main parts should be clear. You will find a few things that are new. First, we build the model in the function itself with

```
model = build_model(number_neurons, learning_rate)
```

Additionally, we evaluate the accuracy on the train and dev datasets and return the values to the caller. In this way we can run a loop for several values of the number of neurons in the hidden layer and get the accuracies. Note that this time the function has an additional input parameter: `number_neurons`. We need to pass this number to the function that builds the model.

Let's suppose we choose the following parameters: mini-batch size = 50, we train for 100 epochs, learning rate $\lambda = 0.001$, and we build our model with 15 neurons in the hidden layer.

We then run the model

```
train_accuracy, test_accuracy = train_model(15, 0.001, 100, 50)
print(train_accuracy)
print(test_accuracy)
```

For the train dataset, we get 0.8549 and for the dev dataset, we get 0.8396 accuracy. Can we do better? Well, we can surely do a grid search to start with

```
nn = [1, 5, 10, 15, 25, 30, 50, 150, 300, 1000, 3000]
for nn_ in nn:
    train_accuracy, test_accuracy = train_model(nn_, 0.001, 100, 50)
    print('Number of neurons:', nn_, 'Acc. Train:', acc_train, 'Acc. Test',
    acc_test)
```

Keep in mind that this will take a lot of time. 3000 neurons is a large number, so be warned in case you want to try. We will get the results shown in Table 6-1.

Table 6-1. *Overview of the Accuracy on the Training and Test Datasets for a Different Number of Neurons*

Number of Neurons	Accuracy on the train Dataset	Accuracy on the test Dataset
1	0.1000	0.1000
5	0.8242	0.8123
10	0.8476	0.8323
15	0.8549	0.8373
25	0.8604	0.8444
30	0.8577	0.8413
50	0.8661	0.8470
150	0.8677	0.8513
300	0.8694	0.8512
1000	0.8740	0.8555
3000	0.8754	0.8569

Not surprisingly, more neurons deliver better accuracy, with no signs of overfitting of the train dataset, since the accuracy on the dev dataset is almost equal to the one on the train dataset. Figure 6-26 shows a plot of the accuracy on the test dataset vs. the number of neurons in the hidden layer. Note that the x-axis uses a logarithmic scale, to make the changes more evident.

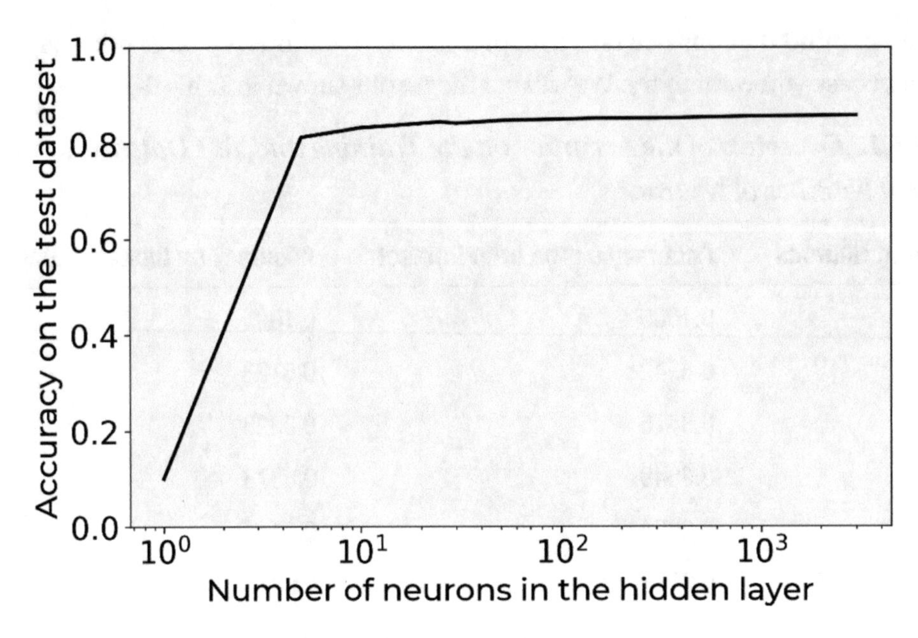

Figure 6-26. *Accuracy on the test dataset vs. the number of neurons in the hidden layer*

If your goal is to reach 80% accuracy, you could well stop here. But there are a few things to consider: first, we may be able to do better, and second, training the network with 3,000 neurons takes a lot of time. On a medium-performance laptop roughly 35 minutes. We should see if we can get the same result in a fraction of the time. We want a model that trains as fast as possible! Let's try a slightly different approach. Since we want to be faster let's consider a model with four layers. We could also tune the number of layers, but let's stick to four for this example and tune the other parameters. We will try to find the optimal value for the learning rate, the mini-batch size, the number of neurons in each layer, and the number of epochs. We will use random search. For each parameter, we will randomly select ten values:

- Number of neurons: Between 35 and 60
- Learning rate: We will use the search on the logarithmic scale between 10^{-1} and 10^{-3}
- Mini-batch size: Between 20 and 80
- Number of epochs: Between 40 and 100

We can create arrays with the possible values using this code

```
neurons_ = np.random.randint(low = 35, high = 60.0, size = (10))
r = - np.random.random([10])*3.0 - 1
learning_ = 10**r
mb_size_ = np.random.randint(low = 20, high = 80, size = 10)
epochs_ = np.random.randint(low = 40, high = 100, size = (10))
```

Note that we will not try all possible combinations. We consider only ten possible combinations: the first value of each array, the second value of each array, and so on. Let's see how efficient random search can be with just ten evaluations. We can test this model with the following loop

```
for i in range(len(neurons_)):
    train_accuracy, test_accuracy = train_model(neurons_[i], learning_[i],
    epochs_[i], mb_size_[i])
    print()
    print('Epochs:', epochs_[i], 'Number of neurons:', neurons_[i],
    'Learning rate:', learning_[i], 'Minibatch size', mb_size_[i],
          'Training accuracy:', train_accuracy, 'Test Accuracy', test_accuracy)
```

You will find that the combinations with 44 epochs, 36 neurons in each layer, a learning rate of 0.07537, and a mini-batch size of 54 gives you an accuracy on the dev dataset of 0.86. Not bad, considering that this run took 1.5 minutes, so 23 times faster than the model with one layer and 3,000 neurons. And it's 2% better. Our naïve initial test gave us an accuracy of 0.84, so with hyper-parameter tuning, we got 2% better than our initial guess. In the deep learning, this is considered a good result. What we did should give you an idea of how powerful hyper-parameter tuning can be if it's done properly. Keep in mind that you should spend quite some time doing it, and especially thinking about how to do it.

Note Always think about how you want to do your hyper-parameter tuning and use your experience or ask for help from someone with experience. It is useless to invest time and resources to try combinations of parameters that you know will not work. For example, it's better to spend time testing learning rates that are very small than to test learning rates around 1. Remember that every training round of your network will take time, even if the results are not useful!

The point of this last section is not to get the best model possible, but to give you an idea about how the tuning process may work. You could continue trying different optimizers (for example Adam), considering wider ranges for the parameters, more parameter combinations, and so on.

A Quick Note about the Radial Basis Function

Before finishing this chapter, let's discuss a minor point about the Radial Basis Function

$$K(x,x_i)=\sigma^2 e^{-\frac{1}{2l^2}\|x-x_i\|^2}$$

It is important that you understand the role of the parameter l. In our examples, we chose $l = 2$ but we did not discuss why. Choosing an l that's too small will make the acquisition function develop very narrow peaks around the points we already have, as you can see in the left plot of Figure 6-27. Big values of l will have a smoothing effect on the acquisition function, as you can see in the middle and right plots of Figure 6-27.

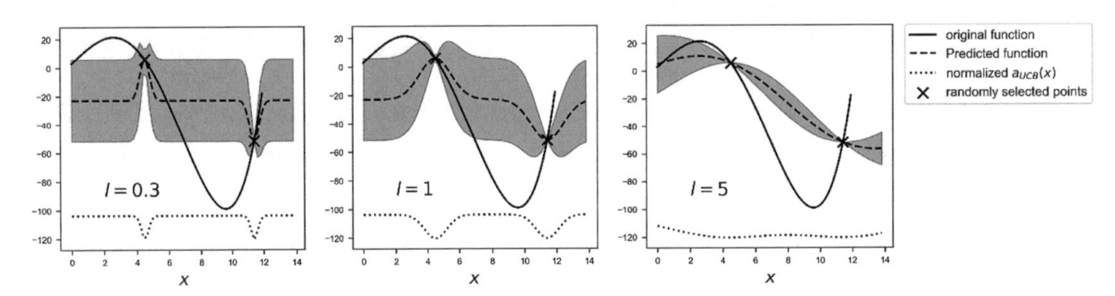

Figure 6-27. *The effect of changing the l parameter in the radial basis function*

Usually it's good practice to avoid values for l that are too small or too big, to be able to have a variance that varies in a smooth way between known points, as in Figure 6-27 for $l = 1$. Having a very small l will make the variance between points almost constant and therefore make the algorithm almost always choose the middle point between points, as you can check from the acquisition function. While choosing an l that's too big will make the variance small, and therefore difficult to use. As you can see for $l = 5$ in Figure 6-27, the acquisition function is almost constant. Typical values are around 1 or 2.

Exercises

EXERCISE 1 (LEVEL: EASY)

Try different optimizers (for example, Adam) for the examples in this chapter and consider wider ranges for the parameters, more parameters combinations, and so on. See how the results change.

EXERCISE 2 (LEVEL: DIFFICULT)

Try to implement Bayesian optimization from scratch, as described in this chapter.

EXERCISE 3 (LEVEL: MEDIUM)

Try to optimize a multiclass classification model like the one you saw in this chapter, but using the MNIST database of handwritten digits (`http://yann.lecun.com/exdb/mnist/`). To download the dataset from TensorFlow, use the following lines of code:

```
from tensorflow import keras
(x_train, y_train), (x_test, y_test) = keras.datasets.mnist.load_data()
```

References

[1] Bergstra, James, and Yoshua Bengio. "Random search for hyper-parameter optimization." *Journal of Machine Learning Research,* 13.2 (2012).

[2] Nadaraya, Elizbar A. "On estimating regression." *Theory of Probability & Its Applications,* 9.1 (1964): 141-142.

[3] Do, Chuong B. "Gaussian processes." Stanford University, Stanford, CA, accessed Dec 5 (2007): 2017.

[4] Knudde, Nicolas, et al. "GPflowOpt: A Bayesian optimization library using TensorFlow." *arXiv preprint arXiv:1711.03845* (2017).

CHAPTER 7

Convolutional Neural Networks

In the previous chapters, we looked at fully connected networks and all the problems you encounter while training them. The network architecture we used, one where each neuron in a layer is connected to all the neurons in the previous and next layers, is not good at many fundamental tasks like image recognition, speech recognition, time series prediction, and many more. *Convolutional Neural Networks* (CNN) and *Recurrent Neural Networks* (RNN) are the most advanced architectures used today. This chapter looks at convolution and pooling, the basic building blocks of CNNs. We also discuss a complete, although basic, implementation of CNNs in Keras. RNNs are discussed, although briefly, in the next chapter.

Kernels and Filters

One of the main components of CNNs are filters, which are square matrices that have dimensions $n_K \times n_K$, where usually n_K is a small number, like 3 or 5. Sometimes filters are also called *kernels*. Let's define four different filters and check later in the chapter their effect when used in convolution operations. For those examples, we will work with 3×3 filters. For the moment just take the following definitions as a reference and you will see how to use them later in the chapter.

- The following kernel will allow the detection of horizontal edges

$$\mathfrak{I}_H = \begin{pmatrix} 1 & 1 & 1 \\ 0 & 0 & 0 \\ -1 & -1 & -1 \end{pmatrix}$$

U. Michelucci, *Applied Deep Learning with TensorFlow 2*, https://doi.org/10.1007/978-1-4842-8020-1_7

- The following kernel will allow the detection of vertical edges

$$\mathfrak{I}_V = \begin{pmatrix} 1 & 0 & -1 \\ 1 & 0 & -1 \\ 1 & 0 & -1 \end{pmatrix}$$

- The following kernel will allow the detection of edges when luminosity changes drastically

$$\mathfrak{I}_L = \begin{pmatrix} -1 & -1 & -1 \\ -1 & 8 & -1 \\ -1 & -1 & -1 \end{pmatrix}$$

- The following kernel will blur edges in an image

$$\mathfrak{I}_B = -\frac{1}{9}\begin{pmatrix} 1 & 1 & 1 \\ 1 & 1 & 1 \\ 1 & 1 & 1 \end{pmatrix}$$

In the next sections, we will apply convolution to a test image with the filters and you will see the effect.

Convolution

The first step to understanding CNNs is to understand convolution. The easiest way is to see it in action is with a few simple cases. First, in the context of neural networks, convolution is done between tensors. The operation gets two tensors as input and produces a tensor as output. The operation is usually indicated with the operator $*$. Let's see how it works. Let's get two tensors, both with dimensions 3×3. The convolution operation is done by applying the following formula

$$\begin{pmatrix} a_1 & a_2 & a_3 \\ a_4 & a_5 & a_6 \\ a_7 & a_8 & a_9 \end{pmatrix} * \begin{pmatrix} k_1 & k_2 & k_3 \\ k_4 & k_5 & k_6 \\ k_7 & k_8 & k_9 \end{pmatrix} = \sum_{i=1}^{9} a_i k_i$$

In this case, the result is simply the sum of each element a_i multiplied by the respective element k_i. In a more typical matrix formalism, this formula could be written with a double sum, as

$$\begin{pmatrix} a_{11} & a_{12} & a_{13} \\ a_{21} & a_{22} & a_{23} \\ a_{31} & a_{32} & a_{33} \end{pmatrix} * \begin{pmatrix} k_{11} & k_{12} & k_{13} \\ k_{21} & k_{22} & k_{23} \\ k_{31} & k_{32} & k_{33} \end{pmatrix} = \sum_{i=1}^{3}\sum_{j=1}^{3} a_{ij} k_{ij}$$

The first version has the advantage of making the fundamental idea very clear: each element from one tensor is multiplied by the correspondent element (the element in the same position) of the second tensor. Then all the values are summed to get the result.

In the previous section we talked about kernels, and the reason is that convolution is usually done between a tensor, which we indicate here with A, and a kernel. Typically, kernels are small, 3×3 or 5×5, while the input tensors A are normally bigger. In image recognition for example, the input tensors A are the images that may have dimensions as high as $1024 \times 1024 \times 3$, where 1024×1024 is the resolution and the last dimension (3) is the number of the color channels, the RGB values. In advanced applications the images may even have higher resolutions. How do we apply convolution when we have matrices with different dimensions? To understand this, let's consider a matrix A that is 4×4

$$A = \begin{pmatrix} a_1 & a_2 & a_3 & a_4 \\ a_5 & a_6 & a_7 & a_8 \\ a_9 & a_{10} & a_{11} & a_{12} \\ a_{13} & a_{14} & a_{15} & a_{16} \end{pmatrix}$$

Let's see how to do convolution with a kernel K that we will take for this example to be 3×3

$$K = \begin{pmatrix} k_1 & k_2 & k_3 \\ k_4 & k_5 & k_6 \\ k_7 & k_8 & k_9 \end{pmatrix}$$

The idea is to start on the top-left corner of the matrix A and select a 3×3 region. In the example that would be

$$A_1 = \begin{pmatrix} a_1 & a_2 & a_3 \\ a_5 & a_6 & a_7 \\ a_9 & a_{10} & a_{11} \end{pmatrix}$$

Or the elements marked in bold here

$$A = \begin{pmatrix} \boldsymbol{a_1} & \boldsymbol{a_2} & \boldsymbol{a_3} & a_4 \\ \boldsymbol{a_5} & \boldsymbol{a_6} & \boldsymbol{a_7} & a_8 \\ \boldsymbol{a_9} & \boldsymbol{a_{10}} & \boldsymbol{a_{11}} & a_{12} \\ a_{13} & a_{14} & a_{15} & a_{16} \end{pmatrix}$$

Then we perform the convolution between this smaller matrix A_1 and K getting (we will indicate the result with B_1)

$$B_1 = A_1 * K = a_1k_1 + a_2k_2 + a_3k_3 + k_4a_5 + k_5a_5 + k_6a_7 + k_7a_9 + k_8a_{10} + k_9a_{11}$$

Then we need to shift the selected 3 × 3 region in matrix A one column to the right and select the elements marked in bold

$$A = \begin{pmatrix} a_1 & \boldsymbol{a_2} & \boldsymbol{a_3} & \boldsymbol{a_4} \\ a_5 & \boldsymbol{a_6} & \boldsymbol{a_7} & \boldsymbol{a_8} \\ a_9 & \boldsymbol{a_{10}} & \boldsymbol{a_{11}} & \boldsymbol{a_{12}} \\ a_{13} & a_{14} & a_{15} & a_{16} \end{pmatrix}$$

That will give us the second sub-matrix A_2

$$A_2 = \begin{pmatrix} a_2 & a_3 & a_4 \\ a_6 & a_7 & a_8 \\ a_{10} & a_{11} & a_{12} \end{pmatrix}$$

We perform the convolution between this smaller matrix A_2 and K

$$B_2 = A_2 * K = a_2k_1 + a_3k_2 + a_4k_3 + a_6k_4 + a_7k_5 + a_8k_6 + a_{10}k_7 + a_{11}k_8 + a_{12}k_9$$

Now we cannot shift our 3 × 3 region anymore to the right, since we have reached the end of the matrix A. So what we do is shift it one row down and start again from the left side. The next selected region would be

$$A_3 = \begin{pmatrix} a_5 & a_6 & a_7 \\ a_9 & a_{10} & a_{11} \\ a_{13} & a_{14} & a_{15} \end{pmatrix}$$

Again, we perform convolution of A_3 with K

$$B_3 = A_3 * K = a_5k_1 + a_6k_2 + a_7k_3 + a_9k_4 + a_{10}k_5 + a_{11}k_6 + a_{13}k_7 + a_{14}k_8 + a_{15}k_9$$

As you might have guessed at this point, the last step is to shift the 3 × 3 selected region to the right one column and perform convolution. Our selected region will now be

$$A_4 = \begin{pmatrix} a_6 & a_7 & a_8 \\ a_{10} & a_{11} & a_{12} \\ a_{14} & a_{15} & a_{16} \end{pmatrix}$$

And the convolution will give the result

$$B_4 = A_4 * K = a_6 k_1 + a_7 k_2 + a_8 k_3 + a_{10} k_4 + a_{11} k_5 + a_{12} k_6 + a_{14} k_7 + a_{15} k_8 + a_{16} k_9$$

Now we cannot shift our 3 × 3 region anymore, neither right nor down. We have calculated four values: B_1, B_2, B_3, and B_4. Those elements will form the resulting tensor of the convolution operation, giving us the tensor B

$$B = \begin{pmatrix} B_1 & B_2 \\ B_3 & B_4 \end{pmatrix}$$

The same process can be applied when the tensor A is bigger. You will simply get a bigger resulting B tensor, but the algorithm to get the elements B_i is the same. Before moving on, there is still a small detail that we need to discuss, and that is the concept of *stride*. In the process above we moved our 3 × 3 region one column to the right and one row down. The number of rows and columns, in this example 1, is called *stride* and is often indicated with s. Stride $s = 2$ means simply that we shift our 3 × 3 region two columns to the right and two rows down.

Something else that we need to discuss is the size of the selected region in the input matrix A. The dimensions of the selected region that we shifted around in the process must be the same as the kernel used. If you use a 5 × 5 kernel, you need to select a 5 × 5 region in A. In general, given a $n_K \times n_K$ kernel, you will select a $n_K \times n_K$ region in A.

In a more formal definition, convolution with stride s in the neural network context is a process that takes a tensor A of dimensions $n_A \times n_A$ and a kernel K of dimensions $n_K \times n_K$ and gives as output a matrix B of dimensions $n_B \times n_B$ with

$$n_B = \frac{n_A - n_K}{s} + 1$$

Where we have indicated with $\lfloor x \rfloor$ the integer part of x (in the programming world, this is often called the floor of x). A proof of this formula would take too long to discuss but is easy to see why it is true (try to derive it). To make things a bit easier, we will suppose that n_K is odd. You will see soon why this is important (although not fundamental). We start explaining the case with a stride $s = 1$. The algorithm generates a new tensor B from an input tensor A and a kernel K according to this formula

$$B_{ij} = \left(A * K\right)_{ij} = \sum_{f=0}^{n_K-1} \sum_{h=0}^{n_K-1} A_{i+f,j+h} K_{i+f,j+h}$$

The formula is cryptic and is very difficult to understand. Let's see some more examples to grasp the meaning better. Figure 7-1 shows how convolution works. Suppose you have a 3 × 3 filter. Then, in the figure, you can see that the top-left nine elements of the matrix A, marked by a square drawn with a black continuous line, are the ones used to generate the first element of the matrix B_1 according to the formula. The elements marked by the square drawn with a dotted line are the ones used to generate the second element B_2, and so on.

To reiterate, the basic idea is that each element of the 3 × 3 square from matrix A is multiplied by the corresponding element of the kernel K and all the numbers are summed. The sum is then the element of the new matrix B. After calculating the value for B_1 you shift the region you are considering in the original matrix of one column to the right (the square indicated in Figure 7-1 with a dotted line) and repeat the operation. You continue to shift your region to the right until you reach the border and then you move one element down and start again from the left. You continue in this fashion until you reach the lower-right angle of the matrix. The same kernel is used for all the regions in the original matrix.

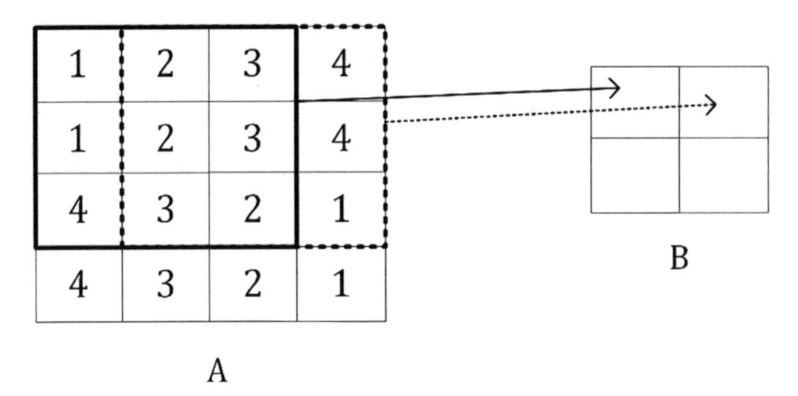

***Figure 7-1.** A visual explanation of convolution*

Given the kernel $\mathfrak{I}_H$ for example you can see in Figure 7-2 which element of A gets multiplied by which element in $\mathfrak{I}_H$ and the result for the element B_1, that is nothing more than the sum of all the multiplications

$$B_{11} = 1\times1 + 2\times1 + 3\times1 + 1\times0 + 2\times0 + 3\times0 + 4\times(-1) + 3\times(-1) + 2\times(-1) = -3$$

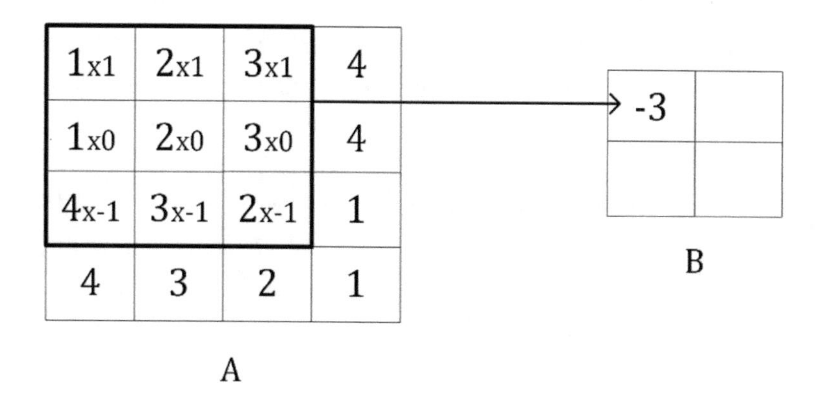

***Figure 7-2.** A visualization of convolution with the kernel $\mathfrak{I}_H$*

Figure 7-3 shows an example of convolution with stride $s = 2$.

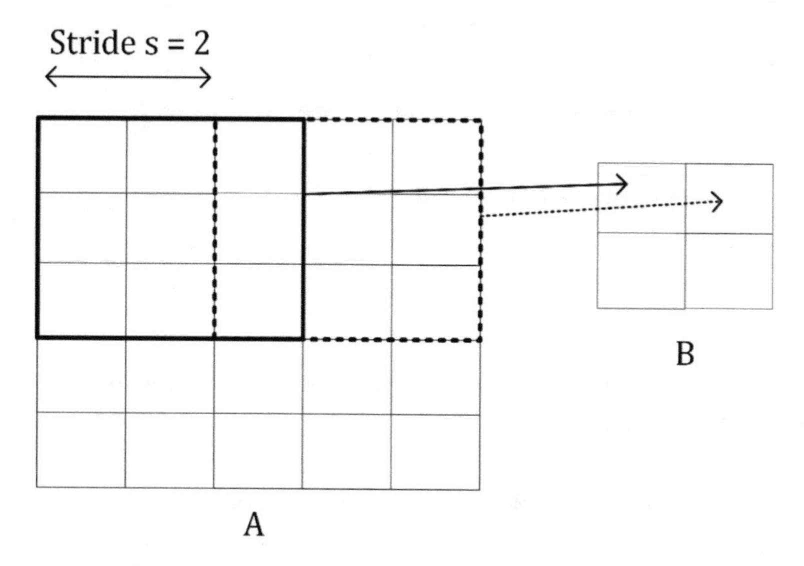

***Figure 7-3.** A visual explanation of convolution with stride $s = 2$*

The reason that the dimension of the output matrix takes only the floor (the integer part) of $\frac{n_A - n_K}{s} + 1$ can be seen in Figure 7-4. If $s > 1$, what can happen, depending on the dimensions of A, is that at a certain point you cannot shift your window on matrix A (the black square you can see in Figure 7-3 for example) anymore, and you cannot cover the matrix A completely. Figure 7-4 shows that you need an additional column on the right of matrix A (marked by many X) to be able to perform the convolution operation. In Figure 7-4, we chose $s = 3$, and since we have $n_A = 5$ and $n_K = 3$, B will be a scalar as a result:

$$n_B = \left\lfloor \frac{n_A - n_K}{s} + 1 \right\rfloor = \left\lfloor \frac{5-3}{3} + 1 \right\rfloor = \left\lfloor \frac{5}{3} \right\rfloor = 1$$

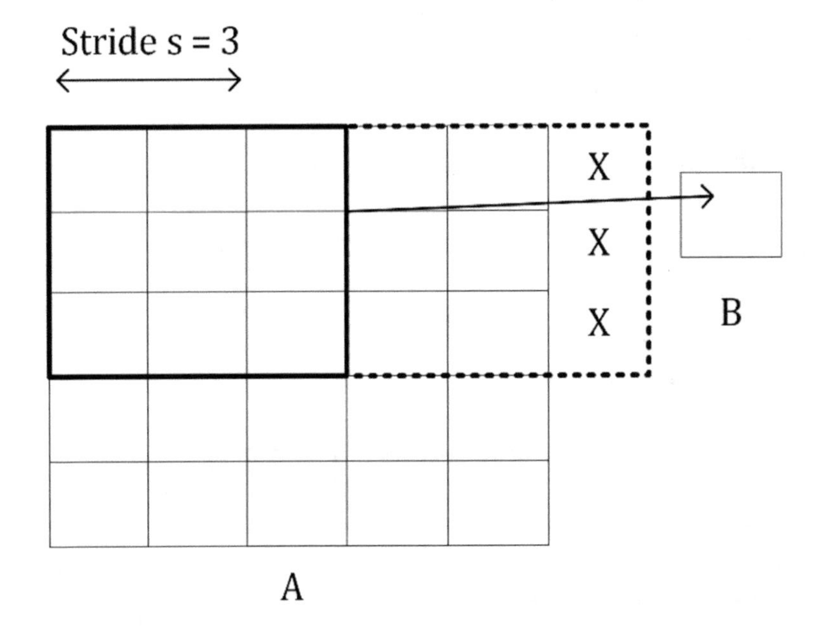

Figure 7-4. *A visual explanation why the floor function is needed when evaluating the resulting matrix B dimensions*

You can see from Figure 7-4 how, with a 3 × 3 region, you can only cover the top-left region of A, since with stride $s = 3$ you would end up outside A. Therefore, you can consider one region only for the convolution operation, thus ending up with a scalar for the resulting tensor B.

Let's now look at a few additional examples to make this formula even clearer. We start with a small matrix 3 × 3

$$A = \begin{pmatrix} 1 & 2 & 3 \\ 4 & 5 & 6 \\ 7 & 8 & 9 \end{pmatrix}$$

and consider this kernel

$$K = \begin{pmatrix} k_1 & k_2 & k_3 \\ k_4 & k_5 & k_6 \\ k_7 & k_8 & k_9 \end{pmatrix}$$

With the stride $s = 1$. The convolution will be given by

$$B = A * K = 1 \cdot k_1 + 2 \cdot k_2 + 3 \cdot k_3 + 4 \cdot k_4 + 5 \cdot k_5 + 6 \cdot k_6 + 7 \cdot k_7 + 8 \cdot k_8 + 9 \cdot k_9$$

and the result B will be a scalar, since $n_A = 3$, $n_K = 3$, therefore

$$n_B = \left\lfloor \frac{n_A - n_K}{s} + 1 \right\rfloor = \left\lfloor \frac{3-3}{1} + 1 \right\rfloor = 1$$

If now consider a matrix A with dimensions 4 × 4, or $n_A = 4$, $n_K = 3$ and $s = 1$, you will get as output a matrix B with dimensions 2 × 2, since

$$n_B = \left\lfloor \frac{n_A - n_K}{s} + 1 \right\rfloor = \left\lfloor \frac{4-3}{1} + 1 \right\rfloor = 2$$

For example, you can verify that given

$$A = \begin{pmatrix} 1 & 2 & 3 & 4 \\ 5 & 6 & 7 & 8 \\ 9 & 10 & 11 & 12 \\ 13 & 14 & 15 & 16 \end{pmatrix}$$

and

$$K = \begin{pmatrix} 1 & 2 & 3 \\ 4 & 5 & 6 \\ 7 & 8 & 9 \end{pmatrix}$$

you have with stride $s = 1$

$$B = A * K = \begin{pmatrix} 348 & 393 \\ 528 & 573 \end{pmatrix}$$

Let's verify one of the elements: B_{11} with the formula we saw before. We have

$$\begin{aligned}
B_{11} &= \sum_{f=0}^{2}\sum_{h=0}^{2} A_{1+f,1+h} K_{1+f,1+h} \\
&= \sum_{f=0}^{2} (A_{1+f,1} K_{1+f,1} + A_{1+f,2} K_{1+f,2} + A_{1+f,3} K_{1+f,3}) \\
&= (A_{1,1} K_{1,1} + A_{1,2} K_{1,2} + A_{1,3} K_{1,3}) + (A_{2,1} K_{2,1} + A_{2,2} K_{2,2} + A_{2,3} K_{2,3}) + \\
&\quad (A_{3,1} K_{3,1} + A_{3,2} K_{3,2} + A_{3,3} K_{3,3}) \\
&= (1 \cdot 1 + 2 \cdot 2 + 3 \cdot 3) + (5 \cdot 4 + 6 \cdot 5 + 7 \cdot 6) + (9 \cdot 7 + 10 \cdot 8 + 11 \cdot 9) \\
&= 14 + 92 + 242 \\
&= 348
\end{aligned}$$

Note that the formula for the convolution works only for stride $s = 1$, but can be easily generalized for other values of s.

This calculation is very easy to implement in Python. The following function can evaluate the convolution of two matrixes easily enough for $s = 1$ (you can do it in Python with existing functions, but it is instructive to see how to do it from scratch)

```
import numpy as np

def conv_2d(A, kernel):

    output = np.zeros([A.shape[0]-(kernel.shape[0]-1), A.shape[1]-(kernel.
    shape[0]-1)])

    for row in range(1, A.shape[0]-1):
        for column in range(1, A.shape[1]-1):
            output[row-1, column-1] = np.tensordot(A[row-1:row+2,
            column-1:column+2], kernel)

    return output
```

Note that the input matrix A does not even need to a square one, but it is assumed that the kernel is and that its dimension n_K is odd. The previous example can be evaluated with the following code

```
A = np.array([[1,2,3,4],[5,6,7,8],[9,10,11,12],[13,14,15,16]])
K = np.array([[1,2,3],[4,5,6],[7,8,9]])
print(conv_2d(A,K))
```

This gives the result

```
[[ 348. 393.]
[ 528. 573.]]
```

Examples of Convolution

We'll now apply the kernels we defined to a test image and see the results. As a test image let's create a chessboard that's 160 × 160 pixels with this code

```
chessboard = np.zeros([8*20, 8*20])
for row in range(0, 8):
    for column in range (0, 8):
        if ((column+8*row) % 2 == 1) and (row % 2 == 0):
            chessboard[row*20:row*20+20, column*20:column*20+20] = 1
        elif ((column+8*row) % 2 == 0) and (row % 2 == 1):
            chessboard[row*20:row*20+20, column*20:column*20+20] = 1
```

Figure 7-5 shows the chessboard.

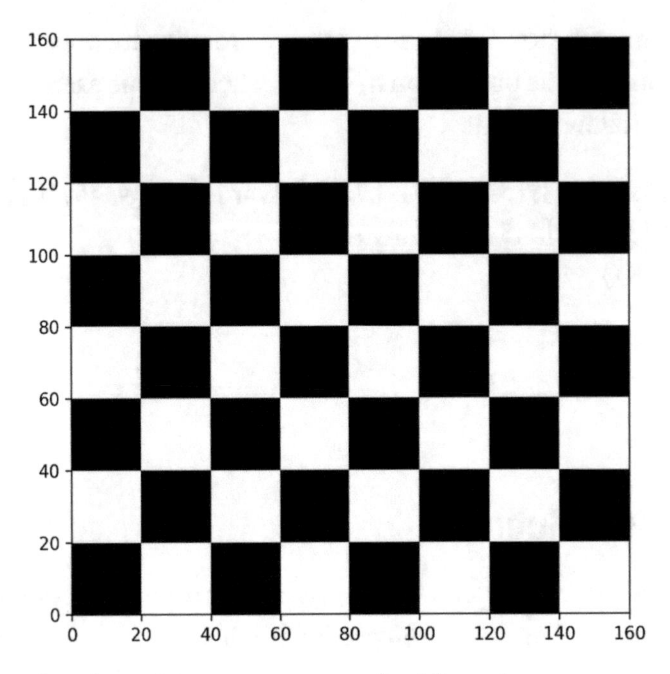

***Figure 7-5.** The chessboard image generated with code*

Now let's try to apply convolution to this image with the different kernels and with stride $s = 1$.

Using the kernel $\mathfrak{I}_H$ will detect the horizontal edges. This can be applied with the code

```
edgeh = np.matrix('1 1 1; 0 0 0; -1 -1 -1')
outputh = conv_2d (chessboard, edgeh)
```

Figure 7-6 shows the output.

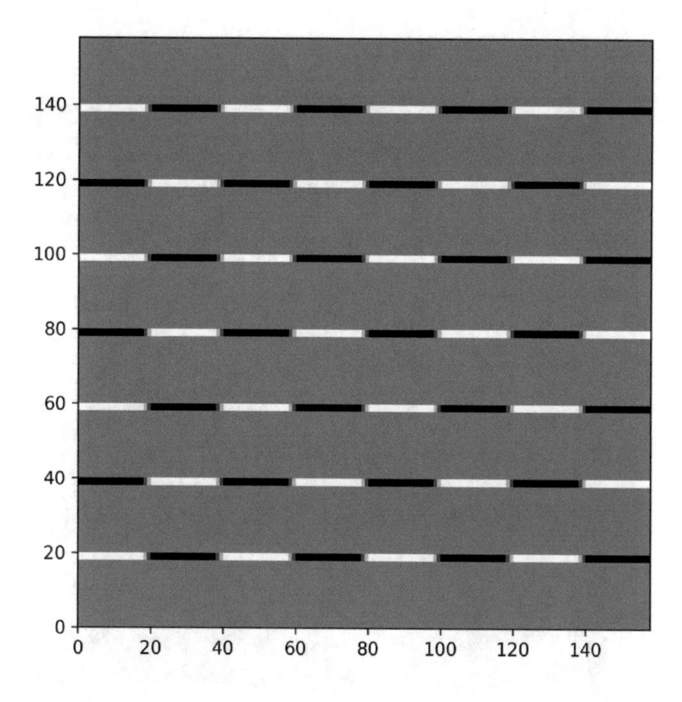

Figure 7-6. *The result of performing a convolution between the kernel $\mathfrak{I}_H$ and the chessboard image*

Now you can understand why this kernel detects horizontal edges. Additionally, this kernel detects when you go from light to dark or vice versa. Note this image is only 158 × 158 pixels, as expected, since

$$n_B = \left\lfloor \frac{n_A - n_K}{s} + 1 \right\rfloor = \left\lfloor \frac{160-3}{1} + 1 \right\rfloor = \left\lfloor \frac{157}{1} + 1 \right\rfloor = \lfloor 158 \rfloor = 158$$

Now let's apply $\mathfrak{I}_V$ with the code

```
edgev = np.matrix('1 0 -1; 1 0 -1; 1 0 -1')
outputv = conv_2d (chessboard, edgev)
```

This gives the result shown in Figure 7-7.

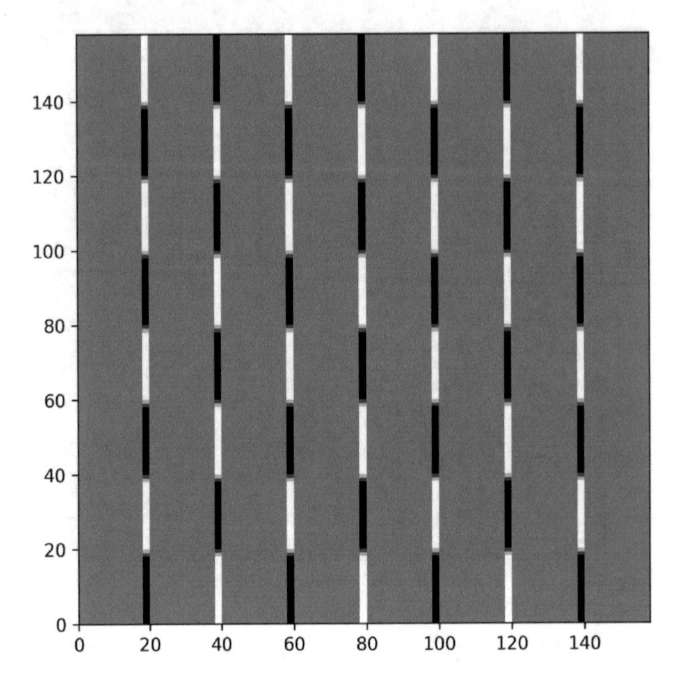

***Figure 7-7.** The result of performing a convolution between the kernel $\mathfrak{I}_V$ and the chessboard image*

Now we can use the kernel $\mathfrak{I}_L$

```
edgel = np.matrix ('-1 -1 -1; -1 8 -1; -1 -1 -1')
outputl = conv_2d (chessboard, edgel)
```

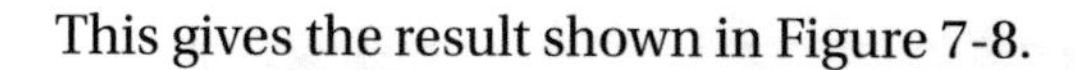

This gives the result shown in Figure 7-8.

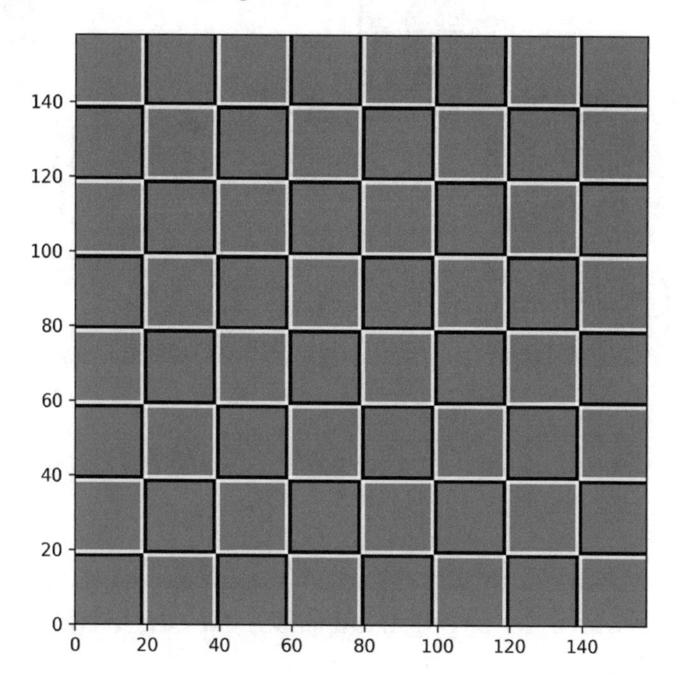

Figure 7-8. *The result of performing a convolution between the kernel $\mathfrak{I}_L$ and the chessboard image*

And finally, we can apply the blurring kernel $\mathfrak{I}_B$

```
edge_blur = -1.0/9.0*np.matrix('1 1 1; 1 1 1; 1 1 1')
output_blur = conv_2d (chessboard, edge_blur)
```

Figure 7-9 shows two plots: on the left the blurred image and on the right the original one. The images show only a small region of the original chessboard to make the blurring clearer.

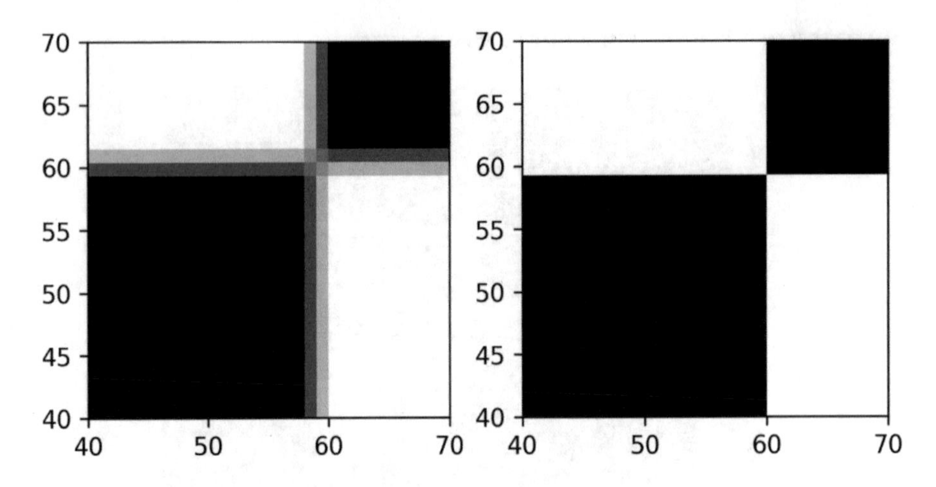

***Figure 7-9.** The effect of the blurring kernel $\Im_B$. On the left is the blurred image and on the right is the original one*

To finish this section, let's try to understand how the edges can be detected. Consider the following matrix with a sharp vertical transition, since the left part is full of 10s and the right part full of 0s.

```
ex_mat = np.matrix('10 10 10 10 0 0 0 0; 10 10 10 10 0 0 0 0; 10 10 10 10 0 0 0 0; 10 10 10 10 0 0 0 0; 10 10 10 10 0 0 0 0; 10 10 10 10 0 0 0 0; 10 10 10 10 0 0 0 0; 10 10 10 10 0 0 0 0')
```

It looks like this

```
matrix([[10, 10, 10, 10,  0,  0,  0,  0],
        [10, 10, 10, 10,  0,  0,  0,  0],
        [10, 10, 10, 10,  0,  0,  0,  0],
        [10, 10, 10, 10,  0,  0,  0,  0],
        [10, 10, 10, 10,  0,  0,  0,  0],
        [10, 10, 10, 10,  0,  0,  0,  0],
        [10, 10, 10, 10,  0,  0,  0,  0],
        [10, 10, 10, 10,  0,  0,  0,  0]])
```

Let's consider the kernel $\Im_V$. We can perform the convolution with the code

```
ex_out = conv_2d (ex_mat, edgev)
```

The result is

```
array([[ 0.,  0., 30., 30.,  0.,  0.],
       [ 0.,  0., 30., 30.,  0.,  0.],
       [ 0.,  0., 30., 30.,  0.,  0.],
       [ 0.,  0., 30., 30.,  0.,  0.],
       [ 0.,  0., 30., 30.,  0.,  0.],
       [ 0.,  0., 30., 30.,  0.,  0.]])
```

Figure 7-10 shows the original matrix (on the left) and the output of the convolution on the right. The convolution with the kernel $\mathfrak{I}_V$ has clearly detected the sharp transition in the original matrix, marking with a vertical black line where the transition from black to white happens. For example, consider $B_{11} = 0$

$$B_{11} = \begin{pmatrix} 10 & 10 & 10 \\ 10 & 10 & 10 \\ 10 & 10 & 10 \end{pmatrix} * \mathfrak{I}_V = \begin{pmatrix} 10 & 10 & 10 \\ 10 & 10 & 10 \\ 10 & 10 & 10 \end{pmatrix} * \begin{pmatrix} 1 & 0 & -1 \\ 1 & 0 & -1 \\ 1 & 0 & -1 \end{pmatrix}$$
$$= 10\times1 + 10\times0 + 10\times-1 + 10\times1 + 10\times0 + 10\times-1 + 10\times1 + 10\times0 + 10\times-1 = 0$$

Note that in the input matrix

$$\begin{pmatrix} 10 & 10 & 10 \\ 10 & 10 & 10 \\ 10 & 10 & 10 \end{pmatrix}$$

there is no transition, as all the values are the same. On the contrary, if you consider B_{13}, you need to consider this region of the input matrix

$$\begin{pmatrix} 10 & 10 & 0 \\ 10 & 10 & 0 \\ 10 & 10 & 0 \end{pmatrix}$$

where there is a clear transition, since the right-most column is made up of 0s, and the rest are 10s. Now you get a different result

$$B_{11} = \begin{pmatrix} 10 & 10 & 0 \\ 10 & 10 & 0 \\ 10 & 10 & 0 \end{pmatrix} * \mathfrak{I}_V = \begin{pmatrix} 10 & 10 & 0 \\ 10 & 10 & 0 \\ 10 & 10 & 0 \end{pmatrix} * \begin{pmatrix} 1 & 0 & -1 \\ 1 & 0 & -1 \\ 1 & 0 & -1 \end{pmatrix}$$
$$= 10\times1 + 10\times0 + 0\times-1 + 10\times1 + 10\times0 + 0\times-1 + 10\times1 + 10\times0 + 0\times-1 = 30$$

As soon as there is a big change in values along the horizontal direction, this is how the convolution will return a high value since the values multiplied by the column with 1 in the kernel will be bigger. When there is a transition from small to high values along the horizontal axis, the elements multiplied by -1 will give a result that is bigger in absolute value. Therefore the final result will be negative and big in absolute value. This is why this kernel can also detect when you pass from a light color to a darker color or vice versa. In fact, if you consider the opposite transition (from 0 to 10) in a hypothetical different matrix A, you would have

$$B_{11} = \begin{pmatrix} 0 & 10 & 10 \\ 0 & 10 & 10 \\ 0 & 10 & 10 \end{pmatrix} * \mathfrak{I}_V = \begin{pmatrix} 0 & 10 & 10 \\ 0 & 10 & 10 \\ 0 & 10 & 10 \end{pmatrix} * \begin{pmatrix} 1 & 0 & -1 \\ 1 & 0 & -1 \\ 1 & 0 & -1 \end{pmatrix}$$

$$= 0\times 1 + 10\times 0 + 10\times -1 + 0\times 1 + 10\times 0 + 10\times -1 + 0\times 1 + 10\times 0 + 10\times -1 = -30$$

Since this time, we move from 0 to 10 along the horizontal direction.

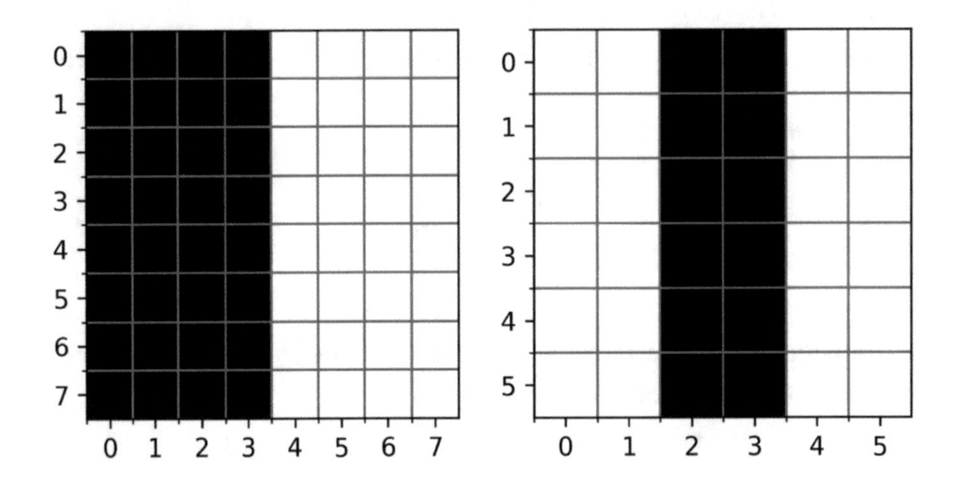

Figure 7-10. *The result of the convolution of the matrix ex_mat with the kernel* $\mathfrak{I}_V$

Note how, as expected, the output matrix is 5 × 5 since the original matrix is 7 × 7 and the kernel is 3 × 3.

Pooling

Pooling is the second operation that is fundamental in CNNs. This operation is much easier to understand than convolution. To understand it, let's again make a concrete example and let's consider what is called *max pooling*. Let's again consider the 4×4 matrix we used in the convolution section

$$A = \begin{pmatrix} a_1 & a_2 & a_3 & a_4 \\ a_5 & a_6 & a_7 & a_8 \\ a_9 & a_{10} & a_{11} & a_{12} \\ a_{13} & a_{14} & a_{15} & a_{16} \end{pmatrix}$$

To perform max pooling, we need to define a region of size $n_K \times n_K$, analogous to what we did for convolution. Let's consider $n_K = 2$. What we need to do is start in the top-left corner of our matrix A and select a $n_K \times n_K$ region, in our case 2×2 from A. Here we would select

$$\begin{pmatrix} a_1 & a_2 \\ a_5 & a_6 \end{pmatrix}$$

or the elements marked in bold face in matrix A here

$$A = \begin{pmatrix} \boldsymbol{a_1} & \boldsymbol{a_2} & a_3 & a_4 \\ \boldsymbol{a_5} & \boldsymbol{a_6} & a_7 & a_8 \\ a_9 & a_{10} & a_{11} & a_{12} \\ a_{13} & a_{14} & a_{15} & a_{16} \end{pmatrix}$$

From the elements selected, a_1, a_2, a_5, and a_6, the max pooling operation selects the maximum value giving a result that we will indicate with B_1

$$B_1 = \max_{i=1,2,5,6} a_i$$

Then we need to shift our 2×2 window two columns, typically the same number of columns the selected region has, to the right, and select the elements marked in bold

$$A = \begin{pmatrix} a_1 & a_2 & \boldsymbol{a_3} & \boldsymbol{a_4} \\ a_5 & a_6 & \boldsymbol{a_7} & \boldsymbol{a_8} \\ a_9 & a_{10} & a_{11} & a_{12} \\ a_{13} & a_{14} & a_{15} & a_{16} \end{pmatrix}$$

Or in other words the smaller matrix

$$\begin{pmatrix} a_3 & a_4 \\ a_7 & a_8 \end{pmatrix}$$

The max pooling algorithm will then select the maximum of the values giving a result that we will indicate with B_2

$$B_2 = \max_{i=3,4,7,8} a_i$$

At this point we cannot shift the 2 × 2 region to the right anymore, so we shift it two rows down and start the process again from the left side of A, selecting the elements marked in bold and getting the maximum and calling it B_3.

$$A = \begin{pmatrix} a_1 & a_2 & a_3 & a_4 \\ a_5 & a_6 & a_7 & a_8 \\ \boldsymbol{a_9} & \boldsymbol{a_{10}} & a_{11} & a_{12} \\ \boldsymbol{a_{13}} & \boldsymbol{a_{14}} & a_{15} & a_{16} \end{pmatrix}$$

The stride s in this context has the same meaning discussed in convolution. It is simply the number of rows or columns you move your region when selecting the elements. Finally, we select the last region 2 × 2 in the bottom-lower part of A, selecting the elements a_{11}, a_{12}, a_{15}, and a_{16}. We then get the maximum, and we call it B_4. With the values we obtain in this process, in the example the four values B_1, B_2, B_3, and B_4, we will build an output tensor

$$B = \begin{pmatrix} B_1 & B_2 \\ B_3 & B_4 \end{pmatrix}$$

In the example, we have $s = 2$. Basically, this operation takes as input a matrix A, a stride s, and a kernel size n_K (the dimension of the region we selected in the example before) and returns a new matrix B with dimensions given by the same formula we discussed for convolution

$$n_B = \left\lfloor \frac{n_A - n_K}{s} + 1 \right\rfloor$$

To reiterate, the idea is to start at the top-left of your matrix A, take a region of dimensions $n_K \times n_K$, apply the max function to the selected elements, then shift the region of s elements toward the right. Then select a new region of dimensions $n_K \times n_K$, apply the function to its values, and so on. Figure 7-11 shows how you select the elements from matrix A with stride $s = 2$.

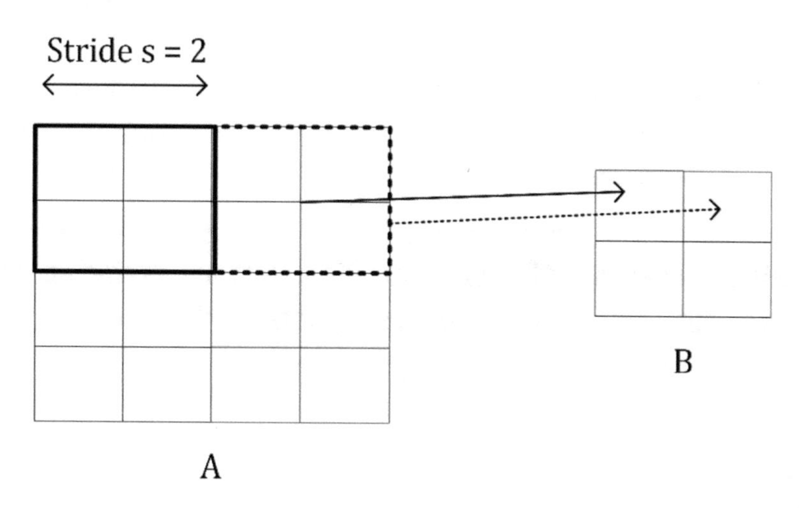

Figure 7-11. *A visualization of pooling with stride s = 2*

For example, applying max pooling to the input A

$$A = \begin{pmatrix} 1 & 3 & 5 & 7 \\ 4 & 5 & 11 & 3 \\ 4 & 1 & 21 & 6 \\ 13 & 15 & 1 & 2 \end{pmatrix}$$

will get you the results (which are very easy to verify)

$$B = \begin{pmatrix} 4 & 11 \\ 15 & 21 \end{pmatrix}$$

Since 4 is the maximum of the values marked in bold

$$A = \begin{pmatrix} \mathbf{1} & \mathbf{3} & 5 & 7 \\ \mathbf{4} & \mathbf{5} & 11 & 3 \\ 4 & 1 & 21 & 6 \\ 13 & 15 & 1 & 2 \end{pmatrix}$$

11 is the maximum of the values marked in bold

$$A = \begin{pmatrix} 1 & 3 & \mathbf{5} & \mathbf{7} \\ 4 & 5 & \mathbf{11} & \mathbf{3} \\ 4 & 1 & 21 & 6 \\ 13 & 15 & 1 & 2 \end{pmatrix}$$

and so on. It's worth mentioning another way of doing pooling, although not as widely used as max pooling, and that's *average pooling*. Instead of returning the maximum of the selected values, it returns the average.

Note The most common pooling operation is max pooling. Average pooling is not as widely used, but can be found in specific network architectures.

Padding

Sometimes, when dealing with images, it's not optimal to get a result from a convolution operation that has dimensions that are different than the original image. In that case, you do what is called "padding." Basically, the idea is very simple: it consists of adding rows of pixels to the top and bottom and to the columns of pixels on the right and on the left of the final images, which are filled with values to make the resulting matrices the same size of the original one. Some strategies are filling the added pixels with zeros, with the values of the closest pixels, and so on. For example, in our example the `ex_out` matrix, zero-padding would look like this

```
array([[ 0., 0., 0., 0., 0., 0., 0., 0.],
       [ 0., 0., 0., 30., 30., 0., 0., 0.],
       [ 0., 0., 0., 30., 30., 0., 0., 0.],
       [ 0., 0., 0., 30., 30., 0., 0., 0.],
       [ 0., 0., 0., 30., 30., 0., 0., 0.],
       [ 0., 0., 0., 30., 30., 0., 0., 0.],
       [ 0., 0., 0., 30., 30., 0., 0., 0.],
       [ 0., 0., 0., 0., 0., 0., 0., 0.]])
```

The use and reasons behind padding goes beyond the scope of this book, but it's important to know that this exists. Only as a reference, if you use padding p (the width of the rows and the columns you use as padding), the final dimensions of the matrix B, in case of convolution and pooling, is given by

$$n_B = \left\lfloor \frac{n_A + 2p - n_K}{s} + 1 \right\rfloor$$

Note When dealing with real images, you always have color images, coded in three channels: RGB. That means that you need to do convolution and pooling in three dimensions: width, height, and color channel. This will add a layer of complexity to the algorithms.

Building Blocks of a CNN

Basically, convolutions and pooling operations are used to build the layers used in CNNs. In CNNs, you can typically find the following layers

- Convolutional layers
- Pooling layers
- Fully connected layers

Fully connected layers are exactly what you have already seen in all previous chapters: a layer where neurons are connected to all neurons of previous and subsequent layers. You know them already. But the first two require some additional explanation.

Convolutional Layers

A convolutional layer takes as input a tensor (it can be three-dimensional, due to the three color channels). For example, an image of certain dimensions. It then applies a certain number of kernels, typically 10, 16, or even more, adds a bias, applies the ReLU activation functions (for example) to introduce non-linearity to the result of the convolution, and produces an output matrix, B. If you remember the notation we used in the previous chapters, the result of the convolution will have the role of $W^{[l]}Z^{[l-1]}$, which we discussed in Chapter 3.

In the previous sections, we saw some examples of applying convolutions with just one kernel. How can you apply several kernels at the same time? Well, the answer is very simple. The final tensor (now we use the word tensor since it will not be a simple matrix anymore) B will have not two dimensions but three. Let's indicate the number of kernels you want to apply with n_c (the c is used since sometimes people talk about channels). *You simply apply each filter to the input independently and stack the results.* So instead of a single matrix B with dimensions $n_B \times n_B$, you get a final tensor $\tilde{B}$ of dimensions $n_B \times n_B \times n_c$. That means that

$$\tilde{B}_{i,j,1} \quad \forall i,j \in [1, n_B]$$

will be the output of convolution of the input image with the first kernel. And

$$\tilde{B}_{i,j,2} \quad \forall i,j \in [1, n_B]$$

will be the output of convolution with the second kernel, and so on. The convolution layer is nothing more than something that transforms the input into an output tensor. But what are the weights in this layer? *The weights, or the parameters that the network learns during the training phase, are the elements of the kernel themselves.* We discussed that we have n_c kernels, each of $n_K \times n_K$ dimensions. That means that we have $n_K^2 n_c$ *parameters in a convolutional layer.*

Note The number of parameters that you have in a convolutional layer, $n_K^2 n_c$, is independent of the input image size. This fact helps reduce overfitting, especially when dealing with large input images.

Sometimes this layer is indicated with the word *CONV* and then a number. In our case, we could call this layer CONV1. Figure 7-12 shows a representation of a convolutional layer. The input image gets transformed by applying convolution with n_c kernels in a tensor of dimensions $n_A \times n_A \times n_c$.

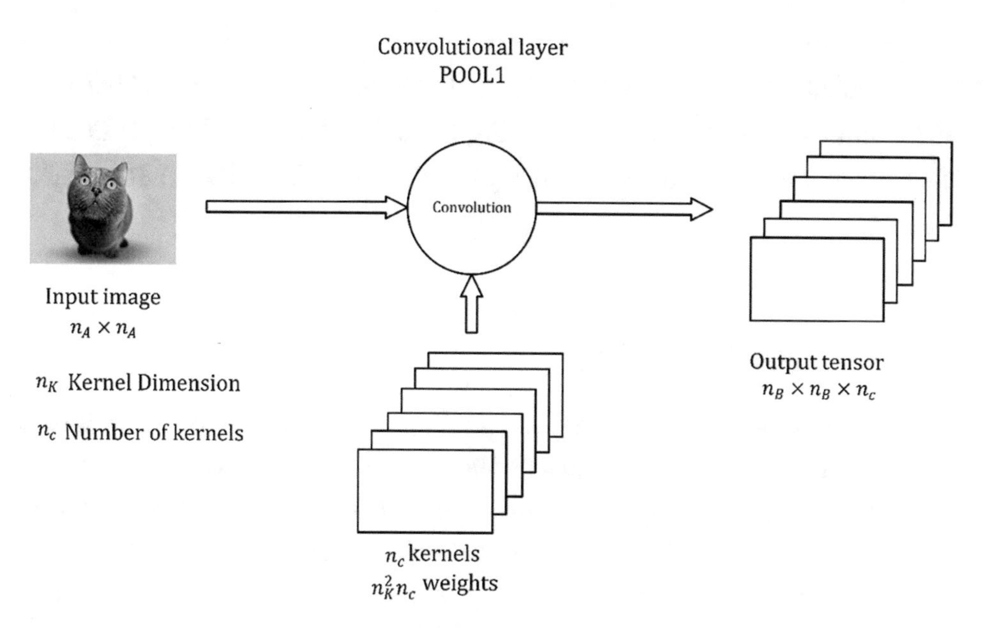

Figure 7-12. *A representation of a convolutional layer*[1]

Of course, a convolutional layer must not necessarily be placed immediately after the inputs. A convolutional layer may get as input the output of any other layer of course. Keep in mind that your input image will usually have dimensions $n_A \times n_A \times 3$, since an image in color has three channels: Red, Green, and Blue. A complete analysis of the tensors involved in a CNN when considering color images goes beyond the scope of this book. Very often in diagrams, the layer is simply indicated as a cube or a square.

Pooling Layers

A pooling layer is usually indicated with *POOL* and a number: for example, POOL1. It takes as input a tensor and gives as output another tensor after applying pooling to the input.

[1] Cat image source: `https://www.shutterstock.com/`

Note A pooling layer has no parameter to learn, but it introduces additional hyper-parameters: n_K and stride s. Typically, in pooling layers you do not use padding, since one of the reasons to use pooling is to reduce the dimensionality of the tensors.

Stacking Layers Together

In CNNs, you often stack convolutional and pooling layers together, one after the other. Figure 7-13 shows a convolutional and a pooling layer stack. A convolutional layer is always followed by a pooling layer. Sometimes the two together are called a layer. The reason is that a pooling layer has no learnable weights and therefore it is simply seen as a simple operation that is associated with the convolutional layer. So be aware when you read papers or blogs and be sure you understand their intentions.

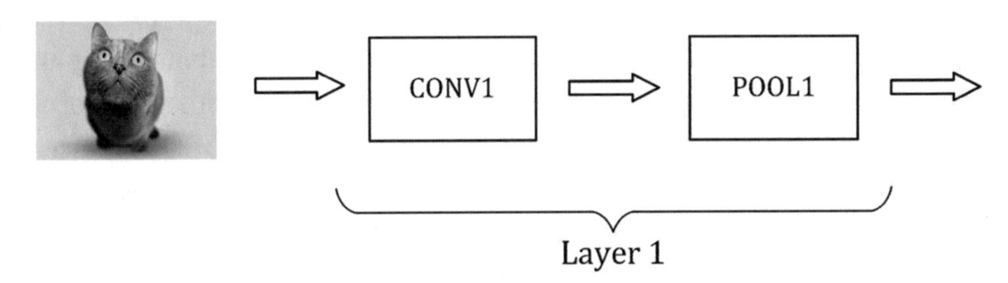

***Figure 7-13.** A representation of how to stack convolutional and pooling layers*

To conclude this part on CNNs, Figure 7-14 shows an example of a CNN. Figure 7-14 is an example like the very famous LeNet-5 network[1]. You have the inputs, then two convolution-pooling layers, then three fully connected layers, and then an output layer. This is where you may have your softmax function, if, for example, you perform multiclass classification. We included some numbers in the figure to give you an idea of the size of the different layers.

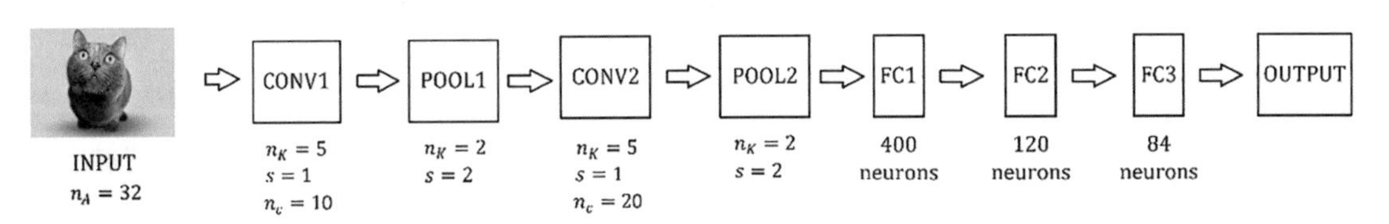

***Figure 7-14.** A representation of a CNN similar to the LeNet-5 network*

An Example of a CNN

Let's try to build a network to give you a feeling of how the process works and what the code looks like. We will not do any hyper-parameter tuning or optimization to keep the section understandable. We will build the following architecture with the following layers, in this order:

- Convolution layer 1 (CONV1): Six filters 5×5, stride $s = 1$.
- We then apply ReLU to the output of the previous layer.
- Max pooling layer 1 (POOL1) with a window 2×2, stride $s = 2$.
- Convolution layer 2 (CONV2): 16 filters 5×5, stride $s = 1$.
- We then apply ReLU to the output of the previous layer.
- Max pooling layer 2 (POOL2) with a window 2×2, stride $s = 2$.
- Fully Connected Layer with 128 neurons with activation function ReLU.
- Fully Connected Layer with ten neurons for classification of the Zalando dataset.
- Softmax output neuron.

We import the Zalando dataset

```
((trainX, trainY), (testX, testY)) = fashion_mnist.load_data()
```

Check there if you do not remember the dataset's details. Let's prepare the data (reshaping the samples and one-hot encoding the labels):

```
labels_train = np.zeros((60000, 10))
labels_train[np.arange(60000), trainY] = 1

data_train = trainX.reshape(60000, 28, 28, 1)
```

and

```
labels_test = np.zeros((10000, 10))
labels_test[np.arange(10000), testY] = 1

data_test = testX.reshape(10000, 28, 28, 1)
```

Note that in this case, we use as network's inputs tensors of dimensions (number_of_images, image_height, image_width, color_channels). Since the Zalando dataset is made up of gray values images, the color_channels will be equal to 1. Each observation is in a row (since feed-forward neural networks take as input *flattened* tensors). If you check the dimensions with the code

```
print('Dimensions of the training dataset: ', data_train.shape)
print('Dimensions of the test dataset: ', data_test.shape)
print('Dimensions of the training labels: ', labels_train.shape)
print('Dimensions of the test labels: ', labels_test.shape)
```

You will get the results

```
Dimensions of the training dataset:  (60000, 28, 28, 1)
Dimensions of the test dataset:  (10000, 28, 28, 1)
Dimensions of the training labels:  (60000, 10)
Dimensions of the test labels:  (10000, 10)
```

We need to normalize the data

```
data_train_norm = np.array(data_train/255.0)
data_test_norm = np.array(data_test/255.0)
```

We can now start to build our network. With Keras, creating and training a CNN model is straightforward; the following function defines the network's architecture

```
def build_model():
  # create model
  model = models.Sequential()
  model.add(layers.Conv2D(6, (5, 5), strides = (1, 1),
            activation = 'relu', input_shape = (28, 28, 1)))
  model.add(layers.MaxPooling2D(pool_size = (2, 2),
            strides = (2, 2)))
  model.add(layers.Conv2D(16, (5, 5), strides = (1, 1),
            activation = 'relu'))
  model.add(layers.MaxPooling2D(pool_size = (2, 2),
            strides = (2, 2)))
  model.add(layers.Flatten())
  model.add(layers.Dense(128, activation = 'relu'))
```

```
    model.add(layers.Dense(10, activation = 'softmax'))
    # compile model
    model.compile(loss = 'categorical_crossentropy',
                  optimizer = 'adam',
                  metrics = ['categorical_accuracy'])
    return model
```

Remember that the convolutional layer will require the two-dimensional image, and not a flattened list of gray values of the pixels.

Note One of the biggest advantages of CNNs is that they use the two-dimensional information contained in the input image, this is why the input of convolutional layers are two-dimensional images, and not a flattened vector.

When building CNNs in Keras, a single line of code (and a Keras method) will correspond to a different layer. The `build_model` function creates a CNN stacking `Conv2D` (which builds a convolutional layer) and `MaxPooling2D` (which builds a max pooling layer) layers. The stride is a tuple since it gives the stride in different dimensions (for rows and columns). In our examples we have gray images, but we could also have RGB, for example. That would mean having more dimensions: the three color channels.

Let's display the architecture of the model so far, using `model.summary()`:

```
Model: "sequential"
_________________________________________________________________
Layer (type)                 Output Shape              Param #
=================================================================
conv2d (Conv2D)              (None, 24, 24, 6)         156
_________________________________________________________________
max_pooling2d (MaxPooling2D) (None, 12, 12, 6)         0
_________________________________________________________________
conv2d_1 (Conv2D)            (None, 8, 8, 16)          2416
_________________________________________________________________
max_pooling2d_1 (MaxPooling2 (None, 4, 4, 16)          0
```

```
_________________________________________________________________
flatten (Flatten)            (None, 256)               0
_________________________________________________________________
dense (Dense)                (None, 128)               32896
_________________________________________________________________
dense_1 (Dense)              (None, 10)                1290
=================================================================
Total params: 36,758
Trainable params: 36,758
Non-trainable params: 0
_________________________________________________________________
```

Note that the output of every convolutional and pooling layer is a 3D tensor of shape (height, width, number_of_filters). The first dimension (i.e., the number of batches), is set to None since the network does not know it yet and thus it can be applied to every set of samples, of any length. The width and height dimensions decrease as you go deeper into the network. The number of output channels for each Conv2D layer is controlled by the first function argument. Typically, as the width and height decrease, you can afford (computationally) to add more output filters to each Conv2D layer.

To complete the model, we added two Dense layers. They take vectors as input (which are 1D), while the current output is a 3D tensor. This is why you first need to flatten the 3D output to 1D, then add one or more Dense layers on top.

Now it's time to train and test our network. We will use mini-batch gradient descent with a batch size of 100 and we will train our network for ten epochs.

```
model.fit(data_train_norm, labels_train, validation_data = (data_test_norm,
labels_test), epochs = 10, batch_size = 100, verbose = 1)
```

If you run this code (it took roughly four minutes on a medium performance laptop), it will start, after just one epoch, with a training accuracy of 76.3%. After ten epochs it will reach a training accuracy of 91% (88% on the dev set). We have trained our network here only for ten epochs. You can get much higher accuracy if you train longer. Additionally, note that we have not done any hyper-parameter tuning so this would get you much better results if you spent time tuning the parameters.

As you may have noticed, every time you introduce a convolutional layer, you will introduce new hyper-parameters for each layer:

- Kernel size
- Stride
- Padding

Those will need to be tuned to get optimal results. Typically, researchers tend to use existing architectures for specific tasks that have been already optimized by other practitioners and are well documented in papers.

Conclusion

You should now have a basic understanding of how CNN networks work, and on what principles they work on. Convolutional neural networks are used extensively in multiple forms for various tasks, from classification (as you have seen here) to object localization, object segmentation, instance segmentation, and much more. This chapter just scratched the surface. But you should understand the building blocks of CNNs and should be able to understand how more complex architecture is built and structured.

Exercises

EXERCISE 1 (CONVOLUTION) (LEVEL: EASY)

Try to apply the different convolution operators like the ones you saw in this chapter, but to different images, such as the handwritten digits of the MNIST database (`http://yann.lecun.com/exdb/mnist/`). To download the dataset from TensorFlow, use the following lines of code:

```
from tensorflow import keras
(x_train, y_train), (x_test, y_test) = keras.datasets.mnist.load_data()
```

EXERCISE 2 (CNN) (LEVEL: EASY)

Try to build a multiclass classification model like the one you saw in this chapter, but using the MNIST database of handwritten digits instead.

EXERCISE 3 (CNN)(LEVEL: MEDIUM)

Try to change the network's parameters to see if you can get a better accuracy. Change kernel size, stride, and padding.

References

[1] `https://goo.gl/hM1kAL`, last accessed 19.07.2021.

[2] `https://goo.gl/FodLp5`, last accessed 21.07.2021.

[3] `https://goo.gl/8Ja3n2`, last accessed 21.07.2021.

CHAPTER 8

A Brief Introduction to Recurrent Neural Networks

In the last chapter, we looked at *convolutional neural networks* (CNNs). Another network architecture that is widely used (for example, in natural language processing) is the recurrent one. Networks with this architecture are called *recurrent neural networks,* or RNNs. This chapter is a superficial description of how RNNs work, with one small application that should help you better understand their inner workings. A full explanation of RNNs would require multiple books, so the goal of this chapter is to give you a very basic understanding of how they work. It's useful for machine learning engineers to have at least a basic understanding of RNNs. I discuss only the very basic components of RNNs to elucidate the very fundamental aspects. I hope you find it useful. At the end of the chapter, I suggest further reading in case you find the subject interesting and want to better understand RNNs.

Introduction to RNNs

RNN are very different from CNNs and are typically used when dealing with sequential information. In other words, for data in which the order matters. The typical example given is a series of words in a sentence. You can easily understand how the order of words in a sentence can make a big difference. For example, saying "the man eats the rabbit" has a different meaning than "the rabbit eats the man." The order of the words changes and that changes who gets eaten by whom.

You can use RNNs to predict, for example, the next word in a sentence. Take for example the phrase "Paris is the capital of." It is easy to complete the sentence with

U. Michelucci, *Applied Deep Learning with TensorFlow 2*, https://doi.org/10.1007/978-1-4842-8020-1_8

"France," and that means that there is information about the final word of the sentence encoded in the previous words. That information is what RNNs exploit in order to predict the next terms in a sequence. The name *recurrent* comes from how they work: the network applies the same operation on each element of the sequence, accumulating information about the previous terms. To summarize:

- RNNs use sequential data and the information encoded in the order of the terms in a sequence.
- RNNs apply the same kind of operation to all terms in a sequence and build a memory of the previous terms in the sequence to predict the next term.

Before exploring how they work in more depth, let's consider a few important use cases. These examples show the range of applications possible.

- **Generating text**: Predicting the probability of words, given a previous set of words. For example, you can easily generate text that looks like Shakespeare with RNNs, as A. Karpathy has done in his blog [2].
- **Translation**: Given a set of words in a language, you predict words in a different language.
- **Speech recognition**: Given a series of audio signals (words), you want to predict the sequence of letters forming the spoken words.
- **Generating image labels**: With CNNs, RNNs can be used to generate labels for images. Check out the paper "Deep Visual-Semantic Alignments for Generating Image Descriptions" by A. Karpathy on the subject [3]. Be aware that this is a rather advanced paper that requires a mathematical background.
- **Chatbots**: When a sequence of words is given as input, RNNs try to generate answers to the input.

As you can imagine, to solve those problems you need sophisticated architectures that are not easy to describe in a few sentences and that require a deeper (pun intended) understanding of how RNNs work. These are topics that go beyond the scope of this chapter and book.

Notation

Consider the sequence: "Paris is the capital of France." This sentence will be fed to a RNN one word at a time: first "Paris," then "is," then "the," and so on.

- "Paris" will be the first word of the sequence: `w1 = 'Paris'`
- "is" will be the second word of the sequence: `w2 = 'is'`
- "the" will be the third word of the sequence: `w3 = 'the'`
- "capital" will be the fourth word of the sequence: `w4 = 'capital'`
- "of" will be the fifth word of the sequence: `w5 = 'of'`
- "France" will be the sixth word of the sequence: `w6 = 'France'`

The words will be fed into the RNN in the following order: `w1`, `w2`, `w3`, `w4`, `w5`, and then `w6`. The different words will be processed by the network one after the other, or at different time points. If word `w1` is processed at time t, then `w2` is processed at time $t + 1$, `w3` at time $t + 2$, and so on. The time t is not related to the real time but is meant to suggest the fact that each element in the sequence is processed sequentially and not in parallel. The time t is also not related to computing time or anything related to it. And the increment of 1 in $t + 1$ does not have any meaning, it simply indicates the next element in the sequence. You may see the following notations when reading papers, blogs, or books:

- x_t: The input at time t. For example, `w1` could be the input at time 1 x_1, `w2` at time 2 x_2, and so on.
- s_t: The notation with which the internal memory, which we have not defined yet, at time t is indicated. This quantity s_t will contain the accumulated information of the previous terms in the sequence we discussed previously. An intuitive understanding of it will have to suffice, since a mathematical definition requires a very detailed explanation.
- o_t: The output of the network at time t, or in other words after all the elements of the sequence until t, including the element x_t, have been fed into the network.

The Basic Idea of RNNs

Typically, a RNN is indicated in the literature as the leftmost part of Figure 8-1. The notation is indicative and has the goal of simply indicating the different elements of the network: x is the inputs, s is the internal memory, W is one set of weights, and U is another set of weights. In reality, this schematic representation is simply a way of depicting the real structure of the network, which you can see on the right side of Figure 8-1. This is sometimes called the *unfolded* version of the network.

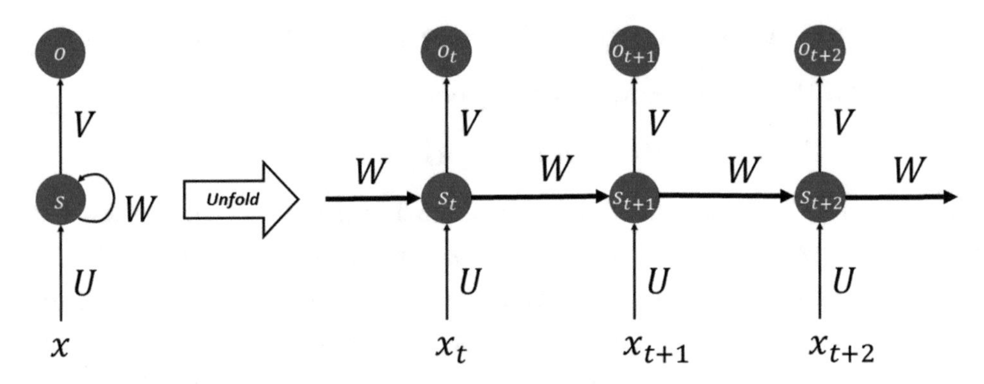

Figure 8-1. *A schematic representation of an RNN*

The right side of Figure 8-1 should be read left to right. The first neuron in the figure does its evaluation at an indicative time t, produces an output o_t, and creates an internal memory state s_t. The second neuron, which does its evaluation at a time $t + 1$, after the first neuron, gets as input both the next element in the sequence x_{t+1} and the previous memory state s. The second neuron then generates an output o_{t+1} and a new internal memory state s_{t+1}. The third neuron (the one at the extreme right of Figure 8-1) gets as input the new element of the sequence x_{t+2} and the previous internal memory state s_{t+1}. The process proceeds this way for a finite number of neurons. You can see in Figure 8-1 that there are two sets of weights: W and U. One set (indicated with W) is used for the internal memory states and one (U) for the sequence element. Typically, each neuron will generate the new internal memory state with a formula that will look like this

$$s_t = f\left(Ux_t + Ws_{t-1}\right)$$

where we indicate with $f()$ one of the activation functions we have seen as ReLU or `tanh`. Additionally, the previous formula will be of course multi-dimensional. s_t can be understood as the memory of the network at time t. The number of neurons (or time steps) that can be used is a new hyperparameter that needs to be tuned, depending on the problem. Research has shown that when this number is too big, the network has problems during training.

Something very important to note is that at each time step, the weights don't change. We are performing the same operation at each step, simply changing the inputs every time we perform an evaluation. Additionally, in Figure 8-1 we have for every step an output in the diagram (o_t, o_{t+1}, and o_{t+2}) but typically this is not necessary. In the example where we wanted to predict the final word in a sentence, we may just need the final output.

Why the Name Recurrent

We need to discuss very briefly why the networks are called *recurrent*. We have mentioned that the internal memory state at a time t is given by the following

$$s_t = f\left(Ux_t + Ws_{t-1}\right)$$

The internal memory state at time t is evaluated using the same memory state at time $t-1$, the one at time $t-1$ with the value at time $t-2$ and so on. This is at the origin of the name recurrent.

Learning to Count

To give you an idea of the power of such networks, this section shows a very basic example of something RNNs are very good at, and that standard fully connected networks, as the one you saw in the previous chapters, are really bad at. Let's try to teach a network to count.

The problem we want to solve is the following: given a certain vector made of 15 elements containing just 0s and 1s, we want to build a neural network that can count the amount of 1s there are. This is a difficult problem for a standard network, but why? Consider the problem we analyzed of distinguishing the 1 and 2 digits in the MNIST dataset. In that case, the learning happens because the 1s and the 2s have black pixels in fundamentally different positions. A digit 1 will always differ in (at least in the MNIST dataset) the same way from the digit 2, and the network will identify those differences. As soon as they are detected, a clear identification can be made. In this case, that is not possible.

Consider for example a simpler case of a vector with just five elements. Consider the case when a 1 appears exactly one time. We have five possible cases: `[1,0,0,0,0]`, `[0,1,0,0,0]`, `[0,0,1,0,0]`, `[0,0,0,1,0]`, and `[0,0,0,0,1]`. There is no discernable pattern to be detected here. There is no easy weight configuration that could cover those cases at the same time. In an image, this problem is similar to the problem of detecting the position of a black square in a white image. We can build a network in TensorFlow and check how good such networks are. Due to the introductory nature of this chapter, there is no hyperparameter discussion, metric analysis, and so on. We simply look at a basic network that can count.

Let's start by creating the vectors. We will create 10^5 vectors that we will split into `training` and dev sets.

```
import numpy as np
import tensorflow as tf
from random import shuffle
from tensorflow import keras
from tensorflow.keras import layers
```

Now we will create the list of vectors. The code is a slightly more complicated, so we look at it in a bit more detail.

```
nn = 15
ll = 2**15
train_input = ['{0:015b}'.format(i) for i in range(ll)]
# consider every number up to 2^15 in binary format
shuffle(train_input) # shuffle inputs
train_input = [map(int, i) for i in train_input]
ti  = []
for i in train_input:
  temp_list = []
  for j in i:
    temp_list.append([j])
  ti.append(np.array(temp_list))
train_input = ti
```

We want to have all possible combinations of 1 and 0 in vectors of 15 elements. So, an easy way to do that is take all numbers up to 2^{15} in binary format. To understand why, suppose you want to do this with only four elements. You want all possible combinations of four 0s and 1s. Consider all the numbers up to 2^4 in binary that you can get with this code

```
['{0:04b}'.format(i) for i in range(2**4)]
```

The code simply formats all the numbers that you get with the `range(2**4)` function from `0` to `2**4` in binary format with `{0:04b}`, which limits the number of digits to four. The result is the following:

```
['0000',
 '0001',
 '0010',
 '0011',
 '0100',
 '0101',
 '0110',
 '0111',
 '1000',
 '1001',
 '1010',
 '1011',
 '1100',
 '1101',
 '1110',
 '1111']
```

As you can easily verify, you have all possible combinations in the list. You have all possible combinations of the 1 appearing one times (`[0001]`, `[0010]`, `[0100]` and `[1000]`), of the 1s appearing two times, and so on. For this example, we will simply do it with 15 digits, which means we will do that with numbers up to 2^{15}. The rest of the code is there to simply transform a string like `'0100'` to a list `[0,1,0,0]` and then concatenate all the lists with all the possible combinations.

If you check the dimension of the output array, you will notice that you get (32768, 15, 1). Each observation is an array of dimensions (15, 1). Then you prepare the target variable, a one-hot encoded version of the counts. That means that if you have an input with four 1s in the vector, the target vector will look like

[0,0,0,0,1,0,0,0,0,0,0,0,0,0,0,0]. As expected, the train_output array will have the dimensions (32768, 16). Now let's split the set into a train and a dev set, as we have done several times. We will do it here in a dumb way

```
NUM_EXAMPLES = ll - 2000
test_input = train_input[NUM_EXAMPLES:]
test_output = train_output[NUM_EXAMPLES:] # everything beyond 10,000

train_input = train_input[:NUM_EXAMPLES]
train_output = train_output[:NUM_EXAMPLES] # till 10,000
```

Remember that this will work since we shuffled the vectors at the beginning, so we should have a random distribution of cases. We will use 2,000 cases for the dev set and the rest (roughly 30000) for the training set. The train_input will have dimensions (30768, 15, 1) and the dev_input will have dimensions (2000, 16).

Now you can build a network with this code, and you should be able to understand almost all of it by now

```
model = keras.Sequential()

model.add(layers.Embedding(input_dim = 15, output_dim = 15))

# Add a LSTM layer with 128 internal units.
model.add(layers.LSTM(24, input_dim = 15))

# Add a Dense layer with 10 units.
model.add(layers.Dense(16, activation = 'softmax'))

model.compile(loss = 'categorical_crossentropy', optimizer = 'adam',
metrics = ['categorical_accuracy'])
```

Let's train the network

```
# we need to convert the input and output to numpy array to be used by
the network
train_input = np.array(train_input)
train_output = np.array(train_output)

test_input = np.array(test_input)
test_output = np.array(test_output)
```

```
model.fit(train_input, train_output, validation_data = (test_input,
test_output), epochs = 10, batch_size = 100)
```

For performance reasons and to show how efficient RNNs are, we use an LSTM kind of neuron. They have a special way of calculating the internal state. A discussion goes well beyond the scope of the book. For the moment, you should focus on the results and not on the code itself. If you let the code run, you will get the following result

```
Epoch 1/10
308/308 [==============================] - 4s 9ms/step - loss: 1.9441 -
categorical_accuracy: 0.3063 - val_loss: 1.1784 - val_categorical_
accuracy: 0.6840
Epoch 2/10
308/308 [==============================] - 2s 7ms/step - loss: 0.7472 -
categorical_accuracy: 0.8332 - val_loss: 0.4515 - val_categorical_
accuracy: 0.9270
Epoch 3/10
308/308 [==============================] - 2s 7ms/step - loss: 0.3311 -
categorical_accuracy: 0.9554 - val_loss: 0.2360 - val_categorical_
accuracy: 0.9630
Epoch 4/10
308/308 [==============================] - 2s 7ms/step - loss: 0.1921 -
categorical_accuracy: 0.9658 - val_loss: 0.1530 - val_categorical_
accuracy: 0.9675
Epoch 5/10
308/308 [==============================] - 2s 7ms/step - loss: 0.1306 -
categorical_accuracy: 0.9760 - val_loss: 0.1071 - val_categorical_
accuracy: 0.9775
Epoch 6/10
308/308 [==============================] - 2s 7ms/step - loss: 0.0937 -
categorical_accuracy: 0.9824 - val_loss: 0.0778 - val_categorical_
accuracy: 0.9870
Epoch 7/10
308/308 [==============================] - 2s 7ms/step - loss: 0.0696 -
categorical_accuracy: 0.9905 - val_loss: 0.0586 - val_categorical_
accuracy: 0.9930
```

```
Epoch 8/10
308/308 [==============================] - 2s 7ms/step - loss: 0.0533 -
categorical_accuracy: 0.9921 - val_loss: 0.0446 - val_categorical_
accuracy: 0.9945
Epoch 9/10
308/308 [==============================] - 2s 7ms/step - loss: 0.0422 -
categorical_accuracy: 0.9924 - val_loss: 0.0367 - val_categorical_
accuracy: 0.9960
Epoch 10/10
308/308 [==============================] - 2s 7ms/step - loss: 0.0346 -
categorical_accuracy: 0.9943 - val_loss: 0.0301 - val_categorical_
accuracy: 0.9955
<tensorflow.python.keras.callbacks.History at 0x7f6b7b3bd990>
```

After just ten epochs, the network is right in 99% of the cases. Just let it run for more epochs to reach incredible precision. An instructive exercise is trying to train a fully connected network (as the ones we have discussed so far) to count. You will see how this is not possible.

Conclusion

This chapter was a very brief description of RNNs. You should have the basics down as to how they work and how LSTM neurons are structured. There is a lot more about RNNs to discuss, but that would go beyond the scope of this book and therefore I have chosen to neglect it here. RNNs are an advanced topic and require a bit more know-how to understand. In the next section, I list two sources that are free on the Internet that you can use to kick-start your RNN education.

Further Readings

If you found this chapter intriguing and would like to learn more about RNNs, there is a huge amount of material that you can find on the Internet. Here are two good sources:

- A much more complete and advanced treatment of RNNs can be found at `www.deeplearningbook.org/contents/rnn.html`. Be aware that this is more advanced and requires a much more advanced mathematics background.
- This review paper is full of information and further references that you can track down and read: `https://arxiv.org/pdf/1808.03314.pdf`.

CHAPTER 9

Autoencoders

In this chapter, we look at autoencoders. This chapter is a theoretical one, covering the mathematics and the fundamental concepts of autoencoders. We discuss what they are, what their limitations are, the typical use cases, and then look at some examples. We start with a general introduction to autoencoders, and we discuss the role of the activation function in the output layer and the loss function. We then discuss what the reconstruction error is. Finally, we look at typical applications, such as dimensionality reduction, classification, denoising, and anomaly detection.

Introduction

As you have seen in many of the previous chapters, neural networks are typically used in a supervised setting. Meaning that for each training observation $\boldsymbol{x}_i$, we have one label or expected value, $\boldsymbol{y}_i$. During training, the neural network model will learn the relationship between the input data and the expected labels. Now suppose we have only unlabeled observations, meaning we only have our training dataset S_T, made of the M observations $\boldsymbol{x}_i$ with $i = 1, \ldots, M$

$$S_T = \{x_i \mid i = 1, \ldots, M\} \tag{9.1}$$

Where in general $\boldsymbol{x}_i \in \mathbb{R}^n$ with $n \in \mathbb{N}$. Autoencoders were introduced[1] by Rumelhart, Hinton, and Williams in 1986 with the goal of "learning to reconstruct the input observations $\boldsymbol{x}_i$ with the lowest error possible".[2]

[1] You can check out Rumelhart, D.E., Hinton, G.E. Williams, R.J.: 'Parallel distributed processing: Explorations in the microstructure of cognition," Vol. 1. Chap. "Learning Internal Representations by Error Propagation," pp. 318-162, MIT Press, Cambridge, MA, USA (1986).

[2] This chapter discusses at length what we mean by error here.

U. Michelucci, *Applied Deep Learning with TensorFlow 2*, https://doi.org/10.1007/978-1-4842-8020-1_9

Why would you want to learn to reconstruct the input observations? If you have problems imagining what that means, think of having a dataset made of images. An autoencoder is an algorithm that can give as output an image that is as similar as possible to the input one. You may be confused, as there is no apparent reason to do this. To better understand why autoencoders are useful, we need a more informative (although not fully unambiguous) definition.

An autoencoder is a type of algorithm with the primary purpose of learning an "informative" representation of the data that can be used for different applications[3] by learning to reconstruct a set of input observations well enough.

To better understand autoencoders, we need to refer to their typical architecture, visualized in Figure 9-1. The autoencoders' main components are an encoder, a latent feature representation, and a decoder. The encoder and decoder are simply functions, while the *latent feature representation* typically means a tensor of real numbers (more on that later). Generally speaking, we want the autoencoder to reconstruct the input well enough. Still, at the same time, it should create a latent representation (the output of the encoder part in Figure 9-1) that is useful and meaningful.

For example, latent features on handwritten digits[4] could be the number of lines required to write each number or the angle of each line and how they connect. Learning how to write numbers certainly does not require learning the gray values of each pixel in the input image. We humans do not learn to write by filling in pixels with gray values. While learning, we extract the essential information that will allow us to solve a problem (writing digits, for example). This latent representation (*how* to write each number) can then be very useful for various tasks (for instance, feature extraction that can be then used for classification or clustering) or simply for understanding the essential features of a dataset.

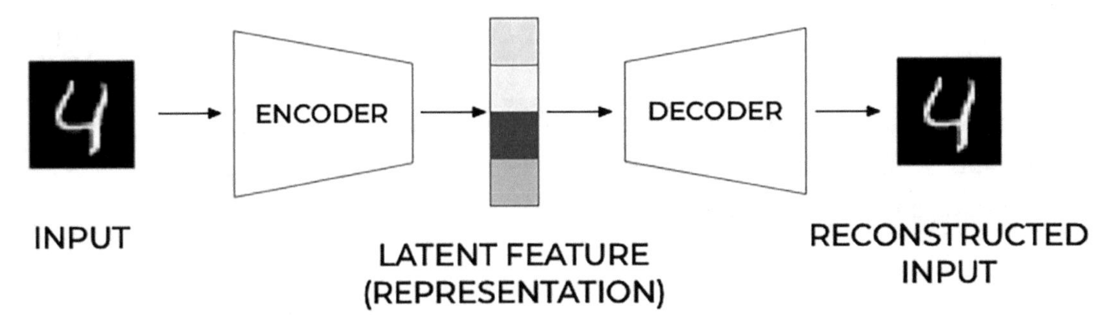

Figure 9-1. *The general structure of an autoencoder*

---

[3] Bank, D., Koenigstein, N., and Giryes, R., Autoencoders, `https://arxiv.org/abs/2003.05991`

[4] Imagine the MNIST dataset that you have seen many times in this book so far.

In most typical architectures, the encoder and the decoder are neural networks[5] (that is the case we will discuss at length in this chapter) since they can be easily trained with existing software libraries such as TensorFlow or PyTorch with backpropagation.

In general, the encoder can be written as a function g that will depend on some parameters

$$\boldsymbol{h}_i = g(\boldsymbol{x}_i)$$

Where $\boldsymbol{h}_i \in \mathbb{R}^q$ (the latent feature representation) is the output of the encoder block in Figure 9-1, when we evaluate it on the input $\boldsymbol{x}_i$. Note that we will have $g : \mathbb{R}^n \to \mathbb{R}^q$.

The decoder (and the output of the network that we indicate with $\tilde{\boldsymbol{x}}_i$) can be written then as a second generic function f of the latent features

$$\tilde{\boldsymbol{x}}_i = f(\boldsymbol{h}_i) = f(g(\boldsymbol{x}_i)).$$

Where $\tilde{\boldsymbol{x}}_i \in \mathbb{R}^n$. Training an autoencoder simply means finding the functions $g(\cdot)$ and $f(\cdot)$ that satisfy

$$\arg\min_{f,g} \langle [\Delta(\boldsymbol{x}_i, f(g(\boldsymbol{x}_i))] \rangle$$

Where Δ indicates a measure of how the input and output of the autoencoder differ (basically our loss function will penalize the difference between the input and output) and $< \cdot >$ indicates the average over all observations. Depending on how you design the autoencoder, it may be possible to find f and g so that the autoencoder learns to reconstruct the output perfectly, thus learning the identity function. This is not very useful, as we discussed at the beginning of the chapter, and to avoid this possibility, two main strategies can be used: creating a bottleneck and adding regularization in some form.

Note We want the autoencoder to reconstruct the input well enough. Still, at the same time, it should create a latent representation (the output of the encoder) that is useful and meaningful.

---

[5] For example, if the encoder and the decoder are linear functions, you get what is called a linear autoencoder. See for more information Baldi, P., Hornik, K.: "Neural networks and principal component analysis: Learning from examples without local minima," *Neural Netw.* **2**(1), 53-58 (1989).

Adding a "bottleneck," (more on that later) is achieved by making the latent feature's dimensionality lower (often much lower) than the input's. That is the case that we look at in detail in this chapter. But before looking at this case, let's briefly discuss regularization.

Regularization in Autoencoders

We will not discuss regularization at length in this chapter, but we should at least mention it. This means enforcing sparsity in the latent feature output. The simplest way of achieving this is to add a ℓ_1 or ℓ_2 regularization term to the loss function. That will look like this for the ℓ_2 regularization term:

$$\text{argmin}_{f,g}\left(\mathbb{E}\left[\Delta(\boldsymbol{x}_i, g(f(\boldsymbol{x}_i))\right] + \lambda\sum_i \theta_i^2\right)$$

In the formula, the θ_i are the parameters in the functions $f(\cdot)$ and $g(\cdot)$ (you can imagine that in the case where the functions are neural networks, the parameters will be the weights). This is typically easy to implement, because the derivative with respect to the parameters is easy to calculate. Another trick that is worth mentioning is to tie the weights of the encoder to the weights of the decoder[6] (in other words, make them equal). Those techniques, and a few others that go beyond the scope of this book, have fundamentally the same effect: to add sparsity to the latent feature representation.

We turn now to a specific type of autoencoders: those that build f and g with feed-forward networks that use a bottleneck. The reason for this choice is that they are very easy to implement and are very effective.

Feed-Forward Autoencoders

A Feed-Forward Autoencoder (FFA) is a neural network made of dense layers[7] with a specific architecture, as shown in Figure 9-2.

[6] For an example in Keras, you can check out the following blog post: `http://toe.lt/19`

[7] A dense layer is simply a set of neurons that gets their inputs from the previous layer. Each neuron in a dense layer gets as input the output of all neurons in the previous layer.

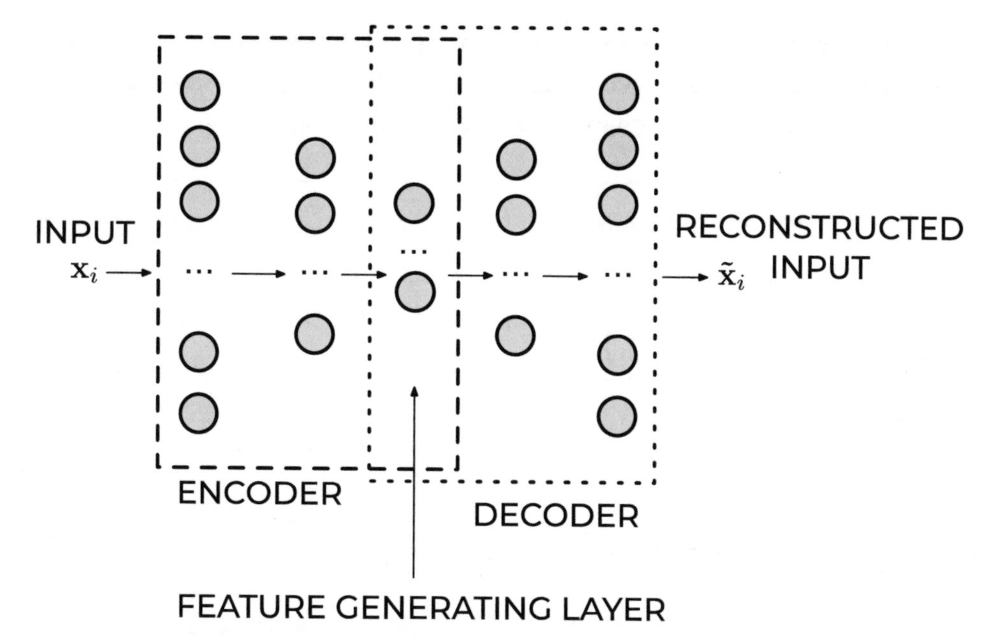

Figure 9-2. *A typical architecture of a Feed-Forward Autoencoder. The number of neurons in the layers at first goes down as we move through the network, until it reaches the middle and then starts to grow again, until the last layer has the same number of neurons as the input dimensions*

A typical FFA architecture (although it's not mandatory) has an odd number of layers and is symmetrical with respect to the middle layer. Typically, the first layer has a number of neurons $n_1 = n$ (the size of the input observation $\boldsymbol{x}_i$). As we move toward the center of the network, the number of neurons in each layer drops in some measure. The middle layer (remember we have an odd number of layers) usually has the smallest number of neurons. The fact that the number of neurons in this layer is smaller than the size of the input is the *bottleneck* mentioned earlier.

In almost all practical applications, the layers after the middle one are a mirrored version of the layers before the middle one. For example, an autoencoder with three layers could have the following numbers of neurons: $n_1 = 10$, $n_2 = 5$ and then $n_3 = n_1 = 10$ (supposing we are working on a problem where the input dimension is $n = 10$). All the layers, up to and including the middle one, make what is called the *encoder*, and all the layers from and including the middle one (up to the output) make what is called the ***decoder***, as you can see in Figure 9-2. If the FFA training is successful, the result will be a good approximation of the input, in other words $\tilde{\boldsymbol{x}}_i \approx \boldsymbol{x}_i$. What is essential to notice is that the decoder can reconstruct the input by using only a much smaller number (q) of features than the input observations initially have (n). The output of the middle layer $\boldsymbol{h}_i$ is also called a *learned representation* of the input observation $\boldsymbol{x}_i$.

Note The *encoder* can reduce the number of dimensions of the input observation (n) and create a learned representation ($\boldsymbol{h}_i$) of the input that has a smaller dimension $q < n$. This learned representation is enough for the decoder to reconstruct the input accurately (if the autoencoder training was successful as intended).

Activation Function of the Output Layer

In autoencoders based on neural networks, the output layer's activation function plays a particularly important role. The most used functions are ReLU and sigmoid. Let's look at both and see some tips on when to use which as well as why you should choose one instead of the other.

ReLU

The *ReLU* activation function can assume all values in the range $[0, \infty]$. As a reminder, its formula is

$$ReLU(x) = \max(0, x).$$

It's a good choice when the input observations $\boldsymbol{x}_i$ assume a wide range of positive values.

If the input $\boldsymbol{x}_i$ can assume negative values, the ReLU is a terrible choice, and the identity function is a much better choice.

Note The ReLU activation function for the output layer is well suited for cases when the input observations $\boldsymbol{x}_i$ assume a wide range of positive, real values.

Sigmoid

The sigmoid function σ can assume all values in the range]0, 1[. As a reminder, its formula is

$$\sigma(x) = \frac{1}{1+e^{-x}}.$$

This activation function can only be used if the input observations $\boldsymbol{x}_i$ are all in the range]0, 1[or if you have normalized them to be in that range. Consider as an example the MNIST dataset. Each value of the input observation $\boldsymbol{x}_i$ (one image) represents the gray values of the pixels that can assume any value from 0 to 255. Normalizing the data by dividing the pixel values by 255 would make each observation (each image) have only pixel values between 0 and 1. In this case, the sigmoid would be a good choice for the output layer's activation function.

Note The sigmoid activation function for the output layer is a good choice in all cases where the input observations assume only values between 0 and 1 or if you have normalized them to assume values in the range]0, 1[.

The Loss Function

As with any neural network model, we need to minimize a loss function. This loss functions should measure the difference between the input $\boldsymbol{x}_i$ and output $\tilde{\boldsymbol{x}}_i$. If you remember the explanations at the beginning, you will realize that this loss function will be

$$\mathbb{E}\left[\Delta(\boldsymbol{x}_i, g(f(\boldsymbol{x}_i))\right].$$

Where for FFAs, g, and f will be the functions that are obtained with dense layers, as discussed in the previous sections. Remember that an autoencoder is trying to learn an approximation of the identity function; therefore, you want to find the weights in the network that give you the smallest difference according to some metric ($\Delta(\cdot)$) between $\boldsymbol{x}_i$

and $\tilde{\boldsymbol{x}}_i$. Two loss functions are widely used for autoencoders: Mean Squared Error (MSE) and Binary Cross-Entropy (BCE). Let's look more in-depth at both since they can only be used when specific requirements are met.

Mean Square Error

Since an autoencoder is trying to solve a regression problem, the most common choice for the loss function is the Mean Square Error (MSE):

$$L_{MSE} = MSE = \frac{1}{M}\sum_{i=1}^{M}\left|\boldsymbol{x}_i - \tilde{\boldsymbol{x}}_i\right|^2$$

The symbol | · | indicates the norm of a vector,[8] and M is the number of the observations in the training dataset. It can be used in almost all cases, independently of how you choose your output layer activation function or how you normalize the input data.

It is easy to show that the minimum of L_{MSE} is found for $\tilde{\boldsymbol{x}}_i = \boldsymbol{x}_i$. To prove it, let's calculate the derivative of L_{MSE} with respect to a specific observation, j. Remember that the minimum is found when this condition

$$\frac{\partial L_{MSE}}{\partial \tilde{x}_j} = 0$$

is met for all $i = 1, \ldots, M$. To simplify the calculations, let's assume that the inputs are one dimensional[9] and let's indicate them with x_i. We can write

$$\frac{\partial L_{MSE}}{\partial \tilde{x}_j} = -\frac{2}{M}\left(x_j - \tilde{x}_j\right) \tag{9.2}$$

Equation (9.2) is satisfied when $x_j = \tilde{x}_j$ as can be easily seen, as we wanted to prove. To be precise, we also need to show that

$$\frac{\partial^2 L_{MSE}}{\partial \tilde{x}_j^2} > 0$$

[8] The norm of a vector is simply the square root of the sum of the square of the components.

[9] If you did not make this assumption, you would have to calculate the gradient of the loss function instead of the simple derivative.

This is easily proved as we have

$$\frac{\partial^2 L_{MSE}}{\partial \tilde{x}_j^2} = \frac{2}{M}$$

This is greater than zero, therefore confirming our assumption that for $x_j = \tilde{x}_j$ we indeed have a minimum.

Binary Cross-Entropy

If the activation function of the output layer of the FFA is a sigmoid function, thus limiting neuron outputs to be between 0 and 1, and the input features are normalized to be between 0 and 1, we can use the binary cross-entropy as a loss function, indicated here with L_{CE}. Note that this loss function is typically used in classification problems, but it works beautifully for autoencoders. The formula for it is

$$L_{CE} = -\frac{1}{M}\sum_{i=1}^{M}\sum_{j=1}^{n}\left[x_{j,i}\log\tilde{x}_{j,i} + \left(1-x_{j,i}\right)\log\left(1-\tilde{x}_{j,i}\right)\right]$$

Where $x_{j,i}$ is the jth component of the ith observation. The sum is over the entire set of observations and over all the components of the vectors. Can we prove that minimizing this loss function is equivalent to reconstructing the input as well as possible? Let's calculate where L_{CE} has a minimum with respect to $\tilde{\boldsymbol{x}}_i$. In other words, we need to find out which values should $\tilde{\boldsymbol{x}}_i$ assume to minimize L_{CE}. As we have done for the MSE, to make the calculations easier, let's consider the simplified case where $\boldsymbol{x}_i$ and $\tilde{\boldsymbol{x}}_i$ are one-dimensional and let's indicate them with x_i and $\tilde{x}_i$.

To find the minimum of a function, as you should know from calculus, we need the first derivative of L_{CE}. In particular, we need to solve the set of M equations

$$\frac{\partial L_{CE}}{\partial \tilde{x}_i} = 0 \quad \text{for} \qquad i = 1,\ldots,M$$

In this case, it is easy to show that the binary cross-entropy L_{CE} is minimized when $x_i = \tilde{x}_i$ for $i = 1, \ldots, M$. Note that strictly speaking, this is true only when x_i is different than 0 or 1 since $\tilde{x}_i$ can be neither 0 nor 1.

To find when the L_{CE} is minimized, we can derive L_{CE} with respect to a specific input $\tilde{x}_j$

$$\frac{\partial L_{CE}}{\partial \tilde{x}_j} = -\frac{1}{M}\left[\frac{x_j}{\tilde{x}_j} - \frac{1-x_j}{1-\tilde{x}_j}\right] = -\frac{1}{M}\left[\frac{x_j(1-\tilde{x}_j) - \tilde{x}_j(1-x_j)}{\tilde{x}_j - \tilde{x}_j^2}\right] = -\frac{1}{M}\left[\frac{x_j - \tilde{x}_j}{\tilde{x}_j - \tilde{x}_j^2}\right] \tag{9.3}$$

Now remember that we need to satisfy the condition

$$\frac{\partial L_{CE}}{\partial \tilde{x}_j} = 0$$

That can happen only if $x_j = \tilde{x}_j$, as can be seen in Equation (9.3). To make sure that this is a minimum, we need to evaluate the second derivative. Since the point for which the first derivative is zero is a minimum only if

$$\frac{\partial^2 L_{CE}}{\partial \tilde{x}_j^2} > 0$$

We can calculate the second derivative at the minimum point $x_j = \tilde{x}_j$ easily

$$\left.\frac{\partial^2 L_{CE}}{\partial \tilde{x}_j^2}\right|_{x_j=\tilde{x}_j} = -\frac{1}{M}\left[\frac{x_j(2\tilde{x}_j - 1) - \tilde{x}_j^2}{(1-\tilde{x}_j^2)\tilde{x}_j^2}\right]_{x_j=\tilde{x}_j} = \frac{1}{M}\left[\frac{\tilde{x}_j(1-\tilde{x}_j)}{(1-\tilde{x}_j^2)\tilde{x}_j^2}\right]$$

Now remember that $\tilde{x}_j \in]0,1[$. We can immediately see that the denominator of the previous formula is greater than zero. The nominator is also clearly greater than zero since $1-\tilde{x}_i > 0$. Dividing two positive numbers gives a positive number, thus we have just proved that

$$\frac{\partial^2 L_{CE}}{\partial \tilde{x}_i^2} > 0$$

The minimum of the cost function is reached when the output is exactly equal to the inputs, as we wanted to prove.

Note An essential prerequisite of using the binary cross-entropy loss function is that the inputs *must* be normalized between 0 and 1 and the activation function for the last layer must be a *sigmoid* or *softmax* function.

The Reconstruction Error

The reconstruction error (RE) is a metric that gives you an indication of how accurately (or poorly) the autoencoder was able to reconstruct the input observation $\boldsymbol{x}_i$. The most typical RE used is the MSE

$$RE \equiv MSE = \frac{1}{M}\sum_{i=1}^{M}\left|x_i - \tilde{x}_i\right|^2 \tag{9.4}$$

That can be easily calculated. The RE is used often when doing anomaly detection with autoencoders, as we explain later. There is an easy explanation of the reconstruction error. When the RE is significant, the autoencoder could not reconstruct the input well, while when it is small, the reconstruction was successful. Figure 9-3 shows an example of big and small reconstruction errors when an autoencoder tries to reconstruct an image.

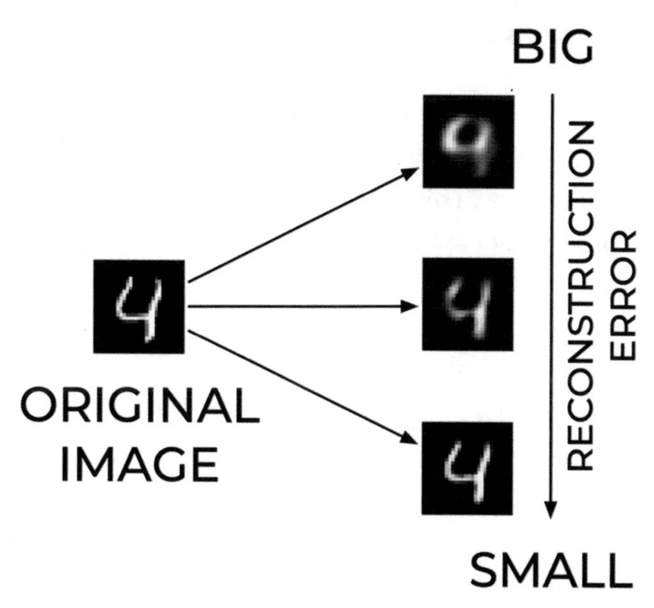

***Figure 9-3.** An example of big and small reconstruction errors when an autoencoder tries to reconstruct an image*

Example: Reconstructing Handwritten Digits

Let's now see how an autoencoder performs with a real example, using the MNIST dataset. This dataset[10] contains 70,000 handwritten digits from 0 to 9. Each image is 28 × 28 pixels with only gray values, which means that we have 784 features (the pixel gray values) as inputs. Let's start with an autoencoder with three layers and with the numbers of neurons in each layer equal to (784,16,784). Note that the first and last layers *must* have a dimension equal to the input dimensions. For this example, we used the Adam optimizer[11] as a loss function the cross-entropy[12] and we trained the model for 30 epochs with a batch size of 256. Figure 9-4 shows two lines of images of digits. The line at the top contains ten random images from the original dataset, while the ones at the bottom are the reconstructed images with the autoencoder.

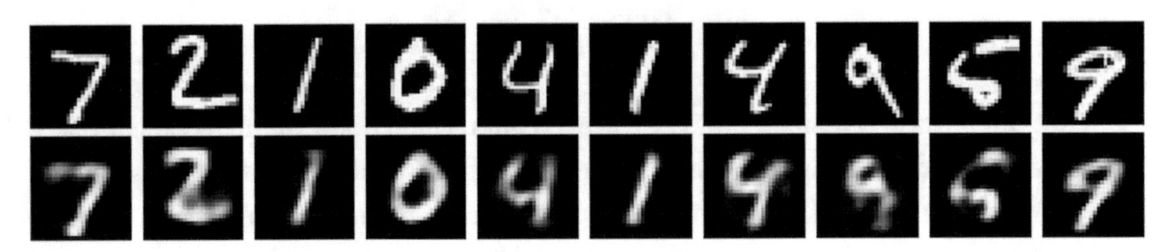

***Figure 9-4.** In the top line, you can see the original digits from the MNIST dataset. In contrast, the bottom line contains the digits reconstructed by the autoencoder with the number of neurons equal to (784, 16, 784)*

It is impressive that, to reconstruct an image with 784 pixels, ten classes, and 70,000 images, only 16 features are needed. The result, although not perfect, allows us to understand almost entirely which digit was used as input. Increasing the middle layer's size to 64 (and leaving all other parameters the same) gets a much better result, as you can see in Figure 9-5.

[10] More information on the dataset can be found at `http://yann.lecun.com/exdb/mnist/`.

[11] You can find the entire code at `https://adl.toelt.ai`.

[12] In this case, we normalized the input features to be between 0 and 1.

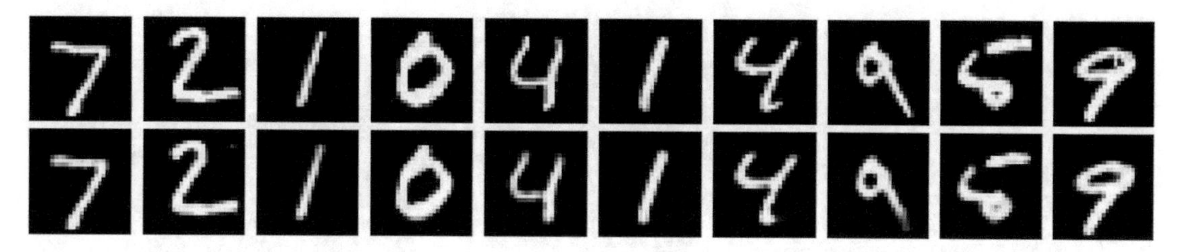

Figure 9-5. *In the top line you can see the original digits from the MNIST dataset. The bottom line shows the digits reconstructed by the autoencoder with the number of neurons equal to (784, 64, 784)*

This tells us that the relevant information on how to write digits is contained in a much lower number of features than 784.

Note An autoencoder with a middle layer smaller than the input dimensions (a bottleneck) can be used to extract the essential features of an input dataset. This creates a learned representation of the inputs given by the function $g(\mathbf{x}_i)$. Effectively an FFA can be used to perform *dimensionality reduction*.

The FFA will not recreate the input digits well if the number of neurons in the middle layer is reduced too much (if the bottleneck is too extreme). Figure 9-6 shows the reconstruction of the same digits with an autoencoder with only eight neurons in the middle layer. With only eight neurons in the middle layer, you can see that some reconstructed digits are wrong. As you can see in Figure 9-6, the 4 is reconstructed as a 9 and the 2 is reconstructed to something that resembles a 3.

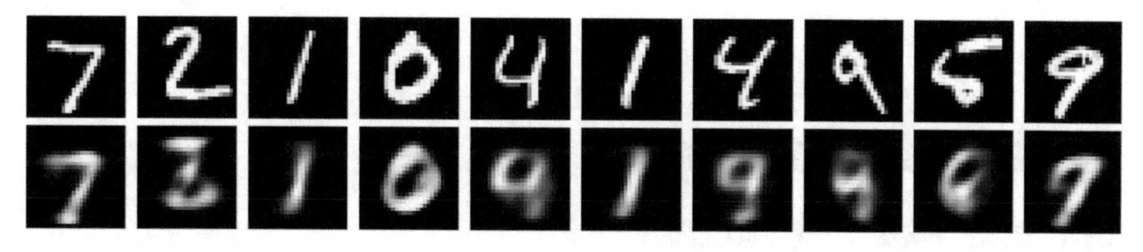

Figure 9-6. *In the top line you can see the original digits from the MNIST dataset. In contrast, the bottom line contains the digits reconstructed by the autoencoder with the number of neurons equal to (784, 8, 784)*

In Figure 9-7, you can compare the reconstructed digits by all the FFAs we have discussed.

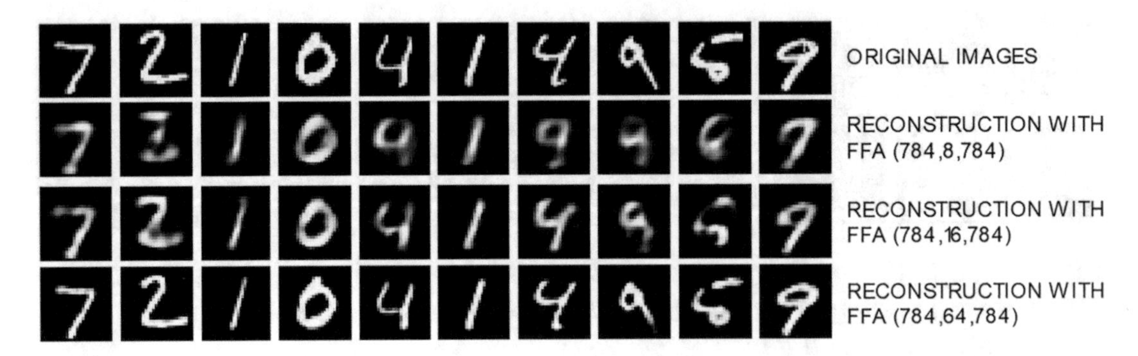

***Figure 9-7.** In the top line, you can see the original digits from the MNIST dataset. The second line of digits contains the digits reconstructed by the FFA (784,8,784). The third line is by the FFA (784,16,784), and the last line is by the FFA (784,64,784)*

From Figure 9-7 you can see how, by increasing the middle layer's size, the reconstruction gets better and better, as we expected.

For these examples, we used the binary cross-entropy as a loss function. But the MSE would have worked also, and results can be seen in Figure 9-8.

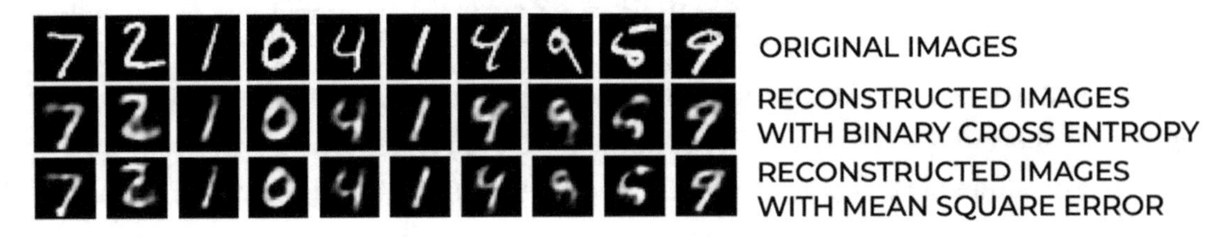

***Figure 9-8.** In the top line, you can see ten random original digits from the MNIST dataset. The second line of digits contains the digits reconstructed with an FFA with 16 neurons in the middle layer and the binary cross-entropy as the loss function. The last line contains images reconstructed with the MSE as a loss function*

Autoencoder Applications

Dimensionality Reduction

As mentioned in this chapter, using the bottleneck method, the latent features will have a dimension q that is smaller than the dimensions of the input observations n. The *encoder* part (once trained) does natural (by design) dimension reduction, thereby producing q real numbers. You can use the latent features for various tasks, such as classification (as you will see in the next section) or clustering.

We would like to point out some of the advantages of dimensionality reduction with an autoencoder compared to a more classical PCA approach. The autoencoder has one main benefit from a computational point of view: it can deal with a very large amount of data efficiently since its training can be done with mini-batches, while PCA, one of the most used dimensionality reduction algorithms, needs to do its calculations using the entire dataset. PCA is an algorithm that projects a dataset on the eigenvectors of its covariance matrix,[13] thus providing a linear transformation of the features Autoencoders are more flexible and consider non-linear transformations of the features. The default PCA method uses $\mathcal{O}\left(d^2\right)$ space for data in $\mathbb{R}^d$. This is, in many cases, not computationally feasible, and the algorithm does not scale up with increasing dataset size. This may seem irrelevant, but in many practical applications, the amount of data and the number of features is so big that PCA is not a practical solution from a computational point of view.

Note The use of an autoencoder for dimensionality reduction has one main advantage from a computational point of view: it can deal with a very large amount of data efficiently since its training can be done with mini-batches.

Equivalence with PCA

This is not well known, but still worth mentioning. An FFA is equivalent to a PCA if the following conditions are met:

- You use a linear function for the encoder $g(\cdot)$
- You use a linear function for the decoder $f(\cdot)$
- You use the MSE for the loss function
- You normalize the inputs to

$$\hat{x}_{i,j} = \frac{1}{\sqrt{M}}\left(x_{i,j} - \frac{1}{M}\sum_{k=1}^{M} x_{k,j}\right)$$

The proof is long and can found in the notes by M.M. Kahpra for the course CS7015 (Indian Institute of Technology Madras) at `http://toe.lt/1a`.

[13] Akshay Balsubramani, Sanjoy Dasgupta, and Yoav Freund. "The fast convergence of incremental pca. In Advances in neural information processing systems," pages 3174–3182, 2013.

Classification

Classification with Latent Features

Let's now suppose that we want to classify our input images of the MNIST dataset. We can simply use all the features, in our case, the 784 pixel values of the images. We can use an algorithm, such as kNN, for illustrative purposes. Doing this with seven nearest neighbors on the training MNIST dataset (with 60,000 images) will take around 16.6 minutes[14] (1000 sec) and gets you an accuracy on the test dataset of 10,000 images of 96.4%. However, what happens if we use this algorithm not with the original dataset, but with the latent features $g(\boldsymbol{x}_i)$? For example, say we consider an FFA with eight neurons in the middle layer and again train a kNN algorithm on the latent features $g(\boldsymbol{x}_i) \in \mathbb{R}^8$. In that case, we get an accuracy of 89% in 1.1 sec. We get a gain of a factor of 1,000 in running time, for a loss of 7.4% in accuracy.[15] See Table 9-1.

Table 9-1. *The Different Accuracies and Running Times When Applying the kNN Algorithm to the Original 784 Features or the Eight Latent Features for the MNIST Dataset*

Input Data	Accuracy	Running Time
Original data $\boldsymbol{x}_i \in \mathbb{R}^{784}$	96.4%	1000 sec. $\approx$16.6 min.
Latent Features $g(\boldsymbol{x}_i) \in \mathbb{R}^8$	89%	1.1 sec.

Using eight features allows us to get very good accuracy in just one second.

We can do the same analysis with another dataset, the Fashion MNIST[16] dataset (a dataset from Zalando very similar to the MNIST one, only with clothing images instead of handwritten digits) for illustrative purposes. The dataset has, as the MNIST one, 60,000 training images and 10,000 test ones. Table 9-2 shows a summary of the results of applying kNN to the testing portion of this dataset.

[14] The examples were run on Google Colab.

[15] You can run these tests yourself by going to `https://adl.toelt.ai`.

[16] `https://research.zalando.com/welcome/mission/research-projects/fashion-mnist/`.

Table 9-2. *The Difference in Accuracy and Running Time When Applying the kNN Algorithm to the Original 784 Features with an FFA with Eight Neurons and with an FFA with 16 Neurons for the Fashion MNIST Dataset*

Input Data	Accuracy	Running Time
Original data $\mathbf{x}_i \in \mathbb{R}^{784}$	85.4%	1040 sec. ≈16.6 min.
Latent Features $enc(\mathbf{x}_i) \in \mathbb{R}^{8}$	79.9%	1.2 sec.
Latent Features $enc(\mathbf{x}_i) \in \mathbb{R}^{16}$	83.6%	3.0 sec.

It is exciting to note that with an FFA with 16 neurons in the middle layer, we reach an accuracy of 83.6% in just three sec. When applying a kNN algorithm to the original features (784), we get an accuracy only 1.8% higher but with a running time of around 330 times longer.

Note Using autoencoders and doing classification with the latent features is a good way to reduce the training time by several orders of magnitude while incurring a minor drop in accuracy.

The Curse of Dimensionality: A Small Detour

Is there any other reason that you would want to do dimensionality reduction before doing the classification? Reducing running time is one reason, but another important one plays a significant role when the input dimension is very large, i.e., the datasets that have a very large number of features: the curse of dimensionality. To understand why, we need to take a quick detour to look at the problem of high dimensionality classification and discuss the *curse of dimensionality*. Let's consider the unit cube $[0, 1]^d$ with d being an integer and m points in it distributed randomly. How big should the length l of the smallest hyper-cube be to contain at least one point? We can easily calculate it as follows

$$l^d \approx \frac{1}{m} \to l \approx \left(\frac{1}{m}\right)^{1/d}$$

We can easily calculate this value of l for various values of d. Let's suppose that we consider $m = 1000$ and summarize the results in Table 9-3.

Table 9-3. *The Length l of the Smallest Hyper-Cube to Contain at Least One Point from a Population of Randomly Distributed m Points*

d	l
2	0.003
10	0.50
100	0.93
1000	0.99

Furthermore, as you can see, the data becomes so sparse in high dimensions that you need to consider the entire hyper-cube to capture one single observation. When the data becomes so sparse, the number of observations you need in order to train an algorithm properly becomes much bigger than the size of existing datasets.

We could look at this differently. Let's consider a small hyper-cube of side $l = 1/10$. How many observations will we find on average in this small portion of the hyper-cube? This is easy to calculate and is given by

$$\frac{m}{10^d}$$

You can see that this number is very small for high values of d. For example, if we consider $d = 100$ it's easy to see that we would need more observations than atoms in the universe[17] to find at least one observation in that small portion of the hyper-cube.

Note Performing dimensionality reduction is a viable method for reducing running time dramatically while incurring a small drop in accuracy. In high-dimensionality datasets, this becomes fundamental due to the curse of dimensionality.

---

[17] `https://www.universetoday.com/36302/atoms-in-the-universe/`

Anomaly Detection

Autoencoders are often used to perform anomaly detection on different datasets. The best way to understand how anomaly detection works with autoencoders is to look at it with a practical example. Let's consider an autoencoder with only three layers with 784 neurons in the first, 64 in the latent feature generation layer, and 784 neurons in the output layers. We will train it with the MNIST dataset and in particular with the 60,000 training portion of it, as we did in the previous sections of this chapter. Now let's consider the Fashion MNIST dataset. Let's choose an image of a shoe (see Figure 9-9) from this dataset.

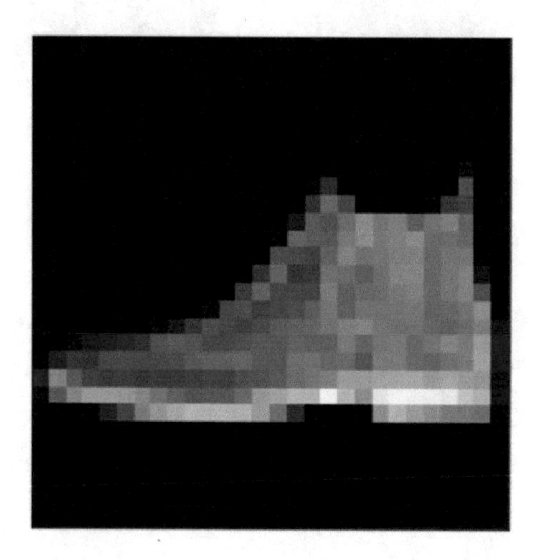

Figure 9-9. *A random image from the Zalando MNIST dataset.*

Let's add it to the testing portion of the MNIST dataset. The original testing portion of MNIST has 10,000 images. With the shoe, we will have 10,001 images. How can we use an autoencoder to find the shoe automatically in those 10,001 images? Note that the shoe is an "outlier," it's an "anomaly" since it is an entirely different image class than handwritten digits. To do that, we will take the autoencoder we trained with the 60,000 MNIST images and calculate the reconstruction error for the 10,001 test images.

The main idea is that, since the autoencoder has only seen handwritten digits images, it will not be able to reconstruct the shoe image. Therefore we expect this image to have the biggest reconstruction error. We can see if that is the case by taking the top two reconstruction errors. For this example, we used the MSE for the reconstruction error. You can check out the code at `https://adl.toelt.ai`. The shoe has the highest reconstruction error: 0.062. The autoencoder cannot reconstruct the image, as shown in Figure 9-10.

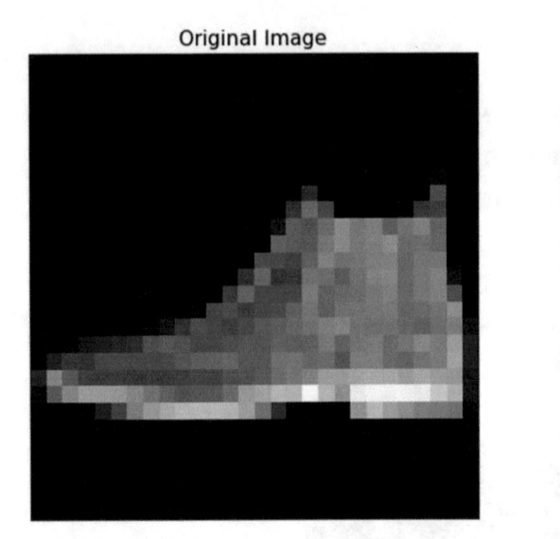

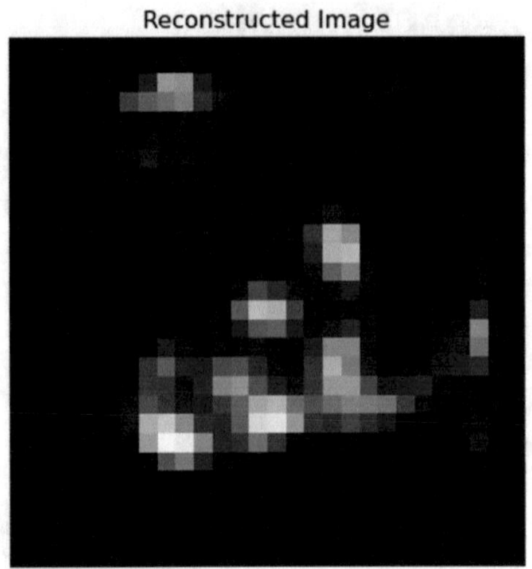

Figure 9-10. *The shoe and the autoencoder's reconstruction trained on the 60,000 handwritten images of the MNIST dataset. This image has the biggest RE in the entire 10,001 test dataset we built, with a value of 0.062*

The second biggest RE is slightly less than one third of that of the shoe: 0.022. This indicates that the autoencoder is doing a good job understanding how to reconstruct handwritten digits. You can see the image with the second biggest RE in Figure 9-11. This image could also be classified as an outlier, as it's not completely clear if it is a 4 or an incomplete 9.

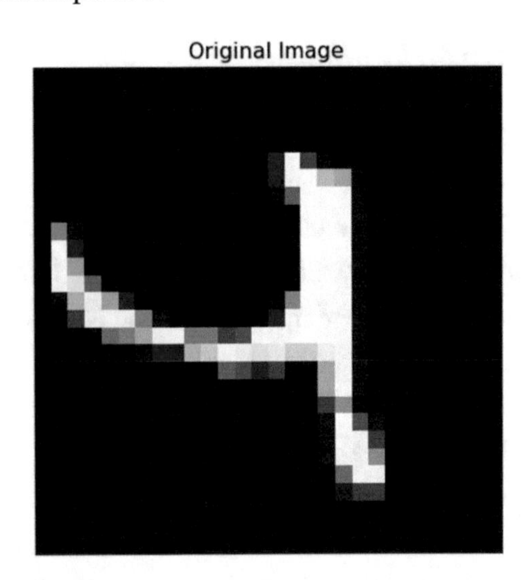

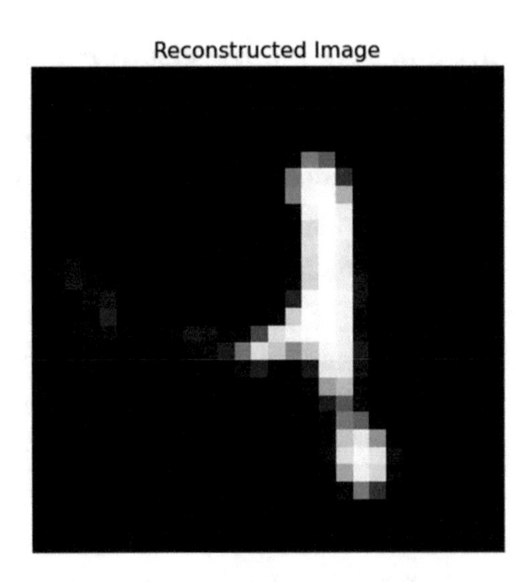

Figure 9-11. *The image with the second biggest RE in the 10,001 test dataset: 0.022*

The readers with the most experience may have noticed that we trained our autoencoders on a dataset without any outliers and applied it to a second dataset with outliers. This is not always possible, as very often the outliers are not known and are lost in a big dataset. In general, you want to find outliers in a single big dataset without any information on how many there are or how they look. Generally speaking, anomaly detection can be done following these main steps.

1. Train an autoencoder on the entire dataset (or if possible, on a portion of the dataset known not to have an outlier).
2. For each observation (or input) of the portion of the dataset known to have the wanted outliers, calculate the RE.
3. Sort the observations by the RE.
4. Classify the observations with the highest RE as outliers. The number of observations that classify as outliers will depend on the problem at hand and requires an analysis of the results (and usually a lot of knowledge of the data and the problem).

Note that if you train the autoencoder on the entire dataset, there is an essential assumption: the outliers are a negligible part of the dataset and their presence will not influence how the autoencoder learns to reconstruct the observations. This is one of the reasons that regularization is so essential. If the autoencoders could learn the identity function, anomaly detection could not be done.

A classic example of anomaly detection is finding fraudulent credit card transactions (the outliers). This case usually presents around 0.1% fraudulent transactions and therefore this would be a case that would allow us to train the autoencoder on the entire dataset. Another is fault detection in an industrial environment.

Note If you train the autoencoder on the entire dataset at disposal, there is an essential assumption: the outliers are a negligible part of the dataset and their presence will not influence how the autoencoder learns to reconstruct the observations.

Model Stability: A Short Note

Note that doing anomaly detection as described in the previous section seems easy, but those methods are prone to overfitting and often give inconsistent results. This means that training an autoencoder with a different architecture may well give different REs and therefore other outliers. There are several ways of solving this problem, but one of the simplest ways of dealing with instability of results is to train different models and then take the average of the REs. Another often used technique involves taking the maximum of the REs evaluated from several models. This kind of approaches are called ensemble methods but go beyond the scope of this book.

Note Anomaly detection done with autoencoders is prone to problems related to overfitting and unstable results. It is essential to be aware of these problems and check the results coming from different models to interpret the results correctly.

Note that this section serves to give you some pointers and is not meant to be an exhaustive overview on how to solve this problem.

Like autoencoders ensembles,[18] more advanced techniques are also used to deal with problems of instable results coming, for example, from small datasets.

Denoising Autoencoders

Denoising autoencoders[19] are developed to auto-correct errors (noise) in the input observations. As an example, imagine the handwritten digits we considered before where we added some noise (for example, Gaussian noise) in the form of randomly changing the gray values of the pixels. In this case, the autoencoders should learn to reconstruct the image without the added noise. As a concrete example, consider the MNIST dataset. We can add to each pixel a random value generated by a normal distribution scaled by a factor (you can check out the code at `https://adl.toelt.ai`).

---

[18] See for example `https://saketsathe.net/downloads/autoencode.pdf`.

[19] Vincent, P., Larochelle, H. Bengio, Y. Manzagol, P.A.: "Extracting and composing robust features with denoising autoencoders." In: *Proceedings of the 25th International Conference on Machine Learning,* ICML '08, pp. 1096-1103. ACM, New York, NY USA (2008).

We can train an autoencoder using the noisy images as the input and the original images as the output. The model should learn to remove the noise, since it is random in nature and has no relationship to the images.

Figure 9-12 shows the results. In the left column, you see the noisy images; in the middle, the original ones; and on the right are the de-noised images. It is quite impressive how well it works. Figure 9-12 was generated by training an FFA autoencoder with three layers and 32 neurons in the middle layer.

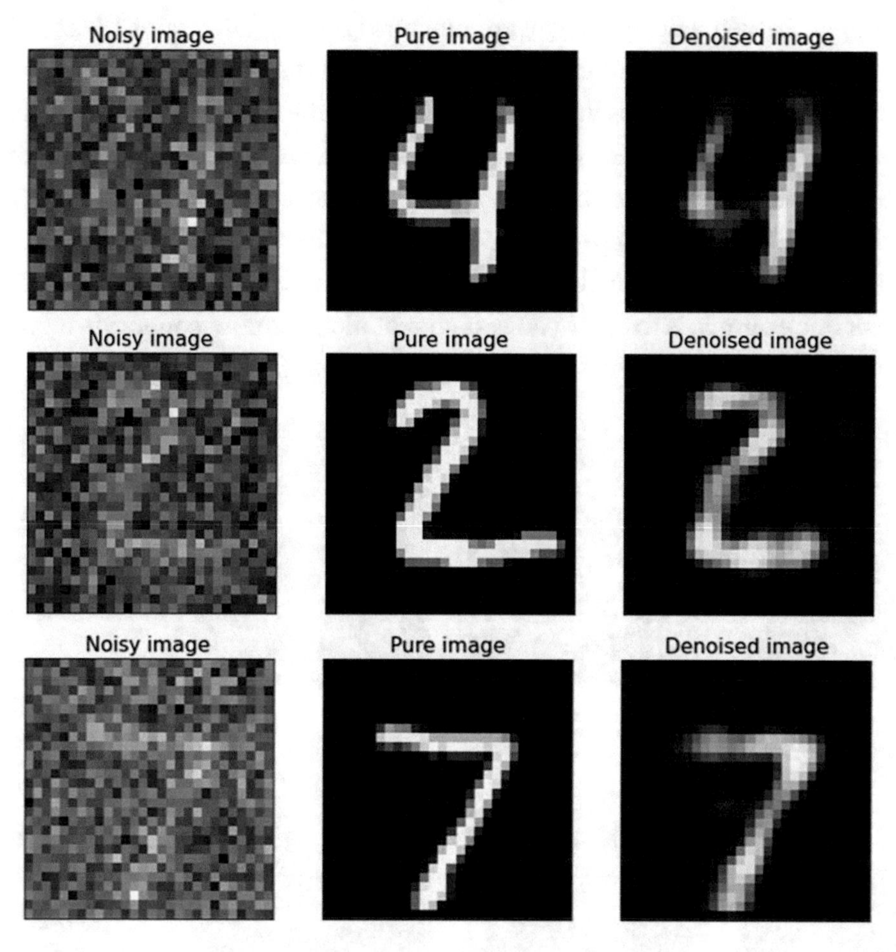

Figure 9-12. *The results of denoising an FFA autoencoder with three layers and 32 neurons in the middle layer. The noise was generated by adding a real number between 0 and 1 taken from a normal distribution. For details, see the code at* *`https://adl.toelt.ai`*

Beyond FFA: Autoencoders with Convolutional Layers

This chapter has described autoencoders with a feed-forward architecture. But autoencoders with convolutional layers works as well, and are often much more efficient (especially when dealing with images). For example, in Figure 9-13, you can see a comparison of the results of an FAA (with architecture 784,32,784) and of a Convolutional Autoencoder (CA) with architecture (28x28), (26x26x64), (24x24,32), (26x26x64), (28x28). Keep in mind the layers are convolutions, so the first two numbers indicate the tensor dimensions and the third indicates the number of kernels, which in this example had a size of 3x3. The two autoencoders have been trained with the same parameters (epochs, mini-batch size, etc.). You can see how a CA gives better results than an FAA, since we are dealing with images. To be fair, note that the feature generating layer is only marginally smaller than the input layer in this example. The purpose of this example is to show you that convolutional autoencoders are a viable solution because they work very well in many practical applications.

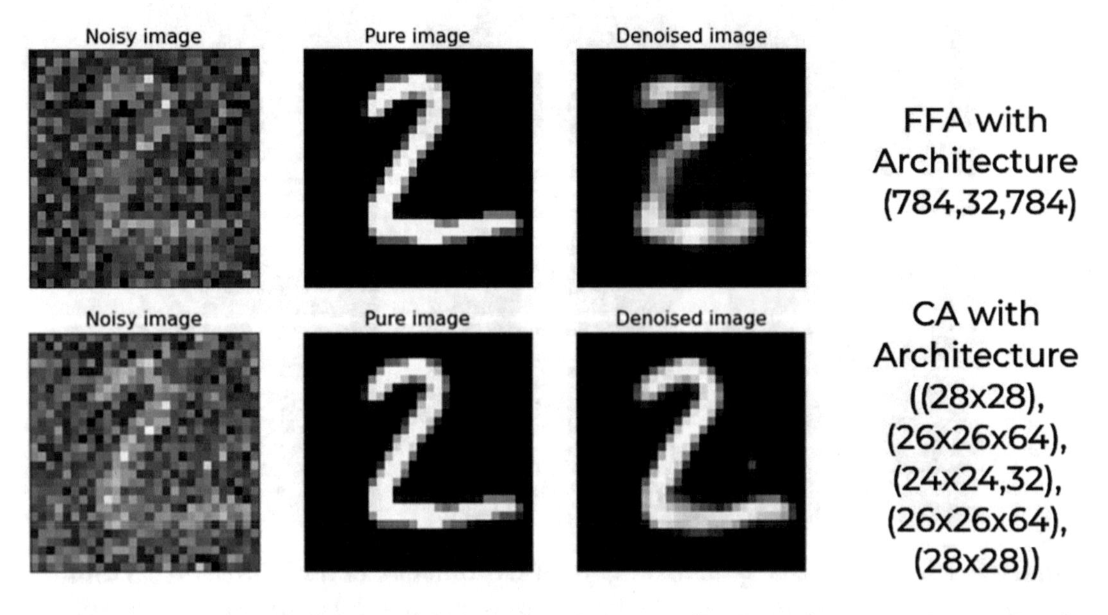

Figure 9-13. *A comparison of the results of an FAA (with architecture 784,32,784) and of a Convolutional Autoencoder (CA) with architecture (28x28), (26x26x64), (24x24,32), (26x26x64), (28x28)*

Another important aspect is that the feature-generating layer can be a convolutional layer but can also be a dense one. There is no fixed rule and testing is required to find the best architecture for your problem. It also depends on how you want to model your latent features: as a tensor (multi-dimensional array) or as a one-dimensional array of real numbers.

Implementation in Keras

Now let's briefly look at how we can implement autoencoders in Keras. It is quite easy so don't worry. You will find many examples on the online version of the book at `https://adl.toelt.ai`. The easiest way of implementing an autoencoder is to use the Keras Functional APIs. As an example, we will consider as input the MNIST dataset, in other words the input shape of the network will be `(784,)` (remember that the MNIST images are 28x28 pixels, and therefore when flattened are vectors with 784 values). The network will consists of two parts: the encoder and the decoder (check out Figure 9-2 if you don't remember exactly which part does what). Let's look at the code and then discuss it.

```
def create_autoencoders(feature_layer_dim = 16):
  input_img = Input(shape = (784,), name = 'Input_Layer')
  # 784 is the total number of pixels of MNIST images

  encoded = Dense(feature_layer_dim, activation = 'relu', name = 'Encoded_
  Features')(input_img)
  decoded = Dense(784, activation = 'sigmoid', name = 'Decoded_Input')
  (encoded)

  autoencoder = Model(input_img, decoded)
  encoder = Model(input_img, encoded)

  encoded_input = Input(shape = (feature_layer_dim,))
  decoder = autoencoder.layers[-1] # Get the last layer
  decoder = Model(encoded_input, decoder(encoded_input))

  return autoencoder, encoder, decoder
```

This is the smallest autoencoder that you can create (at least in terms of the number of layers). It has the input layer (784 input values), the latent feature layer (you decide on its dimension by giving the `feature_layer_dim` parameter to the function), and then the output layer that must have of course the same dimension as the input layer (784). If you want to make the network larger you can of course add as many layers as you want. But the two most important characteristics are that the input and output layer must have the same dimensions and that the latent feature layer should have a smaller dimension than the input/output layer. Once you define this function, you can use it easily to get three models: the autoencoders, an encoder, and a decoder.

```
autoencoder, encoder, decoder = create_autoencoders(16)
```

You can train the autoencoder exactly as you do with Keras with any other neural networks by using the `.fit()` call.

```
history = autoencoder.fit(mnist_x_train, mnist_x_train,
                          epochs = 30,
                          batch_size = 256,
                          shuffle = True,
                          validation_data = (mnist_x_test, mnist_x_test),
                          verbose = 0)
```

Where you can imagine that `mnist_x_train` and `mnist_x_test` are two datasets composed of several flattened MNIST handwritten digits. It is important to note that we have given the dataset `mnist_x_train` as input to the network for the images and for the output. In other words, there are no labels here. The labels are the dataset itself, since we want the output to be as close to the input as possible (remember the previous sections?).

You can easily encode images, or in other words get the latent features of a certain set of inputs by calling `predict()`

```
encoded_imgs = encoder.predict(mnist_x_test)
```

Where `mnist_x_test` is a hypothetical dataset that you want to encode. You can use those encoded_imgs values to do anything you need, for example perform classification or regression. By invoking `encoder.predict`, you basically have performed dimensionality reduction on the input dataset, as we discussed in the previous sections of the chapter. The decoding can be done just as easily

```
decoded_imgs = decoder.predict(encoded_imgs)
```

At `https://adl.toelt.ai` you will find examples of autoencoders, anomaly detection with autoencoders, and denoising with autoencoders, as described in this chapter.

Exercises

EXERCISE 1

List the most useful tasks you can use an autoencoder for. Can you think of an application in your field of work?

EXERCISE 2

Can you explain briefly what a sparse autoencoder is? How is it similar to an autoencoder with a bottleneck?

EXERCISE 3

How do you measure the performance of an autoencoder (which metric do you use)? List the most commonly used metrics that you can use. Can you think of any additional metric, in addition to those discussed in this chapter, that could be used?

EXERCISE 4

Describe how anomaly detection works with autoencoders.

Further Readings

Deep Learning Tutorial from Stanford University

`http://ufldl.stanford.edu/tutorial/unsupervised/Autoencoders/`

Building autoencoders in Keras

`https://blog.keras.io/building-autoencoders-in-keras.html`

Introduction to autoencoders in TensorFlow

`https://www.tensorflow.org/tutorials/generative/autoencoder`

Bank, D., Koenigstein, N., and Giryes, R., "Autoencoders", arXiv e-prints, 2020, `https://arxiv.org/abs/2003.05991`

R. Grosse, University of Toronto, Lecture on autoencoders

`http://www.cs.toronto.edu/~rgrosse/courses/csc321_2017/slides/lec20.pdf`

CHAPTER 10

Metric Analysis

Let's consider a classification problem (similar to one that we have seen so far in the book). While doing all our work we made a strong assumption without explicitly saying it: we assume that all the observations are correctly labelled! We cannot say that with certainty. To perform the labelling, we needed some manual intervention, and therefore there must be a certain number of images that are improperly classified since humans are not perfect. This is an important revelation.

Consider the following scenario: we are supposed to reach 90% accuracy in a classification problem. We could try to get better and better accuracy, but when is it sensible to stop trying? If your labels are wrong in 10% of the cases, the model, as sophisticated as it may be, will never be able to generalize to new data with a very high accuracy since it learned the wrong classes for many images. We spend quite a lot of time checking and preparing the training data and normalizing it, but not much time is typically spent checking the labels themselves. We typically assume that all classes have similar characteristics. (We will discuss later in this chapter what that exactly means; for the moment an understanding of the idea will suffice.) What if the quality of images (suppose our inputs are images) for specific classes is worse than for others? What if the number of pixels whose gray values are different than zero is dramatically different for different classes? What if some images are completely blank. What happens in that case? As you can imagine, we cannot check all the images manually and try to detect such issues. Suppose we have millions of images, a manual analysis is surely not possible.

We need a new weapon in our arsenal to be able to spot such cases and to be able to tell how a model is doing. This new weapon is the focus of this chapter and is what it is typically called "metric analysis." Very often people in this field refer to this array of methods as "error analysis." However, this name is a bit confusing, especially for beginners. Error can refer to too many things: Python code bugs, error in the methods, in the algorithms, errors in the choice of optimizers, and so on.

U. Michelucci, *Applied Deep Learning with TensorFlow 2*, https://doi.org/10.1007/978-1-4842-8020-1_10

You will learn in this chapter how to obtain fundamental information about how your model is doing and how good your data is. You will do this by evaluating your optimizing metric on a set of different datasets that you can derive from your data.

Recall that we discussed, in the case of regression, how in the case of $MSE_{train} \gg MSE_{dev}$ we are in a regime of overfitting. Our metric is the MSE and evaluating it on two datasets, training and dev, and comparing the two values can tell us if the model is overfitting or not. We will expand this methodology in this chapter to be able to extract much more information from the data and the model.

Human-Level Performance and Bayes Error

In most of the datasets that we use for supervised learning, someone has labelled the observations. Take for example a dataset where we have images that are classified. If we ask people to classify all images (imagine this being possible, regardless of the number of images), the accuracy obtained will never be 100%. Some images maybe too blurry to be classified correctly, and people make mistakes. If, for example, 5% of the images are not classified correctly, due for example to how blurry they are, we must expect that the maximum accuracy people can reach will always be less than 95%.

Let's consider a problem of classification. First, let's define what we mean by the word *error*. We associate the word "error" with the following quantity, indicated with ϵ:

$$\epsilon \equiv 1 - Accuracy$$

For example, if we reach an accuracy of 95%, we will have $\epsilon = 1 - 0.95 = 0.05$ or expressed as percent $\epsilon = 5\%$.

A useful concept to understand is the *human-level performance*, which can be defined as "the lowest value for the error ϵ that can be reached by a person performing the classification task." We will indicate it with ϵ_{hlp}.

Let's look at a concrete example. Let's suppose we have a set of 100 images. Now let's suppose we ask three people to classify the 100 images. Let's imagine that they obtain 85%, 83% and 84% accuracy. In this case, human-level performance accuracy will be $\epsilon_{hlp} = 5\%$. Note that someone else maybe much better at this task, and therefore it's always important to consider that the value of ϵ_{hlp} we get is an estimate and should only serve as a guideline.

Now let's complicate things a bit. Suppose we are working on a problem in which doctors classify MRI scans in two classes: ones with signs of cancer and ones without. Now let's suppose we calculate ϵ_{hlp} from the results of untrained students obtaining 15%, from doctors with a few years of experience obtaining 8%, from experienced doctors obtaining 2%, and from experienced groups of doctors obtaining 0.5%. What is ϵ_{hlp} in this case? You should always choose the lowest value you can get, for reasons we will discuss later.

We can now expand the definition of ϵ_{hlp} to a second definition:

Human-level performance is "the lowest value for the error ϵ that can be reached by people or *groups* of people performing the classification task."

Note You do not need to decide which definition is right. Just use the one that gives you the lowest value of ϵ_{hlp}.

Now let's talk a bit about the why we must choose the lowest value we can get for ϵ_{hlp}. The *lowest error that that can be reached by any classifier* is called the *Bayes error* and it's a very important quantity. We will indicate it with ϵ_{Bayes}. Usually, ϵ_{hlp} is very close to ϵ_{Bayes}, at least in tasks where humans excel, like image recognition. *It is commonly said that that human-level performance error is a proxy for the Bayes error.* Normally it's impossible to evaluate ϵ_{Bayes} and therefore practictioners use ϵ_{hlp}, assuming the two are close since the latter is easier (relatively) to estimate.

Now keep in mind that it makes sense to compare the two values and assume that ϵ_{hlp} is a proxy for ϵ_{Bayes} only if people (or groups of people) perform classification in the same way as the classifier. For example, it's okay if both use the same images to do classification. But, in the cancer example, if the doctors use additional scans and analysis to make a diagnosis, the comparison is not fair anymore since human-level performance is not a proxy for Bayes error anymore. Doctors, having more data at their disposal, will clearly be better than the model, which has as the input only the images.

Note ϵ_{hlp} and ϵ_{Bayes} are close to each other only when the classifications by the humans and by the model are done the same way. So always be sure that is the case before assuming that human-level performance is a proxy for the Bayes error.

Something else that you will notice when working on models is that with relatively low effort you can reach a low error, and often (almost) reach ϵ_{hlp}. After passing human-level performance (and in several cases that is possible), progress tends to be very slow, as shown in Figure 10-1.

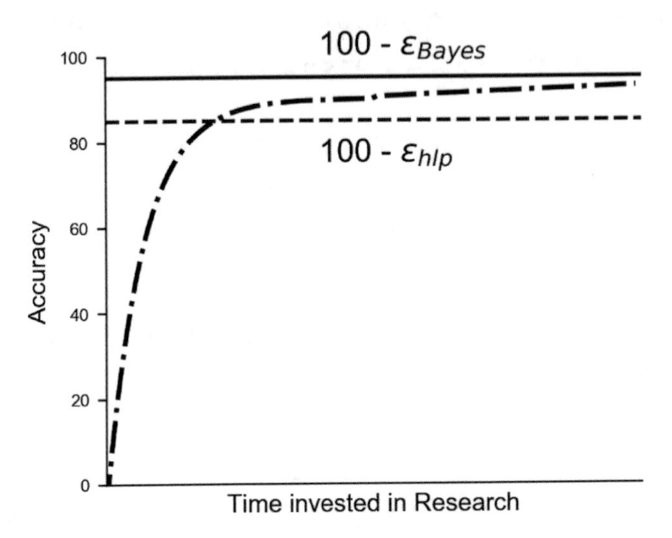

Figure 10-1. *Typical values of accuracy that can be reached vs. amount of time invested. At the beginning, it's very easy with machine learning to get good accuracy, and often reach ϵ_{hlp}. This is indicated by the line in the plot. After that point, progress tends to be very slow*

As long as the error of your algorithm is bigger than ϵ_{hlp} you can use the following techniques to get better results:

- Get better labels from humans or groups, for example from groups of doctors in the case of medical data, as in our example.
- Get more labelled data from humans or groups.
- Do a good metric analysis to determine the best strategy for getting better results. You will learn how to do this in this chapter.

As soon as your algorithm gets better than human-level performance you cannot rely on those techniques anymore. It's important to get an idea of those numbers to be able to decide what to do to get better results. In our example of MRI scans, we could get better labels by relying on other sources that are not related to humans, such as checking the diagnosis a few years after the MRI time point, when it's more clear if the patient developed cancer or not. In the case of image classification, you might decide to label a few thousands of images of specific classes. This is not usually possible, but you must be aware that you can get labels using means other than asking humans to perform the same kind of task that your algorithm is performing.

Note Human-level performance is a good proxy for the Bayes error for tasks that humans excel at, such as image recognition. It can be very far from Bayes error for tasks that humans don't do well.

A Short Story About Human-Level Performance

It is instructive to know about the work that Andrej Karpathy has done, while trying to estimate human-level performance in a specific case. You can read the entire story on his blog post (a long post, but worth reading) [1]. What he did is extremely informative about human-level performance. Karpathy was involved in the ILSVRC contest: ImageNet Large Scale Visual Recognition Challenge in 2014 [2]. The task was made up of 1.2 million images (training set) classified in 1000 categories and included objects like animals, abstract objects like a spiral, scenes, and many others. Results were evaluated on a dev dataset. GoogleLeNet (a model developed by Google) reached an astounding 6.7% error. Karpathy asked himself *how do humans compare*?

The question is a lot more complicated that it may seem at first. Since the images were all classified by humans, shouldn't $\epsilon_{hlp} = 0\%$? Well, not really. In fact, the images were first obtained with a web search, then filtered and labelled by asking people binary questions: is this a hook or not (for example).

The images were collected, as Karpathy mentioned in his blog post, in a binary way. People were not asked to assign to each image a class by choosing from the 1000 available as the algorithms were doing. You may think that this is a technicality, but the difference in how the labelling occurs makes the correct evaluation of ϵ_{hlp} quite a complicated matter. So Karpathy set to work and developed a web interface that consisted of an image on the left and the 1000 classes with examples on the right. You can see an example of the interface in Figure 10-2.

You can try the interface [3] to realize how complicated is such a task. People kept missing classes and making mistakes. The best error that was reached was around 15%. So, he did what every scientist at some point in his career have to do: he bored himself out of his mind and did a careful annotation, sometimes needing 20 minutes for a single image. As he formulated in his blog post, he did it only #forscience. He was able to reach a stunning $\epsilon_{hlp} = 5.1\%$. 1.7% better than the best algorithm at the time. He listed sources of errors that GoogLeNet is more susceptible to than humans, such as problems with multiple objects in an image, and sources of errors that humans are more susceptible to than GoogLeNet, such as problems with classes with a huge granularity (dogs are classified in 120 different subclasses).

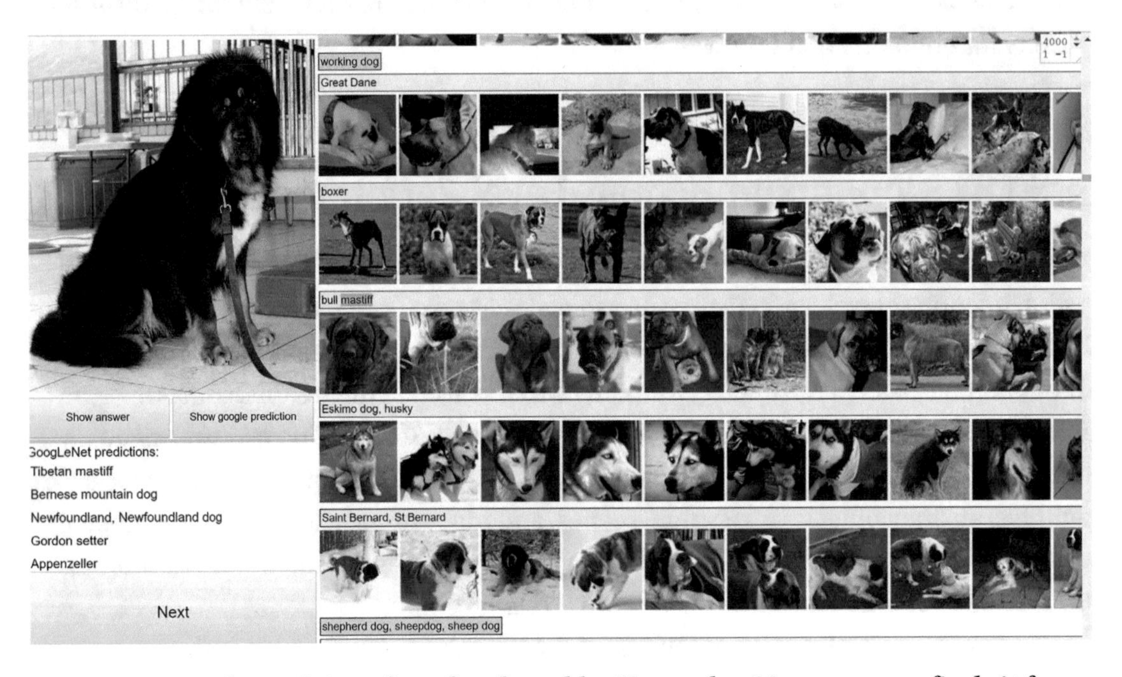

Figure 10-2. *The web interface developed by Karpathy. Not everyone finds it fun to look at 120 breeds of dogs to try to classify the dog on the left (by the way, it's a Tibetan Mastiff)*

If you have a few hours to spare, try it. You will get a whole new appreciation of the difficulties of evaluating human-level performance. Defining and evaluating human-level performance is a very tricky task. It is important to understand that ϵ_{hlp} is dependent on how humans approach the classification task, and is dependent on the time invested, on the patience of the people, and on many factors that are difficult to quantify. The main reason for it being so important, apart from the philosophical aspect of knowing when a machine becomes better than humans, is that it is often taken as a proxy for the Bayes error.

Human-Level Performance on MNIST

Before moving on to the next subject, let's look at another example of human-level performance on a dataset we have analyzed together: the MNIST dataset. Human-level performance has been widely analyzed and has been found to be $\epsilon_{hlp} = 0.2\%$ [4]. Now you may wonder why a human cannot reach a 100% accuracy on simple digits. Look at Figure 10-3 and see if you can say which digits are in the image. You may understand why $\epsilon_{hlp} = 0\%$ is not possible, and why a person cannot reach 100% accuracy. Other reasons may be related to which culture people are coming from. In some countries, the digit 7 is written in a very similar way to 1s are for example, and in some cases, mistakes can be made. In other countries, the digit 7 has a small dash along the vertical bar, making it easier to distinguish from a 1.

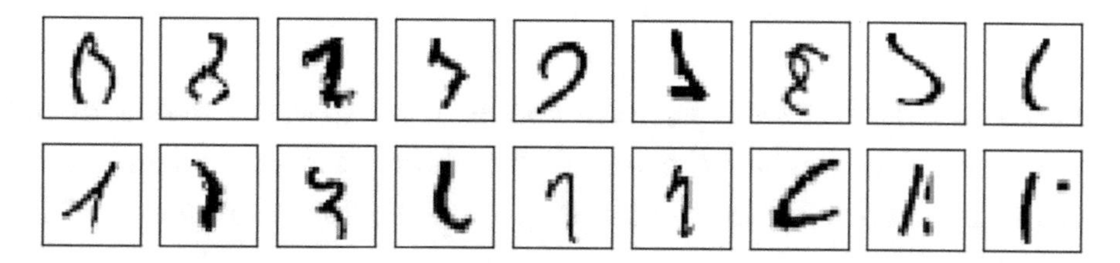

Figure 10-3. *A set of digits from the MNIST dataset almost impossible to recognize. Such examples are one of the reasons that ϵ_{hlp} cannot reach zero*

Bias

Now let's start with metric analysis: a set of procedures that will give you information on how your model is doing—how good or bad your data is—by looking at your optimized metric evaluated on different datasets.

Note Metric analysis consists of a set of procedures that will give you information about how your model is doing—how good or bad your data is—by looking at your optimized metric evaluated on different datasets.

To start, we need to define a third error: the one evaluated on the training dataset, indicated with ϵ_{train}.

The first question we want to answer is if our model is not as flexible or complex as needed to reach human-level performance. Or, in other words, we want to know if our model has a high bias with respect to human-level performance.

To answer the previous question, we can do the following:

- Calculate the error from the model from the training dataset ϵ_{train} and then calculate $|\epsilon_{train} - \epsilon_{hlp}|$. If the number is not small (bigger than a few percent), then we are in the presence of bias (sometimes called avoidable bias). In other words, our model is too simple to capture the real subtleties of our data.

Let's define the following quantity

$$\Delta\epsilon_{Bias} = \left|\epsilon_{train} - \epsilon_{hlp}\right|$$

The bigger $\Delta\epsilon_{Bias}$ is, the more bias in our model. In this case, you want to do better on the training set, since you know you can do better on your training data (we will look at the problem of overfitting in a moment). The following techniques work in reducing bias:

- Bigger networks (more layers or neurons)
- More complex architectures (Convolutional Neural Networks, for example)
- Train your model longer (for more epochs)
- Use better optimizers (like Adam)
- Do a better hyper-parameter search (we looked at it in detail in Chapter 7)

Now there is something else you need to understand. Knowing ϵ_{hlp} and reducing the bias to reach it are two very different things. Suppose you know the ϵ_{hlp} for your problem. This does not mean that you need to reach it. It may well be that you are using the wrong architecture, but you may not have the required skills to be able to develop a network sophisticated enough. It may even be that the effort required to reach that error level is prohibitive (in terms of hardware or infrastructure). Always keep in mind what your problem's requirements are. Always try to understand what is good enough. For an application that recognizes cancer, you may want to invest as much as possible to reach the highest accuracy possible. You do not want to send someone home and discover later that they really do have cancer. On the other hand, if you build a system to recognize cats from web images, you may decide that a higher error than ϵ_{hlp} is completely acceptable.

Metric Analysis Diagram

This chapter looks at different problems you will encounter when developing your models and how to spot them. We looked at the first one: bias, sometimes also called avoidable bias. We saw how this can be spotted by calculating $\Delta\epsilon_{Bias}$. At the end of this chapter, you will have a few of those quantities that you can calculate to spot problems.

To make understanding them easier, let's build a Metric Analysis Diagram (MAD). It is simply a bar diagram, where each bar represents a problem. Let's start building one with (for the moment) the only quantity we have discussed: bias. You can see it in Figure 10-4. At the moment it is a pretty dumb diagram, but you will see how useful it is to keep things under control when you have several problems at the same time.

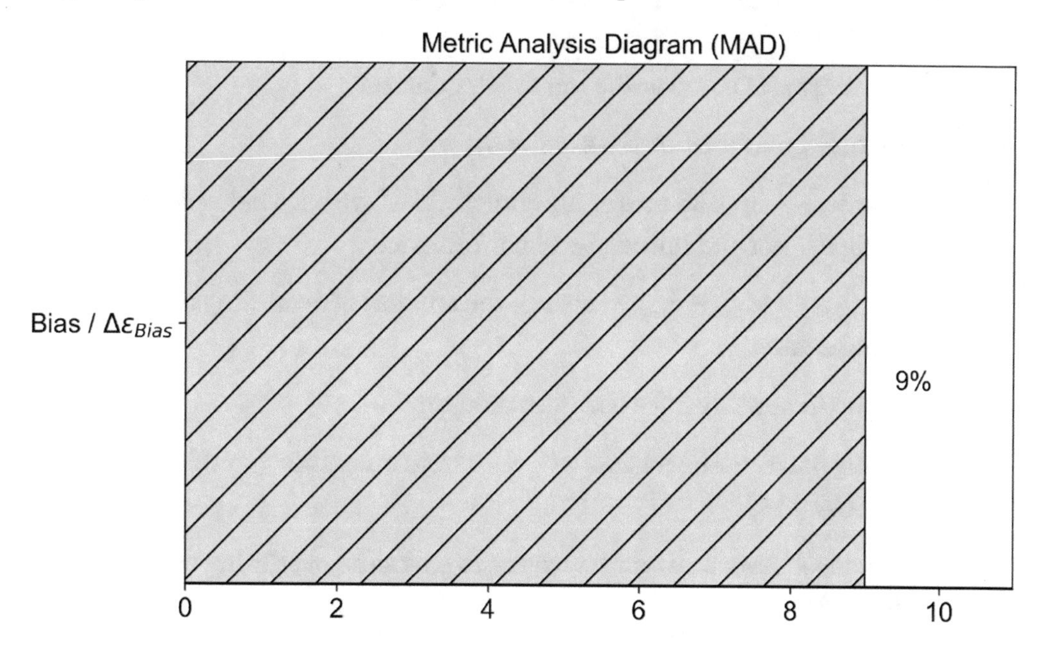

Figure 10-4. *The Metric Analysis Diagram (MAD) with just one of the quantities we encounter in this chapter:* $\Delta\epsilon_{Bias}$

Training Set Overfitting

Another problem we have discussed at length in the previous chapters is overfitting the training data. You will remember that while doing regression you can have overfitting if $MSE_{train} \gg MSE_{dev}$. The same applies in classification problems. Let's indicate, using ϵ_{train}, the error our model has on our training dataset and use ϵ_{dev} for the one on the dev dataset. We can then say we are overfitting the training set if $\epsilon_{train} \gg \epsilon_{dev}$. Let's define a new quantity

$$\Delta\epsilon_{overfitting\ train} = \left|\epsilon_{train} - \epsilon_{dev}\right|$$

With this quantity, we can say we are overfitting the training dataset if $\Delta\epsilon_{overfitting\ train}$ is bigger than a few percent.

Let's summarize what we have defined and discussed so far. We have three errors:

- ϵ_{train}: The error of our classifier on the training dataset
- ϵ_{hlp}: Human-level performance (as discussed in the previous sections)
- ϵ_{dev}: The error of our classifier on the dev dataset

With those three quantities we have defined:

- $\Delta\epsilon_{Bias} = |\epsilon_{train} - \epsilon_{hlp}|$: Measures how much "bias" we have between the training dataset and human-level performance
- $\Delta\epsilon_{overfitting\ train} = |\epsilon_{train} - \epsilon_{dev}|$: Measures the amount of overfitting of the training dataset

In addition, up to now, we have used two datasets:

- Training dataset: The dataset we use to train our model (you should know it by now)
- Dev dataset: A second dataset we use to check the overfitting on the training dataset

Now let's suppose our model has bias and is slightly overfitting the training dataset, meaning we have $\Delta\epsilon_{Bias} = 6\%$ and $\Delta\epsilon_{overfitting\ train} = 4\%$. Our MAD now becomes what is depicted in Figure 10-5.

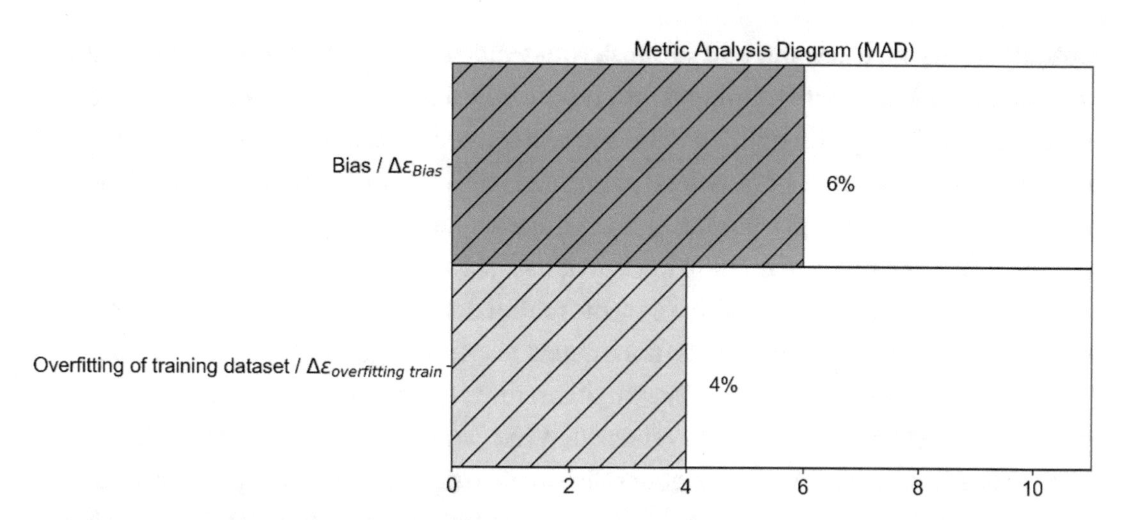

***Figure 10-5.** The MAD diagram for our two problems: bias and overfitting of training dataset*

In Figure 10-5, you can see the relative gravity of the problems we have, and you may decide which one you want to address first.

Usually when you are overfitting the training dataset, it's commonly known as a variance problem. When this happens, you can try the following techniques to minimize this problem:

- Get more data for your training set
- Use regularization (see Chapter 5 for a complete discussion of the subject)
- Try data augmentation (for example, if you are working with images, you can try rotating them, shifting them, etc.)
- Try "simpler" network architectures

As usual, there are no fixed rules, and you must determine which techniques work best on your problem by testing.

Test Set

Now let's quickly mention another problem you may find. Let's recall how you choose the best model in a machine learning project (this is not specific to deep learning by the way). Let's suppose you are working on a classification problem. First you decide which optimizing metric you want, and suppose you decide to use accuracy. Then you build an initial system, feed it with training data, and see how it is doing on the dev dataset to see if you are overfitting your training data. You will remember that in previous chapters we talked often about hyper-parameters: parameters that are not influenced by the learning process. Examples of hyper-parameters are the learning rate, regularization parameter, etc. We have seen many of them in the previous chapters.

Let's say you are working with a specific neural network architecture; you need to search the best values for the hyper-parameters to see how good your model can get. To do that, you train several models with different values of the hyper-parameters and check their performance on the dev dataset. What can happen is that your models work well on the dev dataset but do not generalize at all, since you select only the best values using the dev dataset. You incur in the risk of overfitting the dev dataset with choosing specific values of your hyper-parameters. To see if this is the case, you create a third dataset, called the test dataset, and cut a portion of the observations from your starting dataset. Then you use that test dataset to check the performance of your models.

We must define a new quantity

$$\Delta\epsilon_{overfitting\ dev} = \left|\epsilon_{dev} - \epsilon_{test}\right|$$

Where ϵ_{test} is the error evaluated on the test set. We can add it to the MAD diagram, as shown in Figure 10-6.

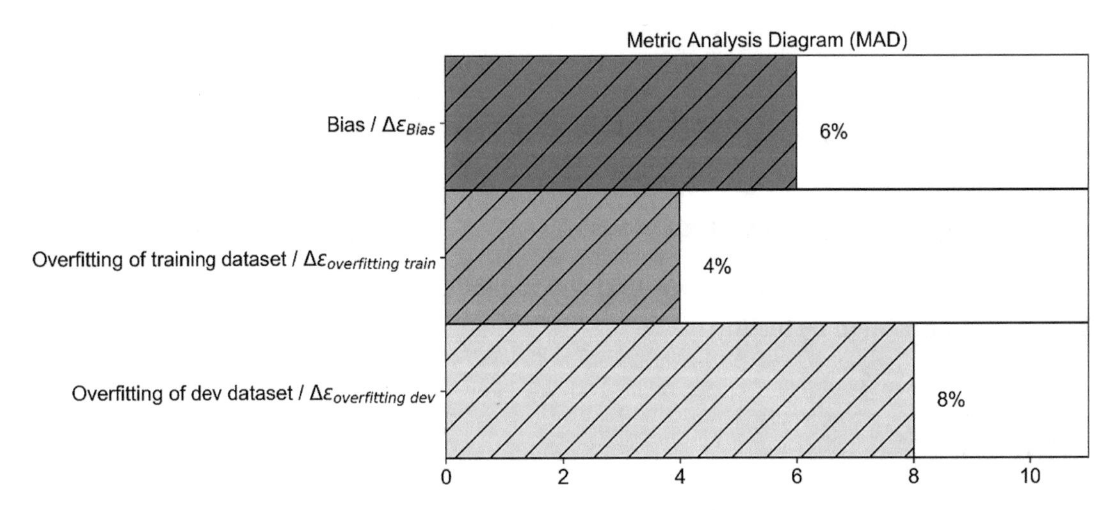

Figure 10-6. *The MAD diagram for the three problems we might encounter: bias, overfitting the training data, and overfitting the dev data*

Note that if you are not doing a hyper-parameter search, you will not need a test dataset. It is only useful when you are doing extensive searches. Otherwise, in most cases, it's useless and takes away observations that you may use for training. What we discussed so far assumes that your dev and test sets observations have the same characteristics. If use high-resolution images from a smartphone for training and the dev dataset, and you use images from the web in low resolution for your test dataset, you may see a big $|\epsilon_{dev} - \epsilon_{test}|$. But that will probably be due to the differences in the images and not because of an overfitting problem. Later in the chapter, we discuss what can happen when different sets come from different distributions (another way of saying that the observations have different characteristics).

How to Split Your Dataset

Now let's briefly discuss how to split the data.

What exactly does *split* mean? Well, as we discussed in the previous section, you will need a set of observations to make the model learn, which you call your training set. You will also need a set of observations that will make your dev set and then a final set that is called the test set. Data is typically split so that 60% for the training set, 20% for the dev set, and 20% for test set. Usually, the kind of split is indicated in this form: 60/20/20, where the first number (60) refers to the percentage of the entire dataset that is in the training set, the second (20) to the percentage of the entire dataset that is in the dev set, and the last (20) to the percentage that is in the test set. You may find in books, blogs, or articles, sentences like this "We split our dataset 80/10/10" for example. This is what it means.

In the deep learning field you will deal with big datasets. For example, if we have $m = 10^6$, we could use a split like 98/1/1. Keep in mind that 1% of 10^6 is 10^4, so still a big number! Remember that the dev/test set must be big enough to give high confidence of the performance of the model, but not unnecessarily big. Additionally, you want to save as many observations as possible for your training set.

Tip When deciding on how to split your dataset, if you have a big number of observations (such as 10^6 or even more), you can split your dataset 98/1/1 or 90/5/5. As soon as your dev and test datasets reach a reasonable size (which depends on your problem), you can stop. When deciding on how to split your dataset, keep in mind how big your dev/test sets must be.

Remember that, as you may know, size is not everything. Your dev and test datasets should be representative of your training dataset and problem. Let's look at an example. Let's consider the ImageNet challenge we described earlier. You want to classify images in 1,000 different classes. To know how your model is performing in your dev and test datasets, you need enough images for each class in each set. If you decide to use only 1000 observations for the dev or test datasets, you are not going to get a reasonable result, since you will have only one observation for each class. You should decide to build your dev and test datasets by choosing for example 100 images for each class at least, thus building two datasets (dev and test) each containing 10^5 observations in total (remember we have 1000 classes). In this case it would not be sensible to go below this number. This is not only relevant in the deep learning context, but in machine learning in general. You should always try to build a dev/test dataset so that it reflects the same distribution of observations you have in your training set. To better understand this concept, take the MNIST dataset for example. Let's load the dataset with the following code (we will see more details about the MNIST data in the following paragraph)

```
import numpy as np
from tensorflow import keras
(x_train, y_train), (x_test, y_test) = keras.datasets.mnist.load_data()
```

Then we can check how often (in percent) each digit appears in the training dataset

```
for i in range(10):
    print ('digit', i, 'makes', np.around(np.count_nonzero(y_train == i)
/60000.0*100.0, decimals = 1), '% of the 60000 observations')
```

This gives us the result

```
digit 0 makes 9.9 % of the 60000 observations
digit 1 makes 11.2 % of the 60000 observations
digit 2 makes 9.9 % of the 60000 observations
digit 3 makes 10.2 % of the 60000 observations
digit 4 makes 9.7 % of the 60000 observations
digit 5 makes 9.0 % of the 60000 observations
digit 6 makes 9.9 % of the 60000 observations
digit 7 makes 10.4 % of the 60000 observations
digit 8 makes 9.8 % of the 60000 observations
digit 9 makes 9.9 % of the 60000 observations
```

Not every digit appears the same number of times in the training dataset. Whenever you build a dev and/or a test dataset, you should check that the data distributions are similar. Otherwise, when applying a model to the dev or test dataset, you could get a result that does not make much sense, since the model learned from a different class distribution. In this case for the sake of clarity we just reasoned on labels to see how the algorithm is working. In real life, you would need to split the features of course. Since the original distribution is almost uniform, you should expect a result that is very similar to the original one. Let's see how often (in percent) each digit appears in the test dataset

```
for i in range(10):
    print ('digit', i, 'makes', np.around(np.count_nonzero(y_test == i)
/10000.0*100.0, decimals = 1), '% of the 10000 observations')
```

This gives us the result

```
digit 0 makes 9.8 % of the 10000 observations
digit 1 makes 11.4 % of the 10000 observations
digit 2 makes 10.3 % of the 10000 observations
digit 3 makes 10.1 % of the 10000 observations
digit 4 makes 9.8 % of the 10000 observations
digit 5 makes 8.9 % of the 10000 observations
digit 6 makes 9.6 % of the 10000 observations
digit 7 makes 10.3 % of the 10000 observations
digit 8 makes 9.7 % of the 10000 observations
digit 9 makes 10.1 % of the 10000 observations
```

You can compare these results to the one from the entire dataset. You will notice that they are very close—not the same, but close enough. In this case, we can proceed without worries. But if this is not the case, be sure you have a similar distribution in every dataset you're going to use.

If the training, dev, and test datasets don't have the same distribution, this can be quite dangerous when checking how the model is doing. Your model may end up learning from a *unbalanced class distribution.*

Note We typically talk about an unbalanced class distribution in a dataset for a classification problem when one or more classes appear a different number of times than others. This becomes a problem in the learning process when the difference is significant. A few percent of difference is often not an issue.

If you have a dataset with three classes, for example, where you have 1000 observations in each class, then the dataset has a perfectly balanced class distribution, but if you have only 100 observations in class 1, 10,000 observations in class 2, and 5,000 in class 3, then this is an unbalanced class distribution. This is not a rare occurrence. Suppose you need to build a model that recognizes fraudulent credit card transactions. It is safe to assume that those transactions are a very small percentage of the entire number of transactions that you have at your disposal!

Tip When splitting your dataset, you must pay great attention not only to the number of observations you have in each dataset, but also to which observations go into each dataset. Note that this problem is not specific to deep learning but is important generally in machine learning.

Details on how to deal with unbalanced datasets are beyond the scope of this book, but is important to understand what kind of consequences this can have. In the next section, you learn what can happen if you feed an unbalanced dataset to a neural network. At the end of that section, we give you a few hints on what to do in such a case.

Unbalanced Class Distribution: What Can Happen

Since we are talking about how to split our dataset to perform metric analysis, it's important to grasp the concept of unbalanced class distribution and know how to deal with it. In deep learning, you will find yourself very often splitting datasets and you should be aware of the problems you may encounter if you do it the wrong way. Let's look at a concrete example of how bad things can go if you do it wrong.

We will use the MNIST dataset [5], and we will do basic logistic regression with a single neuron. The MNIST database is a large database of handwritten digits that we can use to train our model. The MNIST database contains 70,000 images.

"The original black and white (bilevel) images from MNIST were normalized to fit in a 20x20 pixel box while preserving their aspect ratio. The resulting images contain gray levels as a result of the anti-aliasing technique used by the normalization algorithm. The images were centered in a 28x28 image by computing the center of mass of the pixels and translating the image so as to position this point at the center of the 28x28 field" [5].

Our features will be the gray value for each pixel, so we will have 28x28 = 784 features whose values go from 0 to 255 (gray values). The dataset contains all ten digits, from 0 to 9. With the following code, you can prepare the data to use in the following sections. As usual, let's first import the necessary library.

```
# general libraries
import pandas as pd
import numpy as np
import matplotlib
import matplotlib.pyplot as plt
import matplotlib.font_manager as fm

# sklearn libraries
from sklearn.metrics import confusion_matrix

# tensorflow libraries
import tensorflow as tf
from tensorflow import keras
from tensorflow.keras import layers
import tensorflow_docs as tfdocs
import tensorflow_docs.modeling
```

First, we load the data

```
(x_train, y_train), (x_test, y_test) = keras.datasets.mnist.load_data()
```

It is useful to define a function to visualize the digits, to get an idea of how they look.

```
def plot_digit(some_digit):
  plt.imshow(some_digit, cmap = matplotlib.cm.binary, interpolation =
  "nearest")
  plt.axis("off")
  plt.show()
```

For example, we can plot one randomly (see Figure 10-7).

```
plot_digit(x_train[36003])
```

Figure 10-7. *The 36003rd digit in the dataset. It is easily recognizable as a 7*

Now that we have inspected the dataset a bit, here comes the important part. We create a new label. We assign to all observations for the digit 1 the label 0, and to all other digits (0, 2, 3, 4, 5, 6, 7, 8, and 9) the label 1 with this code

```
y_train_unbalanced = np.zeros_like(y_train)
y_train_unbalanced[np.any([y_train == 1], axis = 0)] = 0
y_train_unbalanced[np.any([y_train != 1], axis = 0)] = 1

y_test_unbalanced = np.zeros_like(y_test)
y_test_unbalanced[np.any([y_test == 1], axis = 0)] = 0
y_test_unbalanced[np.any([y_test != 1], axis = 0)] = 1
```

The `y_train_unbalanced` and `y_test_unbalanced` arrays will contain the new labels. Note that the dataset is now heavily unbalanced. Label 0 appears roughly 10% of the time, while label 1 appears 90% of the time.

We then reshape (to provide the data in the right dimension required by the neural network) and normalize the training and the test data

```
x_train_reshaped = x_train.reshape(60000, 784)
x_test_reshaped = x_test.reshape(10000, 784)

x_train_normalised = x_train_reshaped/255.0
x_test_normalised = x_test_reshaped/255.0
```

Then we build our network with a single neuron

```
def build_model():

  # one unit as network's output
  # sigmoid function as activation function
```

```
  # sequential groups a linear stack of layers into a
  # tf.keras.Model
  # activation parameter: if you don't specify anything, no
  # activation
  # is applied (i.e. "linear" activation: a(x) = x).
  model = keras.Sequential([
    layers.Dense(1,
                 input_shape = [len(x_train_normalised[0])],
                 activation = 'sigmoid')
  ])

  # optimizer that implements the Gradient Descent algorithm
  optimizer = tf.keras.optimizers.SGD(momentum = 0.0,
                                      learning_rate = 0.0001)
  # the compile() method takes a metrics argument, which can be a list
  of metrics
  # loss = cross-entropy, metrics = accuracy,
  model.compile(loss = 'binary_crossentropy',
                optimizer = optimizer,
                metrics = ['binary_crossentropy',
                           'binary_accuracy'])

  return model
```

At this point in the book, you should understand this simple model easily, as you have seen it several times. Now we define the function to run the model

```
model = build_model()
```

Let's run the model with the code

```
EPOCHS = 50

history = model.fit(
  x_train_normalised, y_train_unbalanced,
  epochs = EPOCHS, verbose = 0, batch_size = 1000,
  callbacks = [tfdocs.modeling.EpochDots()])
```

And check the accuracy with the code

```
hist = pd.DataFrame(history.history)
hist['epoch'] = history.epoch
hist.tail()
```

	loss	binary_crossentropy	binary_accuracy	epoch
45	0.292092	0.292092	0.887617	45
46	0.290531	0.290531	0.887633	46
47	0.289016	0.289016	0.887633	47
48	0.287542	0.287542	0.887633	48
49	0.286109	0.286109	0.887633	49

You get a very good 88.1% accuracy. Not bad right? But are you sure that the result is that good? Now let's check the confusion matrix[1] for our labels with the code

```
train_predictions = model.predict(x_train_normalised).flatten()
confusion_matrix(y_train_unbalanced, train_predictions > 0.5)
```

When you run the code, you get the following result

```
array([[     0,  6742],
       [     0, 53258]])
```

Slightly better formatted and with some explanatory information, the matrix looks like Table 10-1.

Table 10-1. *Confusion Matrix for the Model Described in the Text*

	Predicted Class 0	Predicated Class 1
Real class 0	0	6742
Real class 1	0	53258

[1] The confusion matrix is, in machine learning classification, a matrix where each column of the matrix represents the number of instances in a predicted class, while each row represents the number of instances in an actual class.

How should we read this table? In the "Predicted Class 0" column, you will see the number of observations that our model predicts of being of class 0 for each real class. 0 is the number of observations our model predicts of being of class 0 and are really in class 0. 0 is the number of observations that our model predicts in class 0 but are really in class 1.

It should be easy to see that our model effectively predicts all observations to be in class 1 (a total of 6742+53258 = 60000). Since we have a total of 60,000 observations in our training set, we get an accuracy of 53258/60000 = 0.887, as our Keras code told us. But not because our model is good, simply because it has effectively classified all observations in class 1. We don't need a neural network in this case to reach this accuracy. What happens is that our model sees observations belonging to class 0 so rarely that they almost don't influence the learning, which is dominated by the observations in class 1.

What at the beginning seemed a nice result, turns out is a really bad one. This is an example of how badly things can go if you don't pay attention to the distributions of your classes. This of course applies not only when splitting your dataset, but in general when you approach a classification problem, regardless of the classifier you want to train (it does not apply only to neural networks).

Tip When splitting your dataset into complex problems, you need to pay close attention not only to the number of observations you have in your datasets, but also to what observations you choose and to the distribution of the classes.

To conclude this section, let me give you a few hints about how to deal with unbalanced datasets:

- **Change your metric:** In our example, you may want to use something else instead of accuracy, since it can be misleading. You could try using the confusion matrix for example, or other metrics like precision, recall, or F1 (review them if you don't know them, as they are very important to know). Another important way of checking how your model is doing, and one that I suggest you learn, is the ROC Curve, which will help you tremendously.

- **Work with an undersampled dataset:** If you have for example 1000 observations in class 1 and 100 in class 2, you may create a new dataset with 100 random observations in class 1 and the 100 you have in class 2. The problem with this method is that you will have a lot less data to feed to your model to train it.
- **Work with an oversampled dataset:** You may try to do the opposite. You may take the 100 observations in class 2 and replicate them ten times to end up with 1000 observations in class 2 (sometimes called sampling with replacement).
- **Get more data in the class with less observations:** This is not always possible. In the case of fraudulent credit card transactions, you cannot go around and generate new data, unless you want to go to jail.

Datasets with Different Distributions

Now I want to discuss another terminology issue, which will lead us to understanding a common problem in the deep learning world. Very often you will hear sentences like: "the sets come from different distributions." This sentence is not always easy to understand. Take for example two datasets formed by images taken with a professional DSLR, and a second one created by images taken with an old smartphone. In the deep learning world, we say that those two sets come from different distributions. But what is the meaning of the sentence? The two datasets differ for various reasons: resolution of images, blurriness due to different quality of lens, amount of colors, how much is in focus, and possibly more. All those differences are what is usually meant by different "distributions."

Let's look at another example. We could consider two datasets: one made of images of white cats and one made of images of black cats. Also, in this case we talk about different distributions. This becomes a problem when you train a model on one set and want to apply it to the other. For example, if you train a model on a set of images of white cats, you probably are not going to do very good on the dataset of black cats, since your model has never seen black cats during training.

Note When talking about datasets coming from different distributions, it is usually meant that the observations have different characteristics in the two datasets: black and white cats, high- and low-resolution images, speech recording in Italian and in German, and so on.

Since data is precious, people often try to create the different datasets (train, dev, etc.) from different sources. For example, you may decide to train your model on a set made of images taken from the web and see how good it is on a set made of images you took with your smartphone. This may seem a nice idea to be able to use as much data as possible, but it may give you many headaches. Let's see what happens in a real case so that you see the consequences of doing something like this.

Let's consider the subset of the MNIST dataset made of the two digits: 1 and 2. We will build a dev dataset coming from a different distribution, shifting a subset of the images ten pixels to the right. We will train our model on the images as they are in the original dataset and apply the model to images shifted ten pixels to the right. Let's first load the data

```
(x_train, y_train), (x_test, y_test) = keras.datasets.mnist.load_data()
```

First select only digits 1 and 2

```
x_train_12 = x_train[np.any([y_train == 1, y_train == 2], axis = 0)]
x_test_12 = x_test[np.any([y_test == 1, y_test == 2], axis = 0)]
y_train_12 = y_train[np.any([y_train == 1, y_train == 2], axis = 0)]
y_test_12 = y_test[np.any([y_test == 1, y_test == 2], axis = 0)]
```

We have 12,700 observations in the training dataset and 2167 in the test dataset (we will call it dev dataset from now on in this chapter).

Then we normalize the features

```
x_train_normalised = x_train_12/255.0
x_test_normalised = x_test_12/255.0
```

And then we transform the matrixes to give them the right dimensions

```
x_train_normalised = x_train_normalised.reshape(x_train_normalised.
shape[0], 784)
x_test_normalised = x_test_normalised.reshape(x_test_normalised.
shape[0], 784)
```

And finally, we shift the labels to have 0 and 1

```
y_train_bin = y_train_12 - 1
y_test_bin = y_test_12 - 1
```

We can check the sizes of the arrays with this code

```
print(x_train_normalised.shape)
print(x_test_normalised.shape)
```

That gives us

```
(12700, 784)
(2167, 784)
```

We have 12,700 observations in our training set and 2167 in the dev set. Now let's duplicate the dev dataset and shift each image to the right ten pixels. We can do this quickly with this code

```
x_test_shifted = np.zeros_like(x_test_normalised)
for i in range(x_test_normalised.shape[0]):
    tmp = x_test_normalised[i,:].reshape(28,28)
    tmp_shifted = np.zeros_like(tmp)
    tmp_shifted[:,10:28] = tmp[:,0:18]
    x_test_shifted[i,:] = tmp_shifted.reshape(784)
```

To make the shift easy, we first reshaped the images in a 28x28 matrix, then simply shifted the columns with `tmp_shifted[:,10:28] = tmp[:,0:18]`. Then we simply reshape the images in a one-dimensional array of 784 elements. The labels remain the same. Figure 10-8 shows a random image from the dev dataset on the left and its shifted version on the right.

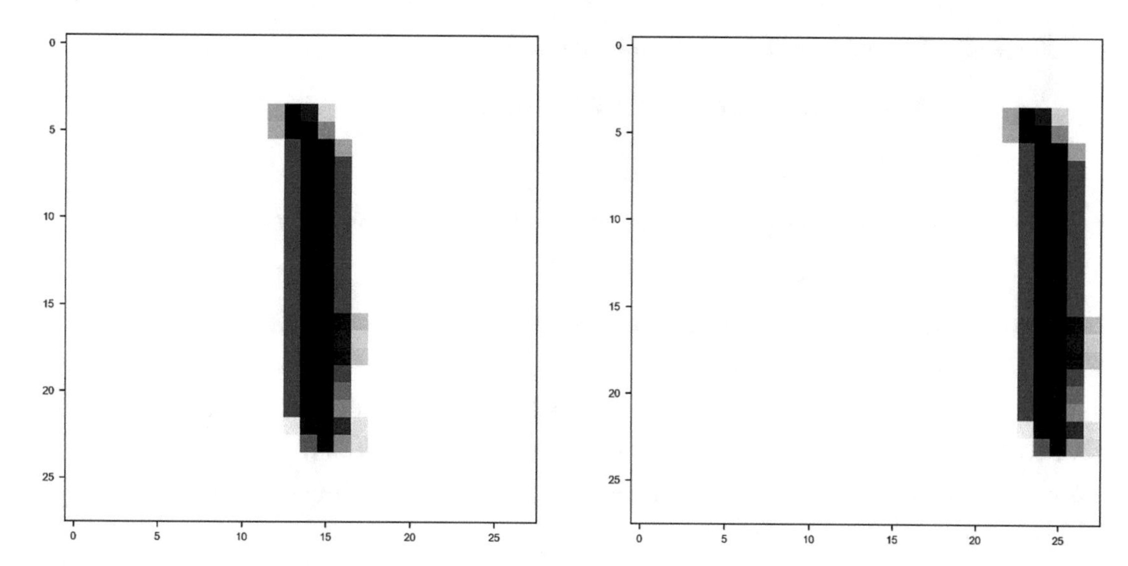

***Figure 10-8.** One random image from the dataset (on the left) and its shifted version (on the right)*

Now let's build a network with a single neuron and See what happens.

```
model = build_model()
EPOCHS = 100

history = model.fit(
  x_train_normalised, y_train_bin,
  epochs = EPOCHS, verbose = 0,
  callbacks = [tfdocs.modeling.EpochDots()])
```

This gives us the output

```
Epoch: 0, binary_accuracy:0.5609,  binary_
crossentropy:0.6886,  loss:0.6886,
.............................................................
```

Let's calculate the accuracy for the three datasets: x_train_normalised, x_test_normalised, and x_test_shifted with the code

```
_, _, train_accuracy = model.evaluate(x_train_normalised, y_train_bin)
_, _, test_accuracy = model.evaluate(x_test_normalised, y_test_bin)
_, _, shifted_test_accuracy = model.evaluate(x_test_shifted, y_test_bin)
```

We get the following results after 100 epochs

- For the training dataset, we get 97%.
- For the dev dataset, we get 98%.
- For the train-dev (you will see later why it's called this), the one with the shifted images, we get 54%. A very bad result.

What has happened is that the model has learned from a dataset where all the images are centered in the box and therefore could not generalize images shifted and not centered.

When training a model on a dataset, you will get good results usually on observations that are like the ones in the training set. But how can you find out if you have such a problem? There is a relatively easy way of doing that, by expanding the MAD diagram. Let's see how to do it.

Suppose you have a training dataset and a dev dataset where the observations have different characteristics (they are coming from different distributions). You create a small subset from the training set, called the train-dev dataset, and end up with three datasets—a training and a train-dev coming from the same distribution (the observations have the same characteristics) and a dev set, where the observations are somehow different, as we discussed previously. Then you train your model on your training set, and evaluate your error ϵ on the three datasets: ϵ_{train}, ϵ_{dev} and $\epsilon_{train-dev}$. If your train and dev sets are coming from the same distributions, so is the train-dev set. In this case you should expect $\epsilon_{dev} \approx \epsilon_{train-dev}$. If we define

$$\Delta\epsilon_{train-dev} = \left|\epsilon_{dev} - \epsilon_{train-dev}\right|$$

we should expect $\Delta\epsilon_{train-dev} \approx 0$. If the train (and train-dev) and the dev set are coming from different distributions (the observations have different characteristics), we should expect $\Delta\epsilon_{train-dev}$ to be big. If we consider the MNIST example, we have in fact $\Delta\epsilon_{train-dev} = 0.46$ or 46%, which is a huge difference.

Let's recap what you should do to find out if your training and your dev (or test) dataset have observations with different characteristics (they are coming from different observations):

- Split your training set in two: one that you will use for training, and that you will call the train set, and a smaller one that you will call "train-dev" set.
- Train your model on the train set.

- Evaluate your error ϵ on the three sets: train, dev, and train-dev.
- Calculate the quantity $\Delta\epsilon_{train-dev}$. If it's big this will give strong evidence that the original training and dev sets are coming from different distributions.

Figure 10-9 is an example of the MAD diagram with the problem we just discussed. Don't look at the numbers; they are just for illustrative purposes (read: I just put them there).

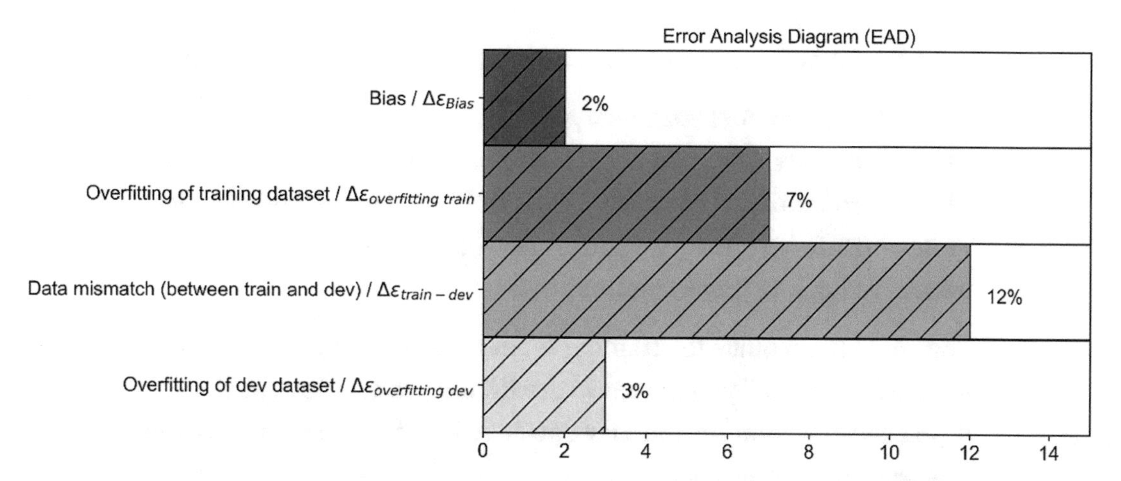

Figure 10-9. *An example of the MAD diagram with the data mismatch problem added*

The MAD diagram in Figure 10-9 can tell us the following things (I highlight only a few ideas; for a more complete list of things you can reread the previous sections):

- The bias (between training and human-level performance) is quite small, so we are not that far from the best we can achieve (let's assume that human-level performance is a proxy for the Bayes error). You could try bigger networks, better optimizers, and so on.
- We are overfitting the datasets, so we could try regularization or get more data.
- We have a strong problem with data mismatch (sets coming from different distributions) between train and dev. You will see what you could do to solve this problem later in the chapter.
- We are also slightly overfitting the dev dataset during our hyper-parameter search.

Note that you don't need to create the bar plot as we have done here. Technically, you just need the four numbers to draw the same conclusions.

Tip Once you have your MAD diagram (or simply the numbers), interpreting it will give you hints about what you should try to get better results, for example a better accuracy.

You can try the following techniques to address data mismatch between sets:

- You can carry manual error analysis to understand the difference between the sets, and then decide what to do (you see an example in the last section of this chapter). This is time consuming and usually quite difficult, because once you know what the difference is, it may be very difficult to find a solution.
- You could try to make the training set more similar to your dev/test sets; for example, if you are working with images and the test/dev sets have a lower resolution you may decide to lower the resolution of the images in the training set.

As usual, there are no fixed rules. Just be aware of the problem and think about the following: your model will learn the characteristics from your training data, so when it's applied to completely different data, it won't usually do well. Always get training data that reflects the data you want your model to work on, not vice versa.

k-fold Cross Validation

This technique is very powerful and should be known by any machine learning practitioner (not only in the deep learning world): k-fold cross validation. The technique is a way of finding a solution to the following two problems:

- What do you do when your dataset is too small to split into a train and dev/test set?
- How do you get information on the variance of your metric?

Let's describe the idea with pseudo-code.

1. Partition your complete dataset in k equally big subsets: $f_1, f_2, \ldots, f_k$. The subsets are also called folds. Normally the subsets are not overlapping, which means that each observation appears in one and only one fold.

2. For i going from 1 to *k*:

   – Train your model on all the folds except f_i.

   – Evaluate your metric on the fold f_i. The fold f_i will be the dev set in iteration *i*.

3. Evaluate the average and variance of your metric on the *k* results.

A typical value for *k* is 10, but that depends on the size of your dataset and on the characteristic of your problem.

Remember that the discussion about how to split a dataset also applies here.

Note When you are creating your folds, you must take care that your folds reflect the structure of your original dataset. If your original dataset has ten classes for example, you must make sure that each of your folds have all the ten classes with the same proportions.

Although this may seem a very attractive technique to deal generally with datasets with less than optimal size, it may be quite complex to implement. But, as you will see shortly, checking your metric on the different folds will give you important information on a possible overfitting of your training dataset.

Let's try this on a real dataset and see how to implement it. Note that you can implement k-fold cross validation easily in `sklearn`, but I will develop it from scratch, to show you what is happening in the background. Everyone (well, almost) can copy code from the web to implement k-fold cross validation in `sklearn`, but not many can explain how it works and understand it, and therefore be able to choose the right `sklearn` method or parameters.

As a dataset we will use the reduced MNIST dataset containing only digits 1 and 2. We will do a simple logistic regression with one neuron to make the code easy to understand and to let you concentrate on the cross-validation part and not on other implementation details that are not relevant here. The goal of this section is to show you how k-fold cross validation works and why it's useful, not to implement it with the smallest amount of code possible.

As before, import the MNIST dataset. Remember that the dataset has 70,000 observations and is made of grayscale images, each 28x28 pixel in size. Then select only digit 1 and 2 and rescale the labels to make sure that digit 1 has label 0 and digit 2 has label 1.

Now we need a small trick. To keep the code simple, we will consider only the training dataset and we will create folds starting from it.

Now let's create ten arrays, each containing a list of indexes that we will use to select images

```
foldnumber = 10
idx = np.arange(0, x_train_12.shape[0])
np.random.shuffle(idx)
al = np.array_split(idx, foldnumber)
```

In each fold, we will have 1270 images (12700/10, 12700 is the total number of observations in the training dataset that have label 1 or 2). Now let's create the arrays containing the images

```
x_train_inputfold = []
y_train_inputfold = []
for i in range(foldnumber):
    tmp = x_train_reshaped[al[i],:]
    x_train_inputfold.append(tmp)
    ytmp = y_train_bin[al[i]]
    y_train_inputfold.append(ytmp)

x_train_inputfold = np.asarray(x_train_inputfold)
y_train_inputfold = np.asarray(y_train_inputfold)
```

If you think this code is convoluted, you are right. There are faster ways of doing it with `sklearn`, but it's very instructive to see how to do it manually step by step. We first create empty lists: `x_train_inputfold` and `y_train_inputfold`. Each element of the list will be a fold, so an array of images or of labels. So, if we want to get all images in fold 2, we simply use `inputfold x_train_inputfold[1]` (remember in Python, indexes start from 0). Those list, converted to the last two lines in NumPy arrays, will have three dimensions, as you can easily see with these statements

```
print(x_train_inputfold.shape)
print(y_train_inputfold.shape)
```

That gives us

```
(10, 1270, 784)
(10, 1270)
```

In `x_train_inputfold` the first dimension indicates the fold number, the second the observation, and the third the gray values of the pixels. In `y_train_inputfold` the first dimension indicates the fold number and the second the label. For example, to get image with index 1234 from fold 0, you would need to use the code

```
x_train_inputfold[0][1234,:]
```

Remember you should check that you still have a balanced dataset in each fold, or in other words that you have as many 1s as 2s. Let's check for fold 0 (you can do the same check for the others):

```
total = 0
for i in range(0,2,1):
    print ("digit", i, "makes", np.around(np.count_nonzero(y_train_
    inputfold[0] == i)/1270.0*100.0, decimals=1), "% of the 1270 observations")
```

That gives us

```
digit 0 makes 51.6 % of the 1270 observations
digit 1 makes 48.4 % of the 1270 observations
```

For our purposes, this is balanced enough. Now we need to normalize the features

```
x_train_inputfold_normalized = np.zeros_like(x_train_inputfold, dtype
= float)
for i in range (foldnumber):
    x_train_inputfold_normalized[i] = x_train_inputfold[i]/255.0
```

You could normalize the data in one shot, but we like to make it evident that we are dealing with folds to make it clear for the reader.

Now we are ready to build our network. We will use a one-neuron network for logistic regression, with the sigmoid activation function. At this point we will need to iterate through the folds. Remember our pseudo-code at the beginning? Select one fold as the dev set and train the model on all other folds concatenated. Proceed in this way for all the folds. The code could look like this (it's a bit long, so take a few minutes to understand it). The code includes comments indicating which step we are talking about, since you will find a numbered list of the steps for explanation next.

```
train_acc = []
dev_acc = []

for i in range(foldnumber): # STEP 1

    # Prepare the folds - STEP 2
    lis = []
    ylis = []
    for k in np.delete(np.arange(foldnumber), i):
        lis.append(X_train[k])
        ylis.append(y_train[k])
        X_train_ = np.concatenate(lis, axis = 0)
        y_train_ = np.concatenate(ylis, axis = 0)
    X_train_ = np.asarray(X_train_)
    y_train_ = np.asarray(y_train_)

    X_dev_ = X_train[i]
    y_dev_ = y_train[i]

    # STEP 3
    print('Dev fold is', i)

    model = build_model()

    EPOCHS = 500

    history = model.fit(
      X_train_, y_train_,
      epochs = EPOCHS, verbose = 0,
      callbacks = [tfdocs.modeling.EpochDots()])
```

```
# STEP 4
_, _, train_accuracy = model.evaluate(X_train_, y_train_)
print('Dev accuracy:', int(train_accuracy*100), '%.')
train_acc = np.append(train_acc, train_accuracy)

_, _, dev_accuracy = model.evaluate(X_dev_, y_dev_)
print('Dev accuracy:', int(dev_accuracy*100), '%.')
dev_acc = np.append(dev_acc, dev_accuracy)
```

The code follows these steps:

1. Do a loop over all the folds (in this case from 1 to 10) iterating with the variable `i` from 0 to 9.
2. For each `i`, use the fold `i` as the dev set and concatenate all the other folds and use the result as the train set.
3. For each `i`, train the model.
4. For each `i`, evaluate the accuracy on the two datasets (train and dev) and save the values in the two lists: `train_acc` and `dev_acc`.

If you run this code, you will get output that looks like this for each fold (you will get ten times the following output, once for each fold):

```
Dev fold is 0

Epoch: 0, binary_accuracy:0.4630,  binary_
crossentropy:0.8690,  loss:0.8690,
..............................................................
Epoch: 100, binary_accuracy:0.9719,  binary_
crossentropy:0.1328,  loss:0.1328,
..............................................................
Epoch: 200, binary_accuracy:0.9801,  binary_
crossentropy:0.0941,  loss:0.0941,
..............................................................
Epoch: 300, binary_accuracy:0.9826,  binary_
crossentropy:0.0785,  loss:0.0785,
..............................................................
```

```
Epoch: 400, binary_accuracy:0.9840,  binary_
crossentropy:0.0697,  loss:0.0697,
358/358 [==============================] - 1s 1ms/step - loss: 0.0640 -
binary_crossentropy: 0.0640 - binary_accuracy: 0.9852
Dev accuracy: 98 %.
40/40 [==============================] - 0s 2ms/step - loss: 0.0576 -
binary_crossentropy: 0.0576 - binary_accuracy: 0.9882
Dev accuracy: 98 %.
```

Notice that you will get slightly different accuracy values for each fold. It is very instructive to study how the accuracy values are distributed. Since we have ten folds, we have ten values to study. In Figure 10-10, you can see the distribution of the values for the train set (left plot) and for the dev set (right plot).

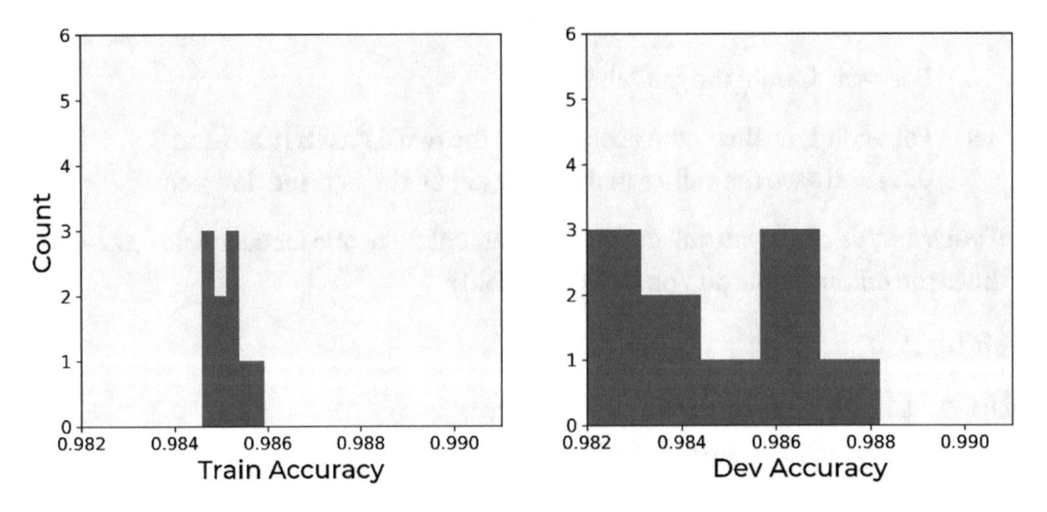

Figure 10-10. *Distribution of the accuracy values for the train set (left plot) and for the dev set (right plot). Note that the two plots use the same scale on both axes*

The image is quite instructive. You can see that the accuracy values for the training set are concentrated around the average, while the ones evaluated on the dev set are much more spread out! This shows how the model on new data behaves less well than on the data it has trained on. The standard deviation for the training data is $3.5 \cdot 10^{-4}$ and for the dev set it's $3.5 \cdot 10^{-3}$, ten times bigger than the value on the train set. In this way you also get an estimate of the variance of your metric when applied to new data, and to how it generalizes.

If you are interested in learning how to do this quickly with `sklearn`, check out the official documentation for the `KFold` method [6]. When you are dealing with datasets with many classes (remember our discussion about how to split your sets?), you must pay attention and do what is called *stratified sampling*. `Sklearn` provides a method to do that too: `stratifiedKFold` [7].

You can now easily find averages and standard deviations. For the training set, we have an average accuracy of 98.5% and a standard deviation of 0.035%, while for the dev set we have an average accuracy of 98.4% with a standard deviation of 0.35%. Now you can even provide an estimate of the variance of your metric. Pretty cool!

Manual Metric Analysis: An Example

We mentioned earlier that sometimes it's useful to do a manual analysis of your data, to check if the results (or the errors) you are getting are plausible. This section has a basic example to shows you a concrete idea of what is meant by this and how complicated it can be. Let's consider the following question: our very simple model (remember we are using only one neuron) can get 98% of accuracy! Is the problem of recognizing digits that easy? Let's try to see if that is the case. First of all, note that our training set does not even have the two-dimensional information about the images. If you remember, each image is converted into a one-dimensional array of values: the gray values of each pixel, starting on the top-left and going row by row from top to bottom. Are the 1s and the 2s easy to recognize? Let's see how the real input for our model looks and start analyzing the digit 1. Let's take an example from fold 0. In Figure 10-11, you can see the image on the left and a bar plot of the gray values of the 784 pixels as they are seen from our model. Remember that as observations, we have a one-dimensional array of the 784 gray values of the pixels in the image.

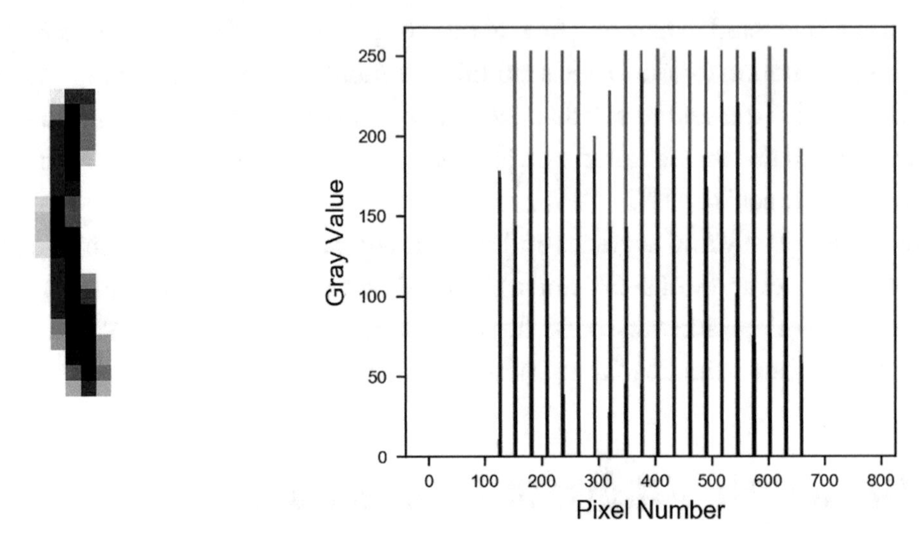

Figure 10-11. *An example from fold 0 for the digit 1. The image on the left and a bar plot of the gray values of the 784 pixels as they are seen from our model on the right. Remember that as inputs we have a one-dimensional array of the 784 gray values of the pixels*

Remember that we reshape our 28x28 image into a one-dimensional array, so when reshaping the digit 1 in Figure 10-11, we find black points roughly each 28 pixels, since the 1 is almost a vertical column of black points. In Figure 10-12, you can see other 1s, and you will notice how, when reshaped as one dimensional, they look all the same: several bars roughly equally spaced. Now that you know what to look for, we can easily say that all the images in Figure 10-12 are of the digit 1.

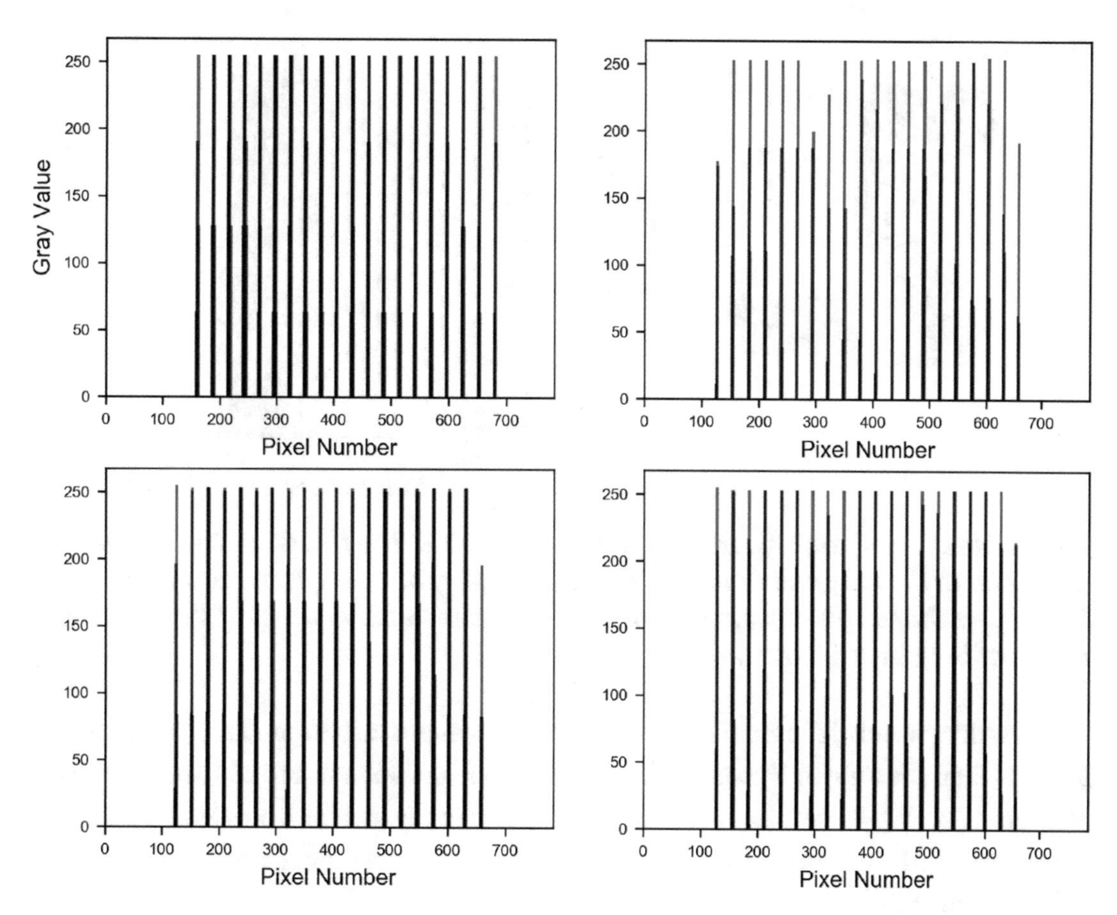

***Figure 10-12.** Four examples of the digit 1, reshaped as one-dimensional arrays. They all look the same: a number of bars roughly equally spaced*

Now let's look at the digit 2. In Figure 10-13, you can see an example, similar to what we had in Figure 10-11.

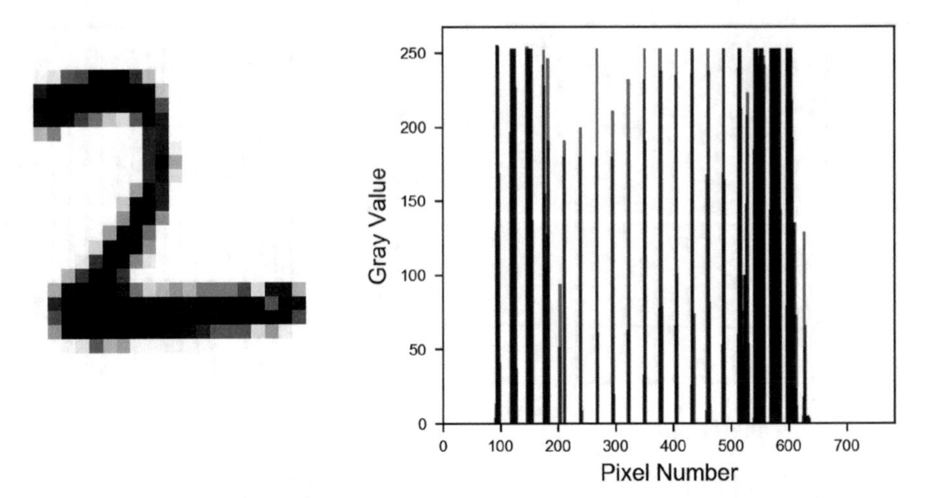

***Figure 10-13.** An example from fold 0 for the digit 2. The image on the left and a bar plot of the gray values of the 784 pixels as they are seen from our model. Remember that as observations we have a one-dimensional array of the 784 gray values of the pixels of the image*

Now things look different. We have two regions where the bars are much more dense in the plot on the right in Figure 10-13. This is between pixels 100 and 200 and especially after pixel 500. Why? Well, the two areas correspond to the two horizontal parts of the image. Figure 10-14 highlights how different parts look when they're reshaped as a one-dimensional array.

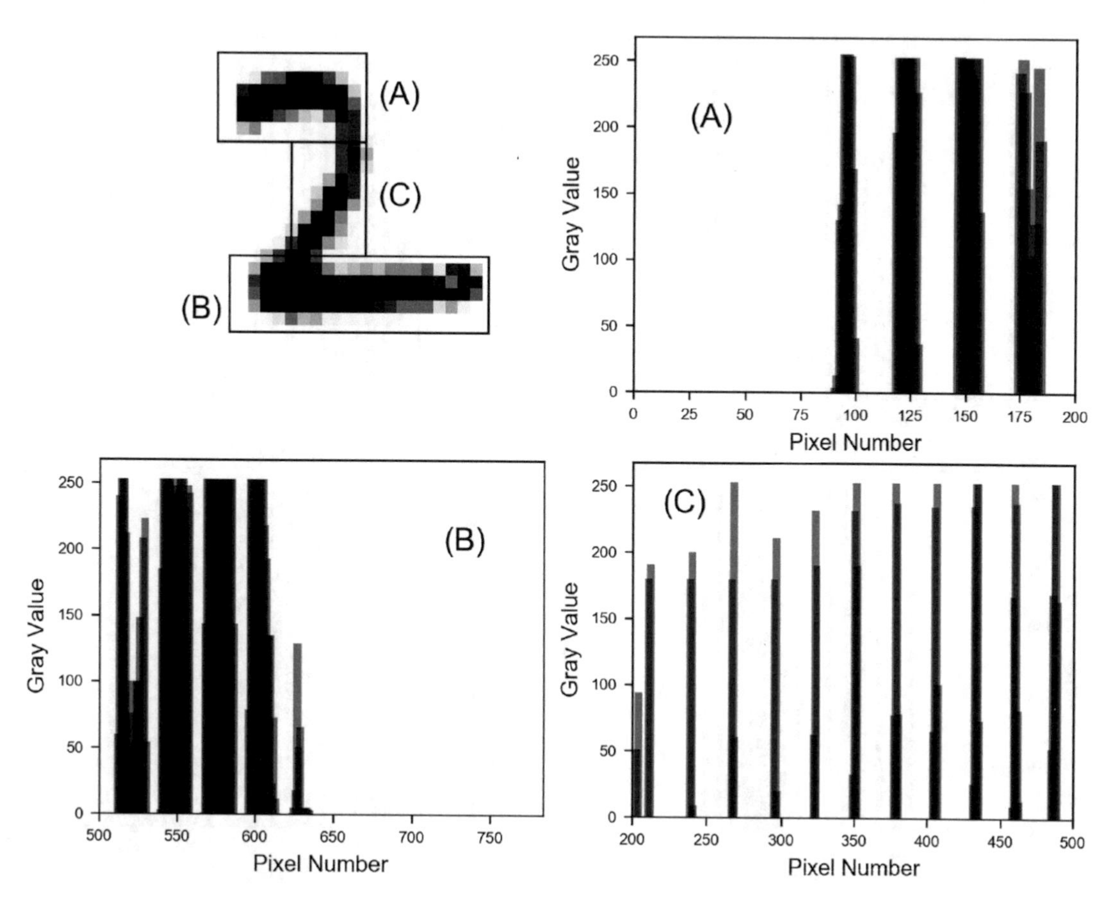

***Figure 10-14.** How different parts of the images look when they're reshaped as a one-dimensional array. Horizontal parts are labelled as (A) and (B), while the more vertical part is labelled as (C)*

Horizontal parts (A) and (B) are clearly different than part (C) when reshaped as a one-dimensional array. The vertical part (C) looks like the digit 1, with many equally spaced bars, as you can check in the lower-right bar plot labelled (C). While the more horizontal parts appear as many bars clustered together in groups, as can be seen in the upper-right and lower-left bar plots labelled (A) and (B). So, when reshaped, if you find those cluster of bars, that means you are looking at a 2. If you see only equally spaced small groups of bars, as in the plot (C) in Figure 10-14, you are looking at a 1. You don't even need to see the two-dimensional image if you know what to look for. Note that this pattern is constant. Figure 10-15 shows four examples of the digit 2, and you can clearly see the wider clusters of bars.

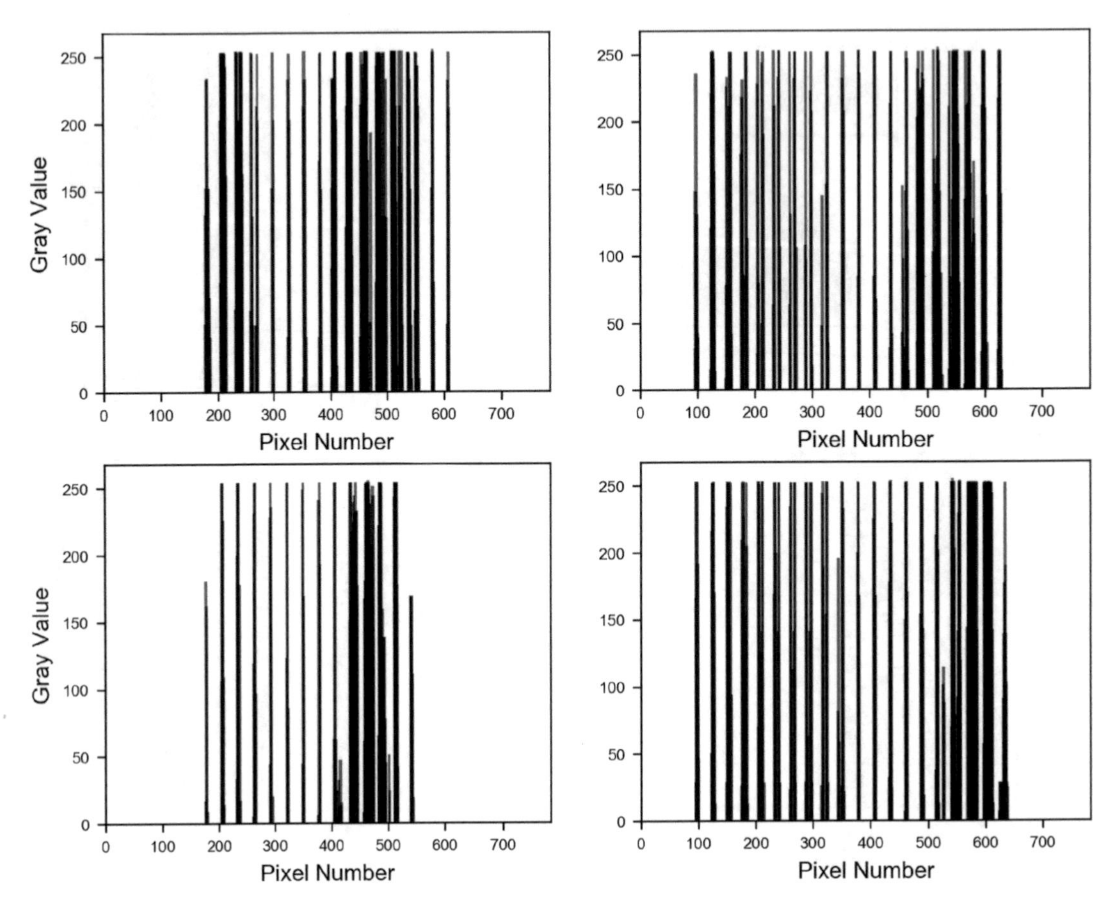

Figure 10-15. *Four examples of the digit 2 reshaped as one-dimensional arrays. The wider clusters of bars can clearly be seen*

As you can imagine, this is an easy pattern to spot for an algorithm, and so it's to be expected that this model works so good. Even a human can spot the images, even when reshaped, without any effort. Such a detailed analysis would not be necessary in a real-life project, but it's really instructive to see what you can learn from your data. Understanding the characteristics of your data may help you design your model or understand why is not working. Advanced architectures, like convolutional networks, will be able to learn exactly those two-dimensional features in a very efficient way.

Let's also check how the network learned to recognize digits. You will remember that the output of our neuron is

$$\hat{y} = \sigma(z) = \sigma\left(w_1 x_1 + w_2 x_2 + \ldots + w_{n_x} x_{n_x} + b\right)$$

Where σ is the sigmoid function, x_i for $i = 1, \ldots, 784$ are the gray values of the pixel of the image, w_i for $i = 1, \ldots, 784$ are the weights and b is the bias. Remember that when $\hat{y} > 0.5$ we classify the image in class 1 (so digit 2), and if $\hat{y} < 0.5$ we classify the image in class 0 (so digit 1). Now if you remember our discussion in Chapter 2 of the sigmoid function you will remember that $\sigma(z) \geq 0.5$ when $z \geq 0$ and $\sigma(z) < 0.5$ for $z < 0$. That means that our network should learn the weights in such a way that for all the 1s, we have $z < 0$ and for all the 2s, $z \geq 0$. Let's see if that is really the case.

In Figure 10-16, you can see a plot for a digit 1 where you can find the weights w_i (as a solid line) of our trained network after 600 epochs (and after reaching an accuracy of 98%) and the gray value of the pixel x_i rescaled to have a maximum of 0.5 (as a dashed line). Look how each time x_i is big, then w_i is negative. And when $w_i > 0$ the x_i are almost zero. Clearly the result $w_1 x_1 + w_2 x_2 + \ldots + w_{n_x} x_{n_x} + b$ will be negative, and therefore $\sigma(z) < 0.5$ and the network will identify the image as a 1. Figure 10-16 is zoomed in to make this behavior more evident.

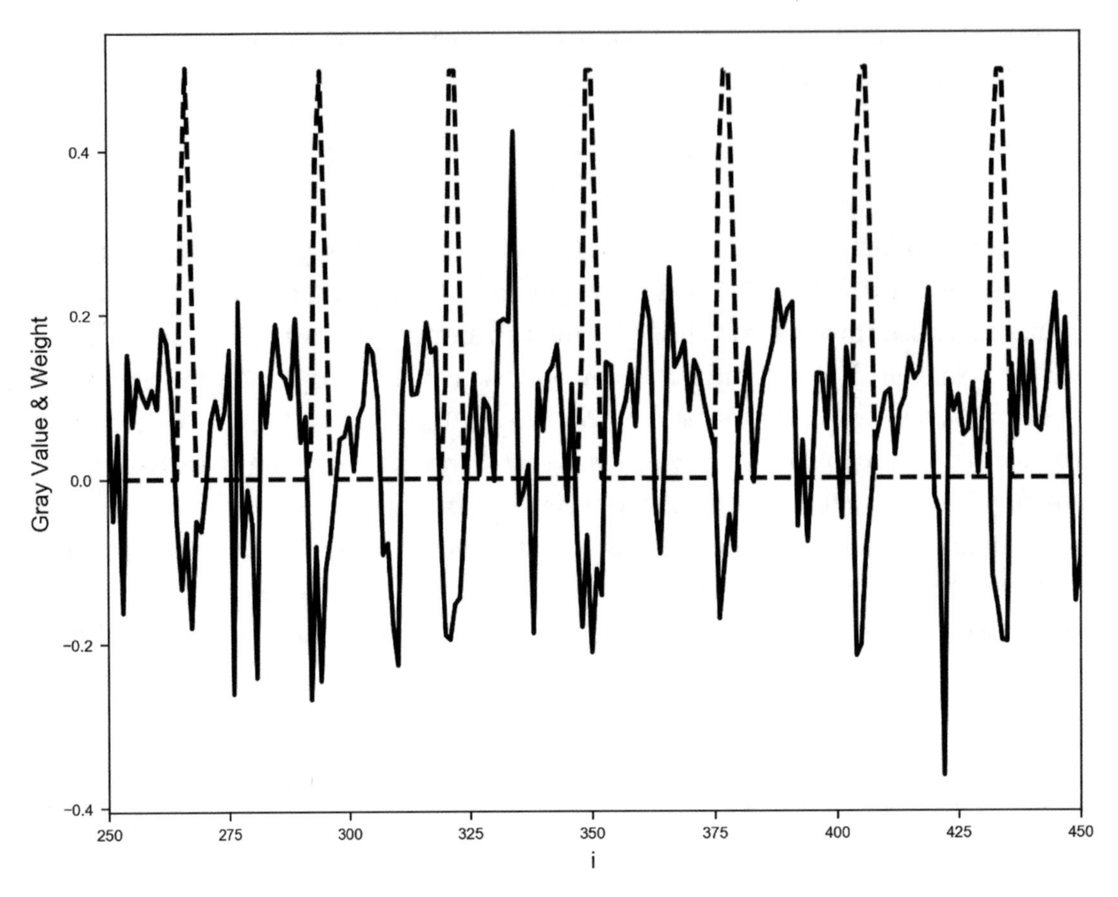

Figure 10-16. *Plot for a digit 1 where you can find the weights w_i (the solid lines) of our trained network after 600 epochs (and after reaching an accuracy of 98%) and the gray value of the pixel x_i rescaled to have a maximum of 0.5 (the dashed lines)*

In Figure 10-17, you can see the same plot for a digit 2. You will remember from our previous discussion that, for a 2, we can see many bars clustered together in groups up to pixel 250 (roughly). Let's look at the weights in that region. Where the pixel gray values are big, the weights are positive, giving then a positive value of z and therefore $\sigma(z) \geq 0.5$. That means the image would be classified as a 2.

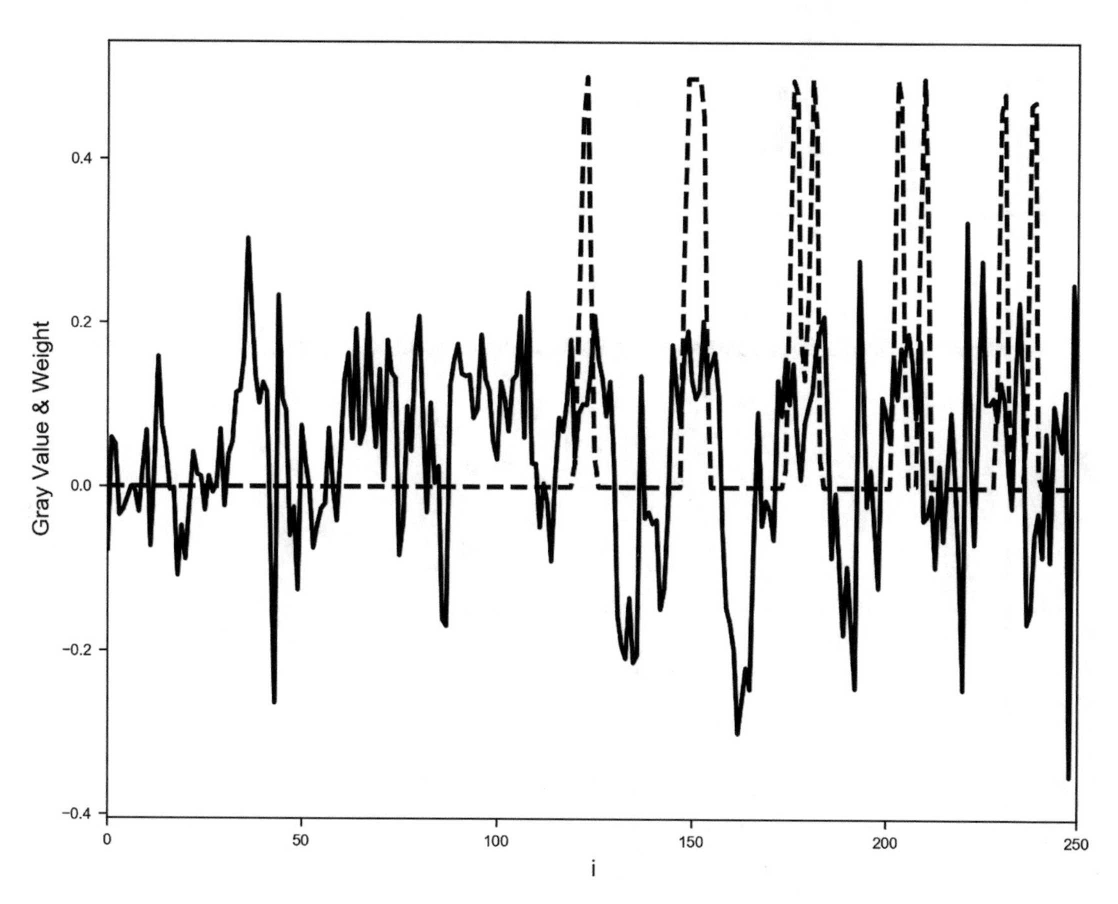

Figure 10-17. *Plot for a digit 2 where you can find the weights w_i (solid lines) of our trained network after 600 epochs (and after reaching an accuracy of 98%) and the gray value of the pixel x_i rescaled to have a maximum of 0.5 (dashed lines)*

As an additional check, Figure 10-18 plots $w_i \cdot x_i$ for all values of i for a digit 1. You can see how almost all the points lie below zero. Note that $b = -0.16$ in this case as well.

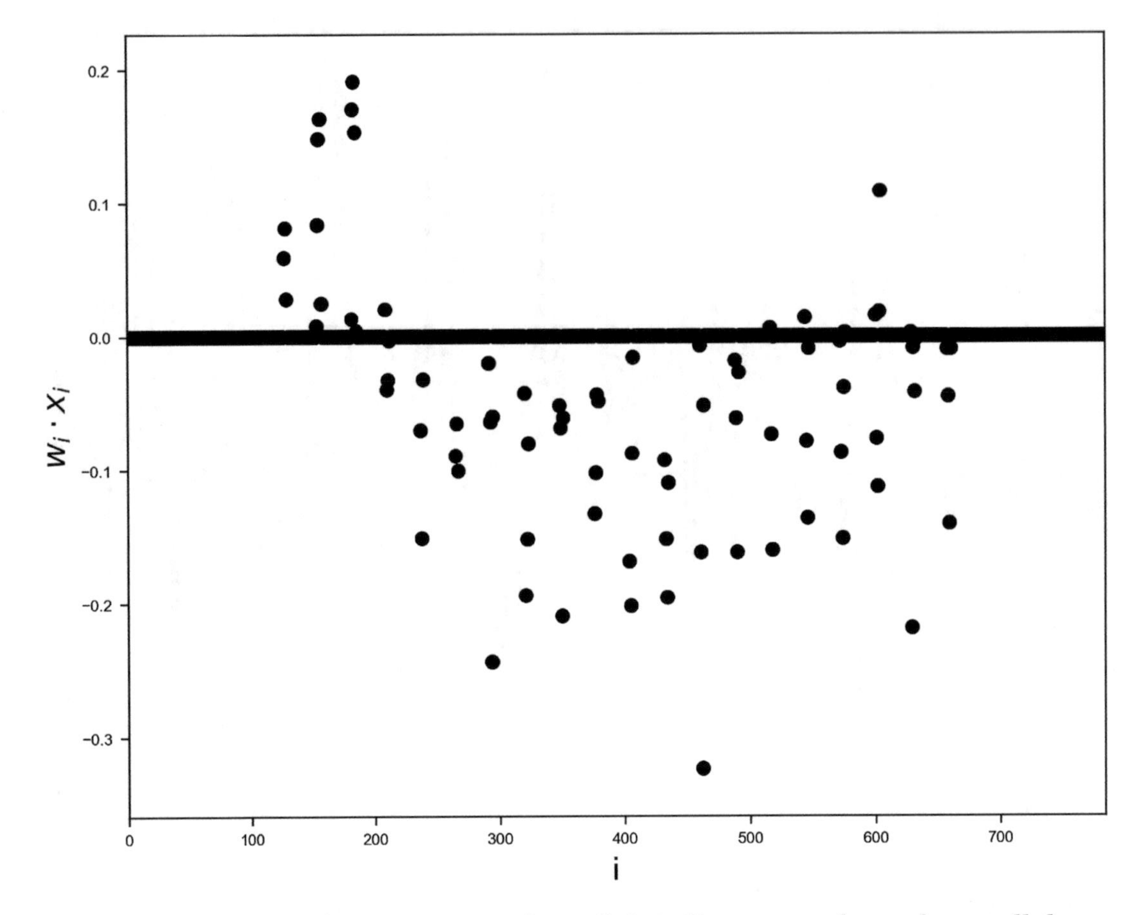

Figure 10-18. *$w_i \cdot x_i$ for i = 1, ..., 784 for a digit 1. You can see how almost all the values lie below zero. The thick line at zero is made of all the points i such that $w_i \cdot x_i = 0$*

As you can see, in very easy cases it is possible to understand how a network learns and therefore it would be much easier to debug strange behaviors. But don't expect this to be possible when dealing with more complex cases. The analysis we did here would not be so easy, for example, if you tried to do the same with digits 3 and 8 instead of 1 and 2.

Exercises

EXERCISE 1 (LEVEL: EASY)

Perform k-fold cross-validation using the built-in function of the `sklearn` library and compare the results.

EXERCISE 2 (LEVEL: MEDIUM)

Look at different performance metrics (for example sensitivity, specificity, ROC curve, etc.) and calculate them for the unbalanced class problem. Try to understand how these can help you in the evaluation of the models.

References

[1] `https://goo.gl/iqCbCO`, last accessed 09.11.2021.

[2] `https://goo.gl/PCHWMJ`, last accessed 09.11.2021.

[3] `https://goo.gl/Rh8S6g`, last accessed 09.11.2021.

[4] Schmidhuber, Jurgen, U. Meier, and D. Ciresan. "Multi-column deep neural networks for image classification." 2012 IEEE Conference on Computer Vision and Pattern Recognition. IEEE Computer Society, 2012 (`https://goo.gl/pEHZVB`, last accessed 09.11.2021).

[5] `http://yann.lecun.com/exdb/mnist/`, last accessed 10.11.2021.

[6] `https://goo.gl/Gq1Ce4`, last accessed 03.12.2021.

[7] `https://goo.gl/ZBKrdt`, last accessed 03.12.2021.

CHAPTER 11

Generative Adversarial Networks (GANs)

Generative Adversarial Networks (GANs) are, in their most basic form, two neural networks that teach each other how to solve a specific task. The idea was invented by Goodfellow and colleagues in 2014.[1] The two networks help each other with the final goal of being able to generate new data that looks like the data used for training. For example, you may want to train a network to generate human faces that are as realistic as possible. In this case, one network will generate human faces as good as it can, and the second network will criticize the results and tell the first network how to improve upon the faces. The two networks learn from each other, so to speak. This chapter looks in detail at how this works and explains how to implement an easy example in Keras.

The goal of this chapter is to give you a basic understanding of how GANs work. Adversarial learning (of which GANs are a specific case) is a vast area of research and starts to be an advanced topic in deep learning. This chapter investigates in detail how a basic GAN system works. We discuss, albeit in a shorter way, how conditional GANs function. Complete examples can be found, as usual, at `https://adl.toelt.ai`.

Introduction to GANs

The best way for you to understand how GANs work is to base this discussion on the diagram in Figure 11-1. After you understand what is going on under the hood, we will look at how to implement GANs in Keras.

[1] Goodfellow, Ian; Pouget-Abadie, Jean; Mirza, Mehdi; Xu, Bing; Warde-Farley, David; Ozair, Sherjil; Courville, Aaron; Bengio, Yoshua (2014)."Generative Adversarial Nets" PDF). *Proceedings of the International Conference on Neural Information Processing Systems* (NIPS 2014). pp. 2672–2680.

U. Michelucci, *Applied Deep Learning with TensorFlow 2*, https://doi.org/10.1007/978-1-4842-8020-1_11

Training Algorithm for GANs

To build a GANs system, we need two neural networks: a *generator* and a *discriminator*. The generator has the goal of producing a fake observation[2] X_{fake}, while the discriminator has the goal of classifying an input X as fake or real.

Imagine the following classical example: the generator (let's call him George) can be an art forger who is trying to produce paintings of some known painter, let's say Vincent van Gogh. And the discriminator (let's call her Anna) is an art critic who scrutinizes the paintings that George has produced to determine if they are genuine or not. They are new to this, so they decide to learn this process together[3]. George produces a painting. Anna examines it and gives some suggestions to George. Every now and then, Anna also trains with some real van Gogh paintings to get better at spotting errors in George's work. This process is repeated many times, until George is so good he can fool Anna. At this point, George can paint like van Gogh, can produce many fakes, and can get rich by selling his fake paintings[4]. This process is depicted in Figure 11-1. Let's see how this story translates into the language of neural networks.

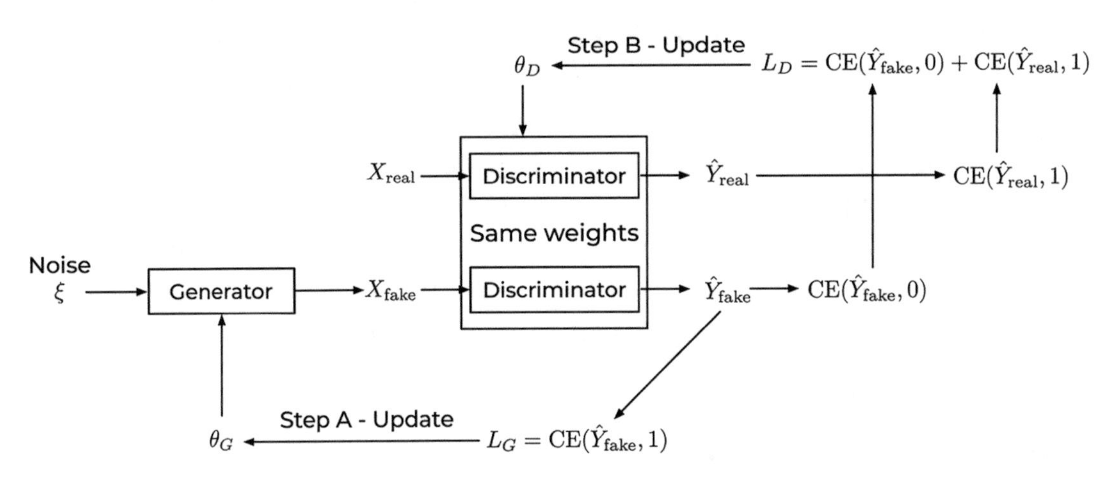

Figure 11-1. *All the components and steps of a GAN setup*

[2] We use the generic term *observation*. They could be fake faces, if you are trying to build a system that generates realistic faces, or an aged version of a face for example. We call an observation one of the inputs in the training dataset.

[3] How unrealistic it is that a forger and a critic would work together is not the point of the story.

[4] Of course, we are not encouraging anyone to be dishonest. It's just a story to help you understand GANs.

The generator gets as input a noise vector $\xi \in \mathbb{R}^k$ taken from a normal distribution. The size of this vector is not fixed and can be chosen based on the problem at hand. In the example we discuss in this chapter, we use $k = 100$. The generator (George) takes the random vector and generates a fake observation X_{fake} (as you can see in Figure 11-1). The output X_{fake} will have the same dimension of the observations contained in the training dataset X_{real} (in this example, van Gogh paintings). If, for example, if X_{real} are 1000x1000 pixels images in color, then X_{fake} will also be a 1000x1000 color image.

Now it's the discriminator's (Anna's) turn. It gets as input a X_{real} (or X_{fake}) and produces a one-dimensional output $\hat{Y}$ (the probability of the input of being real or fake). Basically, the discriminator is performing binary classification.

The steps of the training loop are described here.

1. A vector $\xi \in \mathbb{R}^k$ of k numbers is generated from a normal distribution.

2. Using this ξ the generator gives as output an X_{fake}.

3. The discriminator is used two times: one with real input (X_{real}) and one with the X_{fake} generated in the previous step.

4. Two loss functions are calculated: $L_G = CE(Y_{fake}, 1)$ and $L_D = CE(Y_{real}, 1) + CE(X_{fake}, 0)$.

5. Via an optimizer (Adam, Momentum, etc.), the two loss functions are minimized sequentially (sometimes there is one step for the generator, and multiple steps in updating the weights for the discriminator). Note that minimizing L_G will be done only with respect to the trainable parameters of the generator, while minimizing L_D will be done only with respect to the trainable parameters of the discriminator.

A Practical Example with Keras and MNIST

This section shows a practical example of what we discussed in the previous section implemented with Keras and applied to the MNIST dataset.[5] As usual you can find

[5] At this point in the book, you should know the MNIST dataset very well. In case you don't remember, it is a dataset with 70,000 handwritten digits saved as gray-level images 28x28 pixels in resolution.

the complete code at `https://adl.toelt.ai`, so we will concentrate here only on the relevant parts of the code. In particular, we look at the five steps described in the previous section and see how to implement them. To start, we need to create two neural networks: the generator and the discriminator. This can be done in the usual way. Nothing new here. For example

```
def make_generator_model():
    model = tf.keras.Sequential()
    model.add(layers.Dense(7*7*256, use_bias=False, input_shape=(100,)))
    model.add(layers.BatchNormalization())
    model.add(layers.LeakyReLU())

    model.add(layers.Reshape((7, 7, 256)))
    assert model.output_shape == (None, 7, 7, 256)  # Note: None is the
    batch size

    model.add(layers.Conv2DTranspose(128, (5, 5), strides=(1, 1),
    padding='same', use_bias=False))
    assert model.output_shape == (None, 7, 7, 128)
    model.add(layers.BatchNormalization())
    model.add(layers.LeakyReLU())

    model.add(layers.Conv2DTranspose(64, (5, 5), strides=(2, 2),
    padding='same', use_bias=False))
    assert model.output_shape == (None, 14, 14, 64)
    model.add(layers.BatchNormalization())
    model.add(layers.LeakyReLU())

    model.add(layers.Conv2DTranspose(1, (5, 5), strides=(2, 2),
    padding='same', use_bias=False, activation='tanh'))

    return model
```

The important part of this network is the input shape: `input_shape=(100,)`. Remember that the generator gets as input the random vector ξ that is, in our example, a 100-dimensional vector of random numbers generated from a normal distribution. Figure 11-2 shows a better visualization of the network.

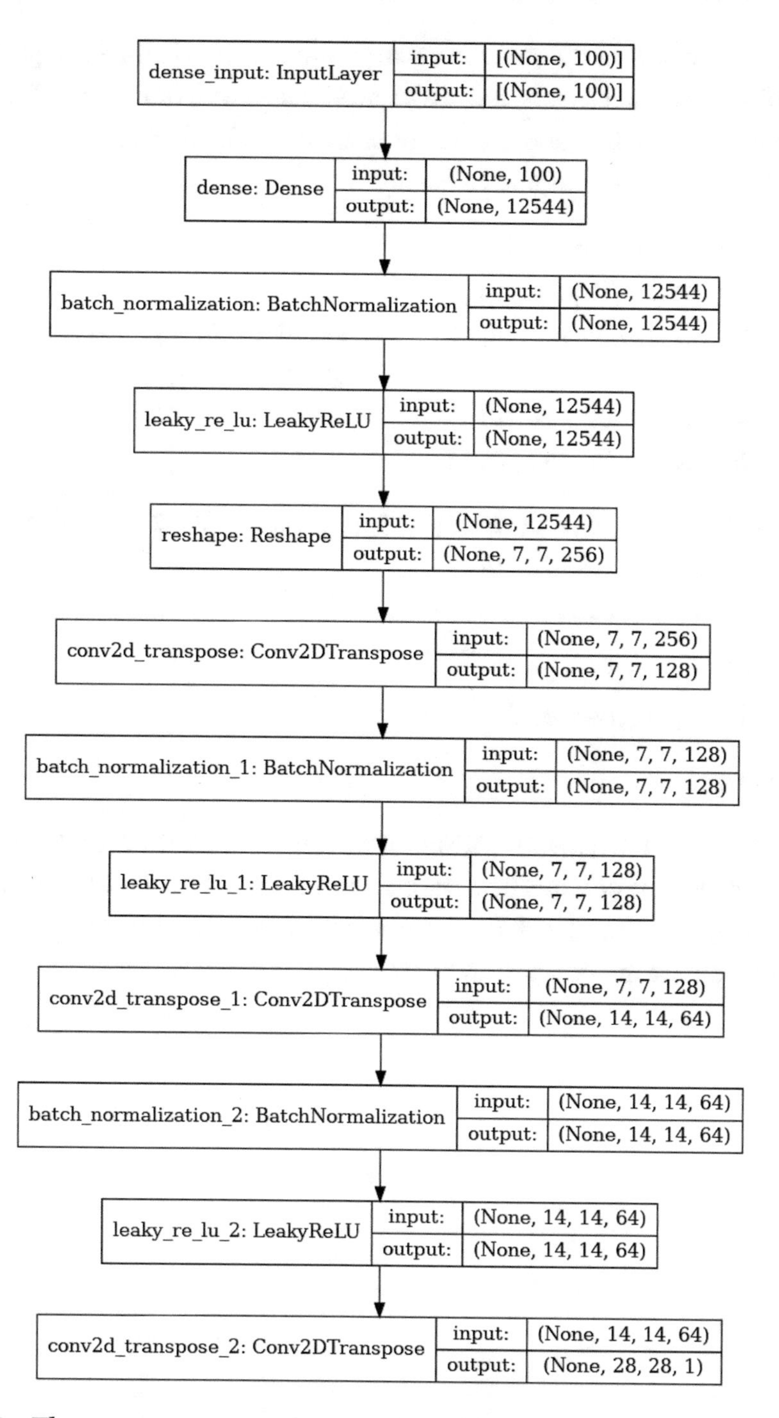

Figure 11-2. *The generator neural network architecture*

Figure 11-2 shows how the random vector is transformed in increasingly larger images, until at the end, the expected 28x28-pixel image with one channel is obtained (this will be the X_{fake} we discussed in the previous section). The discriminator can be created analogously, with standard Keras:

```
def make_discriminator_model():
    model = tf.keras.Sequential()
    model.add(layers.Conv2D(64, (5, 5), strides=(2, 2), padding='same',
                                     input_shape=[28, 28, 1]))
    model.add(layers.LeakyReLU())
    model.add(layers.Dropout(0.3))

    model.add(layers.Conv2D(128, (5, 5), strides=(2, 2), padding='same'))
    model.add(layers.LeakyReLU())
    model.add(layers.Dropout(0.3))

    model.add(layers.Flatten())
    model.add(layers.Dense(1))

    return model
```

That is a rather small network. The input will be an image that's 28x28 pixels in resolution with just one channel (gray levels). The output is the probability that the image is real and is achieved with one neuron called `layers.Dense(1)`.

Figure 11-3 shows the network architecture.

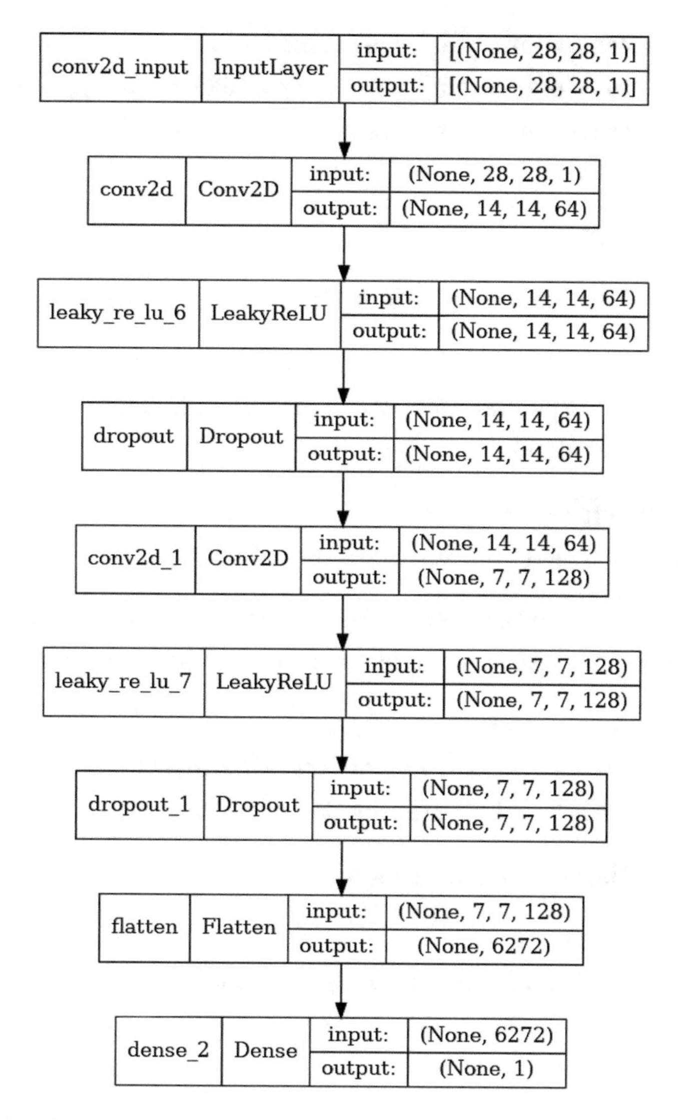

Figure 11-3. *The discriminator neural network architecture*

As discussed, we need to train the two networks in alternate fashion, so you will realize that the standard `compile()/fit()` approach will not be enough and you will need to develop a custom training loop.[6] Before doing that, we need to define the loss functions. This is not difficult, and we can start with the discriminator function L_D:

```
def discriminator_loss(real_output, fake_output):
```

[6] If you have never seen a custom training loop, check out Appendix B, which explains the basics of customizing Keras.

```
    real_loss = cross_entropy(tf.ones_like(real_output), real_output)
    fake_loss = cross_entropy(tf.zeros_like(fake_output), fake_output)
    total_loss = real_loss + fake_loss
    return total_loss
```

After having defined

```
cross_entropy = tf.keras.losses.BinaryCrossentropy(from_logits=True)
```

You will remember that we need the X_{fake} (this will be the fake_output variable) and the X_{real} (the real_output variable) to train the discriminator. The generator loss function L_G is defined analogously

```
def generator_loss(fake_output):
    return cross_entropy(tf.ones_like(fake_output), fake_output)
```

For L_G, as you will remember from the previous section, we only need X_{fake}. At this point, we are almost done. We need to define the optimizers (always using standard Keras functions):

```
generator_optimizer = tf.keras.optimizers.Adam(1e-4)
discriminator_optimizer = tf.keras.optimizers.Adam(1e-4)
```

And now here is the custom training loop

```
def train_step(images):
    # Generation of the xi vector (random noise)
    noise = tf.random.normal([BATCH_SIZE, noise_dim])
    with tf.GradientTape() as gen_tape, tf.GradientTape() as disc_tape:
      # Calculation of X_{fake}
      generated_images = generator(noise, training=True)

      # Calculation of \hat Y_{real}
      real_output = discriminator(images, training=True)

      # Calculation of \hat Y_{fake}
      fake_output = discriminator(generated_images, training=True)

      # Calculation of L_G
      gen_loss = generator_loss(fake_output)
```

```
        # Calculation of L_D
        disc_loss = discriminator_loss(real_output, fake_output)

    # Calculation of the gradients of L_G for backpropagation
    gradients_of_generator = gen_tape.gradient(gen_loss, generator.
    trainable_variables)

    # Calculation of the gradients of L_D for backpropagation
    gradients_of_discriminator = disc_tape.gradient(disc_loss,
    discriminator.trainable_variables)

# Applications of the gradients to update the weights generator_optimizer.
apply_gradients(zip(gradients_of_generator, generator.trainable_variables))

discriminator_optimizer.apply_gradients(zip(gradients_of_discriminator,
discriminator.trainable_variables))
```

Let's summarize the steps:

- We generate X_{fake}: `generated_images = generator(noise, training=True)`
- Then $\hat{Y}_{real}$: `real_output = discriminator(images, training=True)`
- Then $\hat{Y}_{fake}$: `fake_output = discriminator(generated_images, training=True)`
- Then we define L_G: `gen_loss = generator_loss(fake_output)`
- Then we define L_D: `disc_loss = discriminator_loss(real_output, fake_output)`

At this point we can evaluate the gradients:

```
gradients_of_generator = gen_tape.gradient(gen_loss, generator.trainable_
variables)
gradients_of_discriminator = disc_tape.gradient(disc_loss, discriminator.
trainable_variables)
```

and then apply them to update the trainable parameters of the two networks:

```
generator_optimizer.apply_gradients(zip(gradients_of_generator, generator.
trainable_variables))
discriminator_optimizer.apply_gradients(zip(gradients_of_discriminator,
discriminator.trainable_variables))
```

At this point the only thing left is to perform those steps enough times to get the network to learn. By comparing Figure 11-1 with this code, you should be able to immediately see how this GAN is implemented. Figure 11-4 shows examples of digits generated by the generator network. The digits do not exist in the dataset and have been "created" by the neural network.

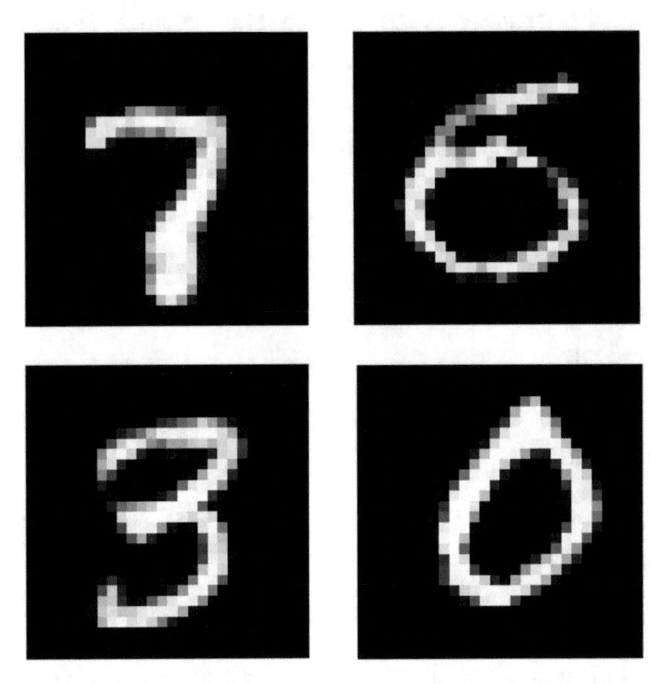

Figure 11-4. *Four examples of digits generated by the generator network. The digits do not exist in the dataset and have been "created" by the neural network*

The only thing you need to do in order to generate images is feed the generator with 100 random numbers. For example, with

```
noise = tf.random.normal([1, 100])
generated_image = generator(noise, training=False)
```

Be aware that due to the dimensions used in the code, if you want to extract the 28x28 image you need to use the code `generated_image[0, :, :, 0]`. You can find the entire code at `https://adl.toelt.ai`. Try different networks, different numbers of epochs, and so on to get a feeling for how such an approach can generate realistic images from a training dataset.

Note that this approach learns from all the classes at the same time. For example, it's not possible to ask the network to generate a specific digit. The generator will simply randomly generate one digit. To be able to do this, we need to implement what is called "conditional" GANs. Those GANs also get as input the class labels and can generate examples from specific classes. If you want to try the code on your laptop, keep in mind that training GANs is rather slow. If you do it on Google Colab and you use a GPU, one epoch may take up to 30 seconds or more. Keep that in mind. On a modern laptop without GPUs, one epoch may take up to 1.5-2 minutes.

A Note on Training

There is an important aspect on why the training is implemented in sequential fashion that we need to discuss. You might wonder why we need to train the two networks in an alternate fashion. Why can't we train the discriminator alone for example until it gets really good at distinguishing fakes and real images? The reason is very simple. Imagine that the discriminator is really good. It will always spot the X_{fake} as fakes, and therefore the generator will never be able to get better, because the discriminator will never make a mistake. Therefore, training in such a situation would never be successful. In practice one of the biggest challenges when training GANs is to make sure that the generator and the discriminator networks remain at approximately the same skill level during the training. This has been shown to be the sweet spot for the training to be efficient and successful.

Conditional GANs

Now let´s turn our attention to conditional GANs (CGANs). CGANs work the same as described in this chapter. The working idea is the same, with the difference that we will be able to specify from which class we want the generator to create an image. In the MNIST example, we could tell the generator that we want one fake digit one, for example. Figure 11-5 shows an updated diagram explaining the training (it's Figure 11-1 updated).

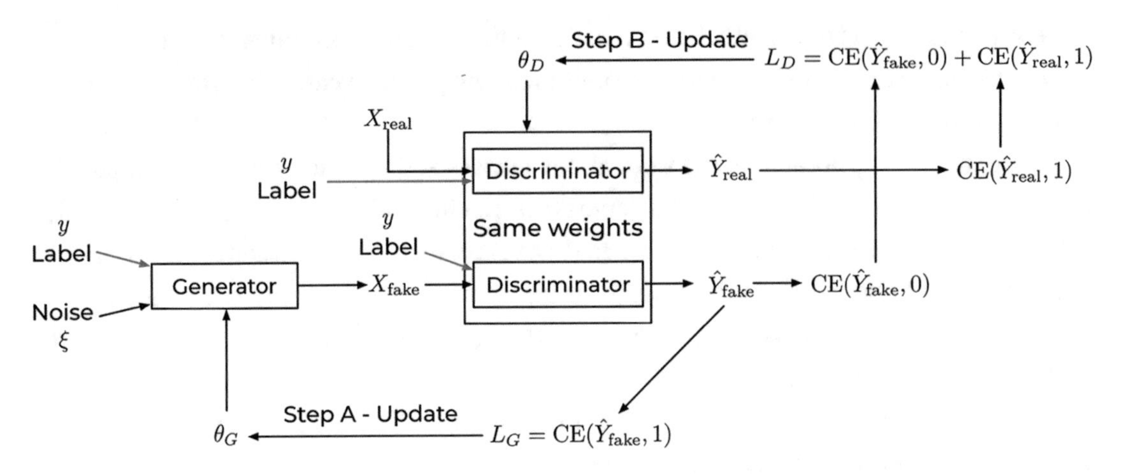

Figure 11-5. *The training of a CGAN system. the red highlights the role of the label that makes it possible for the generator to create fake examples of specific classes*

The main things that we need to change to achieve this are the architectures of the two networks. Figures 11-6 and 11-7 show example architectures of the two networks: a generator and a discriminator, respectively.

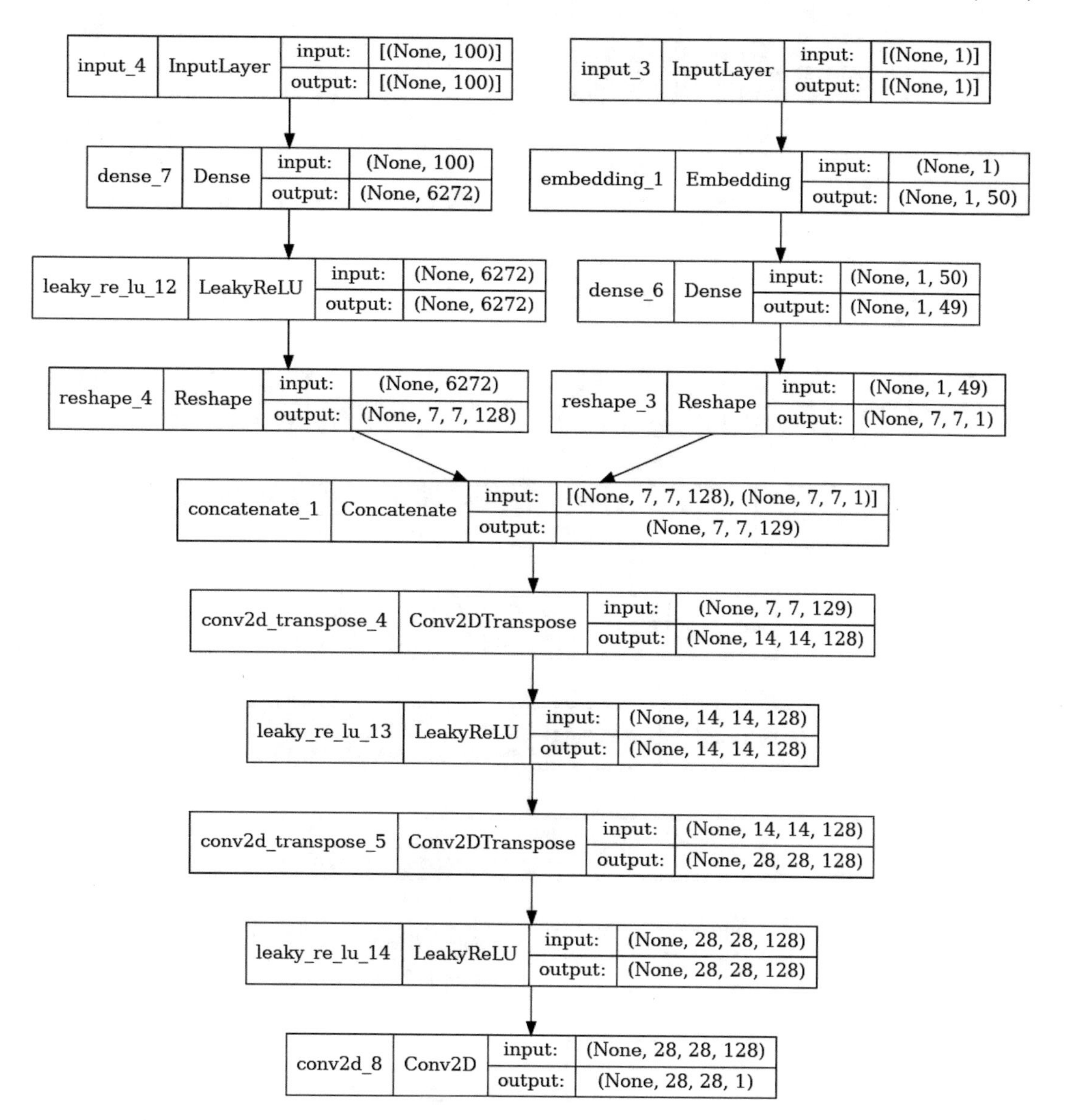

Figure 11-6. *The generator network architecture for a CGAN*

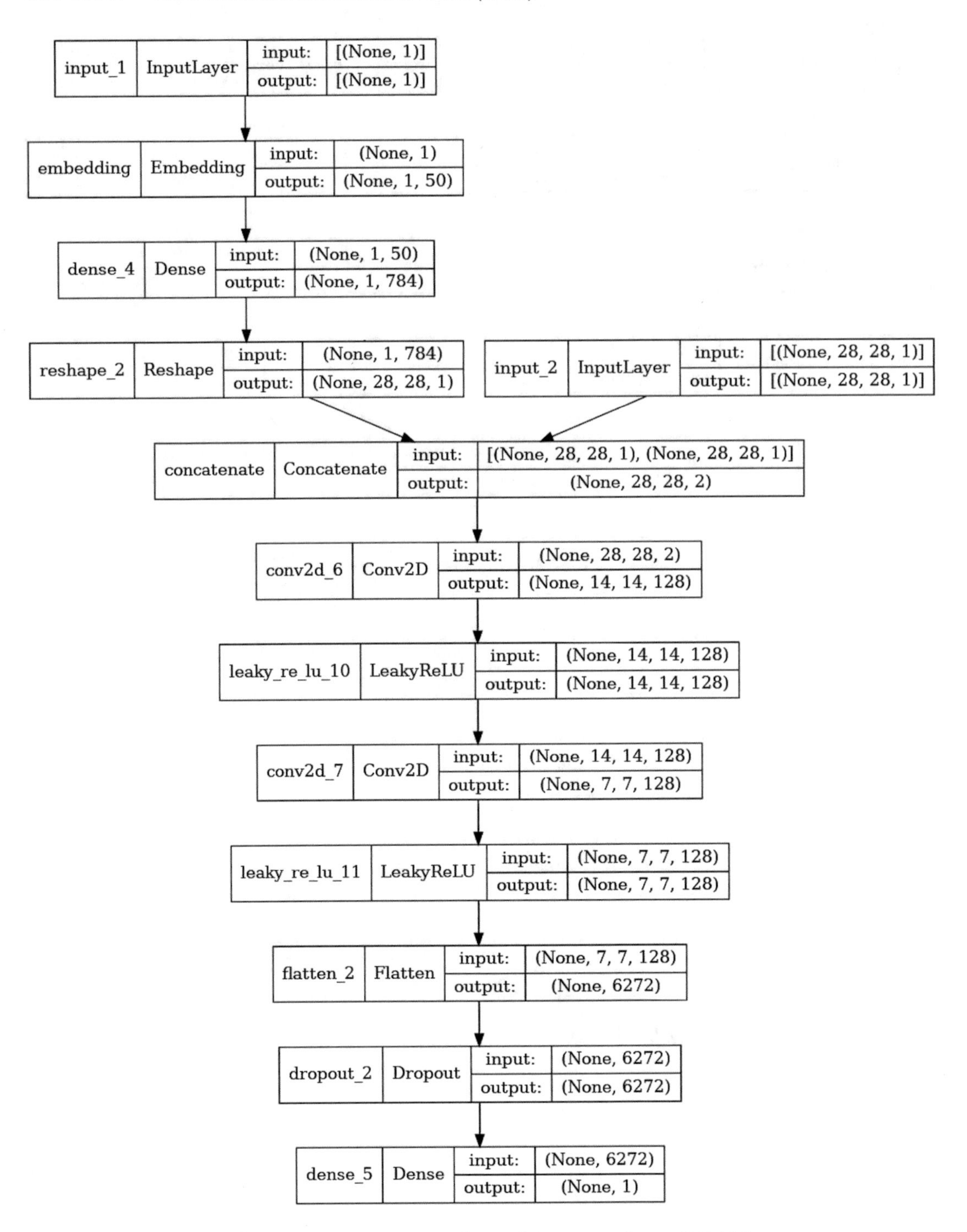

Figure 11-7. *The discriminator network architecture for a CGAN*

From Figures 11-6 and 11-7, you can immediately see that they now have an additional input: a one-dimensional tensor, which will be the class. The kinds of networks that you see here can be easily implemented using the functional Keras API. Just to give you an idea about how to build such networks, here are the first layers of the generator network, up until the merging of the two branches

```
input_label = Input(shape=(1,))
emb = Embedding(n_classes, 50)(input_label)
n_nodes = 7 * 7
emb = Dense(n_nodes)( emb)
emb = Reshape((7, 7, 1))(emb)
in_lat = Input(shape=(latent_dim,))
n_nodes = 128 * 7 * 7
gen = Dense(n_nodes)(in_lat)
gen = LeakyReLU(alpha=0.2)(gen)
gen = Reshape((7, 7, 128))(gen)
merge = Concatenate()([gen, emb])
```

You can see how flexible Keras Functional APIs are. Now, when training the generator network for example, you need to give as input not only a random vector ξ as before, but also a random vector *and* a class label, which you will use to choose the X_{real} to train the discriminator. To generate the random noise and the labels, you use this code

```
latend_dim = 100
x_input = randn(latent_dim * n_samples)
z_input = x_input.reshape(n_samples, latent_dim)
labels = randint(0, n_classes, n_samples)
```

And you will give the generator network the input as `[z_input, labels]`. The `n_samples` variable is simply the batch size you want to use. Analyzing the complete code would make this chapter really long, really difficult to follow, and really boring. Implementing a CGAN starts to be really advanced, and the best way of understanding it is to go through a complete example. As usual, you will find one at `https://adl.toelt.ai` where you can check all the code. In the meanwhile, your best source of knowledge about CGANs are the two papers that you should study to understand what is going on with CGANs.

- Radford, Alec, Luke Metz, and Soumith Chintala. "Unsupervised representation learning with deep convolutional generative adversarial networks." arXiv preprint arXiv:1511.06434 (2015).
- Mirza, Mehdi, and Simon Osindero. "Conditional generative adversarial nets." arXiv preprint arXiv:1411.1784 (2014).

Conclusion

The goal of this chapter was not to go into details about advanced GAN architectures, but to give you an initial understanding of how adversarial learning works. There are many advanced architectures and topics about GANs that we cannot cover in this book, since that would go well beyond the skill level of the average reader. But with this chapter, I hope I have given you an initial understanding of how GANs work and how easy it is to implement them in Keras.

APPENDIX A

Introduction to Keras

Keras is an API designed to make developing neural network easy for humans. It is also the API that we use throughout this book. The goal of this appendix is not to cover all aspects of Keras (surely the space would not be enough), but to give you the minimum amount of information that you need to understand the code in this book and then point you to resources where you can find more. If you want to learn everything about Keras, the most efficient way is to study the book *Deep Learning with Python* by François Chollet and to read the official documentation at `https://keras.io`.

In a few words, citing F. Chollet,[1] "TensorFlow is an infrastructure layer for differentiable programming, dealing with tensors, variables, and gradients. Keras is a user interface for deep learning, dealing with layers, models, optimizers, loss functions, metrics, and more."

If you are looking for custom training loops, custom layers, and so on, check out Appendix B, where we cover those topics, although briefly.

Some History

Keras was developed by F. Chollet, and its first version was made available on March 27, 2015. Up to and including version 2.3, Keras needed a backend, in other words a system that performed the low-level operations needed by your code. At the very beginning Keras did not work with TensorFlow (the first supported backend was Theano). The `tf.keras` package was introduced in TensorFlow 1.10.0. Note that these two imports are very different things:

```
import keras
```

and

```
from tensorflow import keras
```

[1] `https://keras.io/getting_started/intro_to_keras_for_researchers/`

U. Michelucci, *Applied Deep Learning with TensorFlow 2*, https://doi.org/10.1007/978-1-4842-8020-1

After Keras 2.3.0, F. Chollet declared that this release will be in sync with `tf.keras` and that practitioners should use `tf.keras` and not `keras` anymore. After this release, `keras` will not support multiple backends.

Note You should always use `from tensorflow import keras` in your code.

Understanding the Sequential Model

A sequential model is simply a plain stack of layers, where each has one input and one output tensor. The easiest way to create a sequential model is by providing a list of layers to the `keras.Sequential` call. For example

```
model = keras.Sequential(
    [
        layers.Dense(2, activation="relu"),
        layers.Dense(2, activation="relu")
    ]
)
```

In this code, it is assumed you have imported

```
from tensorflow.keras import layers
```

There is an alternative way to create a sequential model and that is by using the `add()` method. The network defined above could be also created with

```
model = keras.Sequential()
model.add(layers.Dense(2, activation="relu"))
model.add(layers.Dense(2, activation="relu"))
```

The two versions of the code are completely equivalent. This second version is a bit more readable than the first when the number of layers is large.

There are situations when the sequential model is not appropriate. For example:[2]

- Your model has multiple inputs or multiple outputs
- Any of your layers have multiple inputs or multiple outputs

---

[2] As taken from the official documentation at `https://keras.io/guides/sequential_model/`.

- You need to do layer sharing
- You want a non-linear topology (e.g., a residual connection, a multi-branch model, etc.)

For those cases, you need the functional APIs, as described later in this appendix.

Understanding Keras Layers

Layers are a fundamental part of Keras. A *layer* includes a state (the weights of the neurons for example) and some computation (implemented in the `call` method). Keras offers a lot of layers that you can use without having to develop your own. The most commonly used (and probably the ones you may have seen so far) are the following:[3]

- Dense: Like the ones we saw in the FFNN discussion
- Conv1D, Conv2D, and Conv3D: Convolutional layers in multiple dimensions
- MaxPooling1D, MaxPooling2D and MaxPooling3D: Max-pooling layers
- AveragePooling1D, AveragePooling2D and AveragePooling3D: Average-pooling layers
- LSTM layers
- Regularization layers as `Dropout`

And many more. Remember that any operation that takes a tensor as input and gives a tensor as output is a layer in the Keras language. For example, flattening a 2D image into a 1D vector is also a layer (see the `Flatten` layer). Also, reshaping an input can be done with a layer (see the `Reshape` layer). Even applying an activation function can be done with a layer (see the `ReLU` layer for example).

Note Remember that any operation that takes a tensor as input and gives a tensor as output is a layer in the Keras language.

---

[3] If you want to see the complete list, consult the official documentation at `https://keras.io/api/layers/`.

In Appendix B, we briefly discuss how to develop your own layers. Note that you can also easily do the following things with layers:

- Retrieve the gradients (see Appendix B)
- Retrieve the weights (see Appendix B)
- Add regularization losses (as discussed in the main part of the book)
- Set the weights to values of your choosing (see Appendix B)
- Use initializers for the weights (for example, He, Glorot, etc.)

Setting the Activation Function

To set the activation function in layers, use the `activation` property. For example, the code

```
layers.Dense(2, activation="relu")
```

creates a layer with two neurons and the `ReLU` activation function. Note that if you don't specify any activation, `none` is used (or, in other words, the identity function is used as the activation function). As usual, Keras offers many activation functions:[4] `relu`, `sigmoid`, `softmax`, `softplus`, `softsign`, `tanh`, `selu`, `elu`, and `exponential function`.

Using Functional APIs

The functional Keras APIs offer a way to create models that are not linear, with shared layers or with multi-input or multi-output layers (or both). The idea behind it is that neural networks are normally a directed acyclic graph, so with the functional APIs you can build graphs of layers (therefore, you can build non-linear architectures). For example, the previous network would be built as follows

```
inputs = keras.Input(shape=(...,))
x = layers.Dense(2, activation="relu")(inputs)
x = layers.Dense(2, activation="relu")(x)
```

[4] As of November 2021.

where we have added an input layer since it is needed when using the functional APIs. As you can see, each layer is "applied" to another one. Or in graph language, it's like drawing a connection between two layers. This model can be graphically plotted with Keras,[5] as shown in Figure A-1.

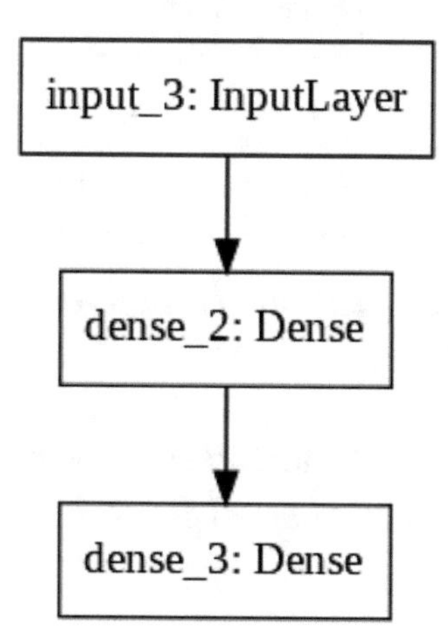

Figure A-1. *A graphical representation of the small network defined in the text as a graph*

For example, the code

```
inputs = keras.Input(shape=(784,))
x1 = layers.Dense(2, activation="relu")(inputs)
x2 = layers.Dense(2, activation="relu")(x1)
y = layers.Dense(2, activation="relu")(x1)
model = keras.Model(inputs, outputs = [x2,y])
```

would give you the architecture shown in Figure A-2.

[5] To plot a model you can use the useful call `keras.utils.plot_model(model, "...")`. Swap the three dots with the filename that you want to use.

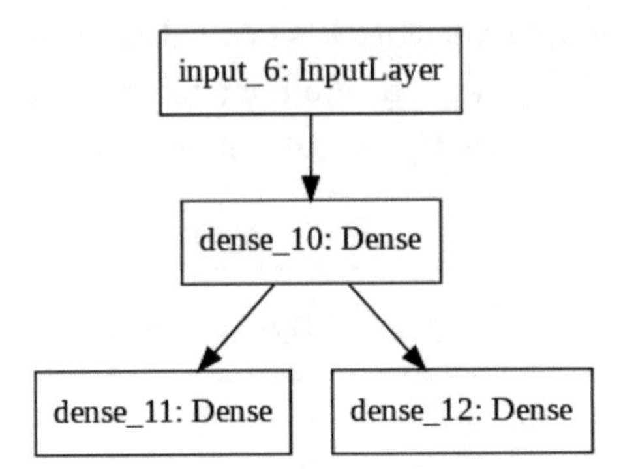

Figure A-2. *A graphical representation of a network with multiple inputs built with the Keras Functional API*

You can get really creative with the architectures you can create. You will find lots of examples in the official documentation at `https://keras.io/guides/functional_api/`.

Specifying Loss Functions and Metrics

To train any neural network, you need of course to specify which loss function you want to minimize and which optimizer you want to use. In Keras, this is achieved by calling the `compile()` method. For example, if you want to use Adam as the optimizer and the MSE as the loss function, you use

```
model.compile(optimizer='Adam', loss='mse')
```

Putting It All Together and Training

The easiest way to train a model in Keras involves three steps:

1. Create the network by specifying the architecture (the number of layers, number of neurons, types of layers, activation functions, etc.). You would do this in the examples with, for example, `keras.Sequential()`.
2. Compile the model with the `compile()` method. This step specifies which loss function and which metrics Keras should use.

3. Train the model by using the `fit()` method. In the `fit()` method you can specify the number of epochs, batch size, and many other parameters.

Here is a minimal example

```
model = keras.Sequential(
    [
        layers.Dense(2, activation="relu"),
        layers.Dense(3, activation="relu"),
        layers.Dense(1),
    ]
)
```

This specifies the network architecture. After that, you compile the model with the `compile()` method.

```
model.compile(optimizer=Adam, loss='mse')
```

where you specify the optimizer Adam and the MSE loss. After that, you can train the model

```
model.fit(x, y, batch_size=32, epochs=10)
```

where `x` indicates the inputs, `y` indicates the labels, the `batch_size` is specified as `32`, and you want to train for ten epochs.

Note The easiest approach to creating and training a model in Keras involves three steps: 1) Create the architecture, 2) Compile the model with `compile()`, and 3) Train the model with the `fit()` method.

The `fit()` method accepts lots of parameters. You can specify:

- How much output you want by specifying the `verbose` parameter (`0` for no output, `1` for a progress bar, and `2` for one line for each epoch).
- Actions at different points during the training by specifying which `callbacks` functions the `fit()` method should use. For more information on callbacks, see the next sections.

- A validation dataset by simply giving a `validation_split` parameter (that is, the fraction of the data that you want to use as the validation dataset). The `fit()` method will give you the metrics for this dataset.

You can specify many more options. As usual, to get a complete overview, check out the official documentation at `https://keras.io/api/models/model_training_apis/`.

Modeling *evaluate*( ) and *predict*( )

Once you have trained your model, you can use the `evaluate()` method. It will return the loss and the metrics applied to the inputs in test mode.6 For example, a call would look like

```
model.evaluate(x,y)
```

And finally you can use the `predict()` method to generate predictions for the input samples. A call would look like

```
model.predict(x)
```

Using Callback Functions

Callback functions are a powerful way to customize training of a model. They can be used with the `fit()`, `evaluate()`, and `predict()` functions.

It is instructive to understand a bit better what Keras callback functions are, since they are used quite often when developing models. From the official documentation[7]

> *A callback is a set of functions to be applied at given stages of the training procedure.*

The idea is that you can pass a list of callback functions to the `.fit()` method of the `Sequential` or `Model` classes. Relevant methods of the callbacks will then be called at each stage of the training. Their use is rather easy. For the `fit()` function, you would use them as

[6] This is relevant when, for example, dealing with a dropout that has a different behavior during training or during testing.

[7] `https://keras.io/callbacks/`

```
model.fit(
    ...,
    callbacks=[Callback()],
)
```

Where `Callback()` is a placeholder name for a callback (you need to change it to the callback function name you want to use). There are callbacks functions that perform many tasks, such as

- `ModelCheckPoint` saves weights and models at specific frequencies
- `LearningRateScheduler` changes the learning rate according to some schedule
- `TerminateOnNaN` stops the training process if NaN appears (so you don't waste time or computing resources)
- And many more

As usual, you can find more information on the official documentation at `https://keras.io/api/callbacks/`. In Appendix B, I discuss how to develop your own custom callback class, since this is one of the best ways to check and control the training process at various stages.

Saving and Loading Models

It is often useful to save a model on disk, so you can continue the training at a later stage or reuse a previously trained model. To learn how to do this, let's consider the MNIST dataset for the sake of giving a concrete example.[8]

You need the following `imports`

```
import os
import tensorflow as tf
from tensorflow import keras
```

Load the MNIST dataset again and take the first 5,000 observations.

[8] The example was inspired by the official Keras documentation at `https://www.tensorflow.org/tutorials/keras/save_and_restore_models`.

```
(train_images, train_labels), (test_images, test_labels) = tf.keras.
datasets.mnist.load_data()
train_labels = train_labels[:5000]
test_labels = test_labels[:5000]
train_images = train_images[:5000].reshape(-1, 28 * 28) / 255.0
test_images = test_images[:5000].reshape(-1, 28 * 28) / 255.0
```

Let's now build a simple Keras model with a Dense layer with 512 neurons, a bit of dropout, and the classical ten-neuron output layer for classification (remember the MNIST dataset has ten classes).

```
model = tf.keras.models.Sequential([
    keras.layers.Dense(512, activation=tf.keras.activations.relu, input_
    shape=(784,)),
    keras.layers.Dropout(0.2),
    keras.layers.Dense(10, activation=tf.keras.activations.softmax)
  ])

model.compile(optimizer='adam',
              loss=tf.keras.losses.sparse_categorical_crossentropy,
              metrics=['accuracy'])
```

We have added a bit of dropout, since this model has 407,050 trainable parameters. You can check this number simply by using `model.summary()`.

What we need to do is first define where we want to save the model on the disk. And we can do that (for example) in this way

```
checkpoint_path = "training/cp.ckpt"
checkpoint_dir = os.path.dirname(checkpoint_path)
```

After that, we need to use a callback (remember what we did in the last section) that will save the weights[9]

```
cp_callback = tf.keras.callbacks.ModelCheckpoint(checkpoint_path,
                                                 save_weights_only=True,
                                                 verbose=1)
```

[9] The `ModelCheckpoint` callback is a standard Keras callback that you can use. You don't need to develop one yourself.

Note that we don't need to define a class as we did in the previous section, since `ModelCheckpoint` inherits from the `Callback` class.

Then we can simply train the model, specifying the correct callback function

```
model.fit(train_images, train_labels,  epochs = 10,
         validation_data = (test_images,test_labels),
         callbacks = [cp_callback])
```

If you check the contents of the folder where your code is running, you should see at least three files:

- `cp.ckpt.data-00000-of-00001`: Contains the weights (if the number of weights is big, you will see many files like this one)
- `cp.ckpt.index`: Contains information about which weights are in which file
- `checkpoint`: Contains information about the checkpoint itself

We can now test our method. This code will give you a model that will reach an accuracy on the validation dataset of roughly 92%.

If we define a second model

```
model2 = tf.keras.models.Sequential([
    keras.layers.Dense(512, activation=tf.keras.activations.relu, input_
    shape=(784,)),
    keras.layers.Dropout(0.2),
    keras.layers.Dense(10, activation=tf.keras.activations.softmax)
  ])

model2.compile(optimizer='adam',
              loss=tf.keras.losses.sparse_categorical_crossentropy,
              metrics=['accuracy'])
```

and we check its accuracy on the validation dataset with

```
loss, acc = model2.evaluate(test_images, test_labels)
print("Untrained model, accuracy: {:5.2f}%".format(100*acc))
```

we will get an accuracy of roughly 8.6%, That was expected, since this model has not been trained yet. But now we can load the saved weights in this model and try again.

```
model2.load_weights(checkpoint_path)
loss,acc = model2.evaluate(test_images, test_labels)
print("Second model, accuracy: {:5.2f}%".format(100*acc))
```

We should get this result

```
5000/5000 [==============================] - 0s 50us/step
Restored model, accuracy: 92.06%
```

That makes again sense, since the new model is now using the weights of the old, trained model. Keep in mind that, to load pretrained weights in a new model, the model needs to have the exact same architecture as the one used to save the weights.

Note To use saved weights with a new model, the model must have the exact same architecture as the one used to save the weights. Using pretrained weights can save you quite a lot of time, since you don't need to waste time in training the network again.

As you will see again and again, the basic idea is to use a callback that will save your weights. Of course, you can customize the callback function. For example, if you want to save the weights every 100 epochs with a different filename each time, so that you could decide to restore a specific check point, you need first to define the filename in a dynamic way as

```
checkpoint_path = "training/cp-{epoch:04d}.ckpt"
checkpoint_dir = os.path.dirname(checkpoint_path)
```

You should use the following callback

```
cp_callback = tf.keras.callbacks.ModelCheckpoint(
    checkpoint_path, verbose=1, save_weights_only=True,
    period=1)
```

Note that `checkpoint_path` can contain named-formatting options (in the name we have `{epoch:04d}`), which will be filled by the values of `epoch` and `logs` (passed in `on_epoch_end`, which you saw in the previous section).[10] You can check the original code

[10] Check out the official documentation at `https://goo.gl/SnKgyQ`.

for `tf.keras.callbacks.ModelCheckpoint` and you will find that the formatting is done in the `on_epoch_end(self, epoch, logs)` method.

```
filepath = self.filepath.format(epoch=epoch + 1, **logs)
```

You can define your filename using the epoch number and the values contained in the `logs` dictionary.

Let's get back to our example. Let's start by saving a first version of the model

```
model.save_weights(checkpoint_path.format(epoch=0))
```

and then we can fit the model as usual

```
model.fit(train_images, train_labels,
          epochs = 10, callbacks = [cp_callback],
          validation_data = (test_images,test_labels),
          verbose=0)
```

Be careful since this will save lots of files. In our example, one every epoch. So for example, the directory content may look like this:

```
checkpoint                          cp-0006.ckpt.data-00000-of-00001
cp-0000.ckpt.data-00000-of-00001    cp-0006.ckpt.index
cp-0000.ckpt.index                  cp-0007.ckpt.data-00000-of-00001
cp-0001.ckpt.data-00000-of-00001    cp-0007.ckpt.index
cp-0001.ckpt.index                  cp-0008.ckpt.data-00000-of-00001
cp-0002.ckpt.data-00000-of-00001    cp-0008.ckpt.index
cp-0002.ckpt.index                  cp-0009.ckpt.data-00000-of-00001
cp-0003.ckpt.data-00000-of-00001    cp-0009.ckpt.index
cp-0003.ckpt.index                  cp-0010.ckpt.data-00000-of-00001
cp-0004.ckpt.data-00000-of-00001    cp-0010.ckpt.index
cp-0004.ckpt.index                  cp.ckpt.data-00000-of-00001
cp-0005.ckpt.data-00000-of-00001    cp.ckpt.index
cp-0005.ckpt.index
```

A last tip before moving on is how to get the latest checkpoint, without bothering to search for the filename. This can be done easily with the following code

```
latest = tf.train.latest_checkpoint('training')
model.load_weights(latest)
```

This will automatically load the weights saved in the latest checkpoint. The variable `latest` is simply a string and contains the last checkpoint filename saved. In this example, that is `training/cp-0010.ckpt`.

Note The checkpoint files are binary files that contain the weights of your model. You will not be able to read them directly, and you should not need to.

Saving Your Weights Manually

Of course, you can simply save your model weights manually when you are done training, without defining a callback function

```
model.save_weights('./checkpoints/my_checkpoint')
```

This command will generate three files, all starting with the string you gave as a name—in this case, `my_checkpoint`. Running this code will generate the three files we described previously:

```
checkpoint
my_checkpoint.data-00000-of-00001
my_checkpoint.index
```

Reloading the weights in a new model is as simple as this:

```
model.load_weights('./checkpoints/my_checkpoint')
```

Keep in mind, that to be able to reload saved weights in a new model, the old model must have the same architecture as the new one. It must be exactly the same.

Saving the Entire Model

Keras also gives us a way to save the entire model on disk: weights, the architecture, and the optimizer. We can re-create the same model by simply moving some files. For example, we could use the following code

```
model.save('my_model.h5')
```

This will save the entire model in one file, called my_model.h5. We can simply move the file to a different computer and re-create the same trained model with

```
new_model = keras.models.load_model('my_model.h5')
```

Note that this model will have the same trained weights as your original model, so it's ready to use. This may be helpful if you want to stop training your model and continue the training on a different machine, for example. Or maybe you must stop the training for a while and continue at a later time.

Conclusion

This appendix presented a very quick and superficial overview of Keras with the goal of giving you enough information to start programming basic neural networks with Keras and to understand the code discussed in this book. I hope this short appendix provided a good overview of the fundamentals concepts and methods of Keras.

APPENDIX B

Customizing Keras

This appendix looks in more detail at the code used to build the GAN. If you studied Chapter 11, you will have realized that we did not use the `compile()/fit()` approach, but instead built a custom training loop. It is important that you understand the fundamental concepts of how this works with Keras. This appendix is here exactly for that reason.

This appendix does not cover custom loss functions, custom layers, or custom activation functions. If you are interested in these topics, you will find plenty of examples in the official documentation.

This appendix is intended as a very short reference that I hope will help you quickly understand how to customize Keras and understand the code in Chapter 11 on GANs. I also added for reference a short section on how to customize callback classes. I hope it is useful.

A complete overview on how to customize Keras would require a book of its own[1] and is not the goal of this book.

Customizing Callback Classes

In Appendix A, you learned what callback functions are. In this section, you will see how you can customize them for your purposes, since this is a really useful thing even when you are using the `compile()/fit()` approach. To do this, you need to understand how the abstract base class `keras.callbacks.Callback` works.

The abstract base class `Callback` can be found (at the moment of this writing) at `tensorflow/python/keras/callbacks.py`.

---

[1] In case you are looking for such a book, a very good introduction is the book by Jojo Moolayil, entitled *Learn Keras for Deep Neural Networks: A Fast-Track Approach to Modern Deep Learning with Python,* and published by Apress.

U. Michelucci, *Applied Deep Learning with TensorFlow 2*, https://doi.org/10.1007/978-1-4842-8020-1

To start customizing, you need simply to define a custom class that inherits from keras.callbacks.Callback. The main methods you want to redefine are the following:

- on_train_begin: Called at the beginning of training
- on_train_end: Called at the end of training
- on_epoch_begin: Called at the start of an epoch
- on_epoch_end: Called at the end of an epoch
- on_batch_begin: Called right before processing a batch
- on_batch_end: Called at the end of a batch

This can be done with the following code

```
from tensorflow import keras
class My_Callback(keras.callbacks.Callback):
    def on_train_begin(self, logs={}):
        # Your code here
       return

    def on_train_end(self, logs={}):
        # Your code here
        return

    def on_epoch_begin(self, epoch, logs={}):
            # Your code here
            return

    def on_epoch_end(self, epoch, logs={}):
            # Your code here
            return

    def on_batch_begin(self, batch, logs={}):
            # Your code here
            return

    def on_batch_end(self, batch, logs={}):
            # Your code here
            self.losses.append(logs.get('loss'))
            return
```

Each of these methods has slightly different inputs that you may use in your class. Let's look at them briefly:

on_epoch_begin, on_epoch_end

```
Arguments:

        epoch: integer, index of epoch.

        logs: dictionary of logs.
```

on_train_begin, on_train_end

```
Arguments:

        logs: dictionary of logs.
```

on_batch_begin, on_batch_end

```
Arguments:

        batch: integer, index of batch within the current epoch.

        logs: dictionary of logs.
```

Let's see with an example how we can use this class.

Example of a Custom Callback Class

Let's again consider the MNIST example. This is the same code you have seen many times by now:

```
import tensorflow as tf
from tensorflow import keras
(train_images, train_labels), (test_images, test_labels) = tf.keras.
datasets.mnist.load_data()

train_labels = train_labels[:5000]
test_labels = test_labels[:5000]

train_images = train_images[:5000].reshape(-1, 28 * 28) / 255.0
test_images = test_images[:5000].reshape(-1, 28 * 28) / 255.0
```

Let's define a `Sequential` model for our example

```
model = tf.keras.models.Sequential([
    keras.layers.Dense(512, activation=tf.keras.activations.relu,
    input_shape=(784,)),
    keras.layers.Dropout(0.2),
    keras.layers.Dense(10, activation=tf.keras.activations.softmax)
  ])

model.compile(optimizer='adam',
              loss=tf.keras.losses.sparse_categorical_crossentropy,
              metrics=['accuracy'])
```

Now let's write a custom callback class, redefining only one of the methods to see what the inputs are. For example, let's see what the variable `logs` contains at the beginning of the training

```
class CustomCallback1(keras.callbacks.Callback):
    def on_train_begin(self, logs={}):
        print (logs)
        return
```

We can then use it with

```
CC1 = CustomCallback1()
model.fit(train_images, train_labels,  epochs = 2,
          validation_data = (test_images,test_labels),
          callbacks = [CC1])  # pass callback to training
```

Remember to always instantiate the class and pass the `CC1` variable, and not the class itself. We will get

```
Train on 5000 samples, validate on 5000 samples
{}
Epoch 1/2
5000/5000 [==============================] - 1s 274us/step - loss: 0.0976 -
acc: 0.9746 - val_loss: 0.2690 - val_acc: 0.9172
Epoch 2/2
5000/5000 [==============================] - 1s 275us/step - loss: 0.0650 -
acc: 0.9852 - val_loss: 0.2925 - val_acc: 0.9114
{}
```

```
<tensorflow.python.keras.callbacks.History at 0x7f795d750208>
```

The logs dictionary is empty, as you can see from the {}. Let's expand the class a bit

```
class CustomCallback2(keras.callbacks.Callback):
    def on_train_begin(self, logs={}):
        print (logs)
        return

    def on_epoch_end(self, epoch, logs={}):
        print ("Just finished epoch", epoch)
        print (logs)
        return
```

Now we train the network with

```
CC2 = CustomCallback2()
model.fit(train_images, train_labels,  epochs = 2,
          validation_data = (test_images,test_labels),
          callbacks = [CC2])  # pass callback to training
```

This will give the following output (reported here for just one epoch for brevity)

```
Train on 5000 samples, validate on 5000 samples
{}
Epoch 1/2
4864/5000 [============================>.] - ETA: 0s - loss: 0.0511 -
acc: 0.9879
Just finished epoch 0
{'val_loss': 0.2545496598124504, 'val_acc': 0.9244, 'loss':
0.05098680723309517, 'acc': 0.9878}
```

Now things are starting to get interesting. The logs dictionary contains a lot more information that we can access and use. In the dictionary, we have val_loss, val_acc, and acc. Let's customize the output a bit. Let's set verbose = 0 in the fit() call to suppress the standard output and generate our own.

Our new class will be

```
class CustomCallback3(keras.callbacks.Callback):
    def on_train_begin(self, logs={}):
```

```
        print (logs)
        return

    def on_epoch_end(self, epoch, logs={}):
        print ("Just finished epoch", epoch)
        print ('Loss evaluated on the validation dataset =',
        logs.get('val_loss'))
        print ('Accuracy reached is', logs.get('acc'))
        return
```

We can train our network with

```
CC3 = CustomCallback3()
model.fit(train_images, train_labels,  epochs = 2,
          validation_data = (test_images,test_labels),
          callbacks = [CC3], verbose = 0)  # pass callback to training
```

Doing this, we will get

```
{}
Just finished epoch 0
Loss evaluated on the validation dataset = 0.2546206972360611
```

The empty {} is simply the empty logs dictionary that on_train_begin received. Of course, you can simply print information every few epochs. For example, by modifying the on_epoch_end() function as

```
def on_epoch_end(self, epoch, logs={}):
        if (epoch % 10 == 0):
          print ("Just finished epoch", epoch)
          print ('Loss evaluated on the validation dataset =',
          logs.get('val_loss'))
          print ('Accuracy reached is', logs.get('acc'))
        return
```

We get the following output if we train the network for 30 epochs

```
{}
Just finished epoch 0
Loss evaluated on the validation dataset = 0.3692033936366439
```

```
Accuracy reached is 0.9932
Just finished epoch 10
Loss evaluated on the validation dataset = 0.3073081444747746
Accuracy reached is 1.0
Just finished epoch 20
Loss evaluated on the validation dataset = 0.31566708440929653
Accuracy reached is 0.9992
<tensorflow.python.keras.callbacks.History at 0x7f796083c4e0>
```

Now you should start to get an idea as to how you can perform several things during training. You can, for example, save accuracy values in lists to plot them later, or simply plot metrics to see how your training is going. The possibilities are almost endless. Callbacks are a great way to customize what happens during training.

Custom Training Loops

The easiest way to train a network with Keras is to use the `compile()/fit()` approach. It makes building and training the network very easy. But the downside of this approach is that you don't have much flexibility as to how the training is implemented. For example, suppose you want to train two networks in alternate fashion (as you learned in Chapter 11 about GANs). In this case, the standard `fit()` call is not enough anymore; you need to implement a custom training loop. Let's see how to do that.

Calculating Gradients

As you know, the `fit()` function will evaluate the gradients of the loss function and, by using the appropriate optimizer, use them to update the weights. The first step in implementing a custom training loop is to understand how to evaluate the gradients of a given function manually. Let's consider the function $y = x^2$. How can we calculate the gradient of it at $x_0 = 2$ with TensorFlow? By manually taking the derivative, we can immediately see that

$$\frac{d}{dx}x^2 = 2x$$

And therefore

$$\left.\frac{d}{dx}x^2\right|_{x=2} = 4$$

With Keras, we can do the same calculation this way

```
x = tf.constant(2.0)
with tf.GradientTape() as g:
  g.watch(x)
  y = x * x
dy_dx = g.gradient(y, x)
```

The dy_dx variable will be a Tensor that will have a single value of 4

```
tf.Tensor(4.0, shape=(), dtype=float32)
```

TensorFlow operations are "recorded" in sequence, like on a tape (hence the name GradientTape), when they are executed within this context manager. TensorFlow checks all the called operations and saves all the gradients of the operations that are evaluated in that context. Note that everything that happens outside that context is ignored. In TensorFlow language, every operation and variable that is recorded is being "watched." Fortunately, when dealing with neural networks, all trainable variables (the weights and bias, typically) are automatically watched. But you can manually ask TensorFlow to watch other tensors by using the watch() call, as we have done in our example with the g.watch(x) code.

To understand what is going on in the background, you need to understand auto-differentiation. But intuitively you can think of the process in this way: when TensorFlow evaluates an operation, it will also save its gradient in memory. By using the gradientTape, you are simply asking TensorFlow to keep the evaluated gradients of specific operations in memory so that they can be used and combined properly to get the right result in the end.

Note that as soon as you call the gradient() function, all the resources held by the GradientTape() are released. If you wanted to calculate the second derivative, for example, you must do it differently than this example. You would need to use the persistent=True parameter in the creation of GradientTape(). For example, suppose you wanted the second derivative of the following function at $x_0 = 2$

$$y = x^3$$

The result is 12 since

$$\frac{d^2}{dx^2} x^3 = 6x.$$

With Keras, the code would look like this

```
x = tf.constant(2.0)
with tf.GradientTape(persistent=True) as g:
  g.watch(x)
  y = x * x * x
dy_dx = g.gradient(y, x)
dy_dx = g.gradient(y, x)
```

That would give the expected result

```
tf.Tensor(12.0, shape=(), dtype=float32)
```

Running the same code without the `persistent=True` parameter will produce an error message when calling the `gradient()` function a second time. There is another important point to note. Consider the following code, where I removed the `g.watch(x)` call.

```
x = tf.constant(2.0)
with tf.GradientTape() as g:
    y = x * x * x
dy_dx = g.gradient(y, x)
print(dy_dx)
```

The result of this code is `None`. No gradient can be evaluated since the variable `x` is not being "watched" by the `GradientTape()`. Now let's see how to implement a custom training loop with a neural network.

Custom Training Loop for a Neural Network

Now consider a very small FFNN with two layers, each having 64 neurons.

```
inputs = keras.Input(shape=(784,), name="digits")
x1 = layers.Dense(64, activation="relu")(inputs)
x2 = layers.Dense(64, activation="relu")(x1)
outputs = layers.Dense(10, name="predictions")(x2)
model = keras.Model(inputs=inputs, outputs=outputs)
```

As a second step, we need to specify an optimizer and a loss function (remember that we will not use the `compile()` function, so we need to use the Keras functions explicitly):

```
optimizer = keras.optimizers.Adam(learning_rate=1e-2)
loss_fn = keras.losses.SparseCategoricalCrossentropy(from_logits=True)
```

Another thing that we need to specify (by using the Keras functions) are the metrics we want to track, because we cannot specify the metrics in the `fit()` call.

```
train_acc_metric = keras.metrics.SparseCategoricalAccuracy()
val_acc_metric = keras.metrics.SparseCategoricalAccuracy()
```

At this point, we have all the ingredients we need. The loop can now be implemented easily

```
epochs = 200
for epoch in range(epochs):
    with tf.GradientTape() as tape:

        # Run the forward pass of the layer.

        logits = model(x_train, training=True)  # Logits for this minibatch

        # Compute the loss funtion
        loss_value = loss_fn(y_train, logits)

    grads = tape.gradient(loss_value, model.trainable_weights)
    optimizer.apply_gradients(zip(grads, model.trainable_weights))

    # Update training metric.
    train_acc_metric.update_state(y_train, logits)

    # Display metrics at the end of each epoch.
    train_acc = train_acc_metric.result()
```

```
    if epoch % 20 == 0:
        print(
            "Training loss (for one batch) at step %d: %.4f"
            % (epoch, float(loss_value))
        )
        print("Training acc over epoch: %.4f" % (float(train_acc),))
```

In the GradientTape() context, we need the following steps:

- The forward pass: Easily done with model(x_train, training=True)
- The loss function: loss_value = loss_fn(y_train, logits)
- Calculate the gradients and apply them to update the weights: grads = tape.gradient(loss_value, model.trainable_weights)
- optimizer.apply_gradients(zip(grads, model.trainable_weights))

The tape.gradient() calculates the gradient of the loss function (that is being watched in the GradientTape context), and apply_gradients() applies the gradients with the optimizers to update the network weights.

After that, we need to keep track of the metrics. The train_acc_metric.update_state(y_train, logits) call updates the metrics we defined, and then the train_acc = train_acc_metric.result() call saves its value in a variable (train_acc) that we can display during training.

This very basic loop shows how you can build your own custom training loops. Of course, there is much more you can do and, as usual, the best place to get all the information is the official Keras documentation.[2] Note that there at least two important things that we have not discussed:

- How to train with mini-batches (the training loop we discussed uses all the data). To do that, we could use

  ```
  train_dataset = tf.data.Dataset.from_tensor_slices
  ((x_train, y_train)
  train_dataset = train_dataset.shuffle(buffer_size=1024).
  batch(batch_size)
  ```

  and then add a loop over the batches with this

  ```
  for step, (x_batch_train, y_batch_train) in enumerate
  (train_dataset):
  ```

- How to speed up the training by using `@tf.function`. This is a more advanced topic that would require a long discussion and goes beyond the scope of this book.

Remember that the goal of this book is to teach you how neural networks work and how easy it is to implement them, not to make you a Keras expert. The goal of this appendix is to give just enough information that you can follow the book. To better understand Keras customization, your best bet is to use the official documentation and work through the examples.

[2] A good place to start is `https://www.tensorflow.org/guide/keras/writing_a_training_loop_from_scratch`.

Index

A

B

U. Michelucci, *Applied Deep Learning with TensorFlow 2*, https://doi.org/10.1007/978-1-4842-8020-1

C

D

E

F

G

H

I, J

K

L

M

N

O

P, Q

R

S

T

U, V, W, X, Y

Z

Printed in the United States
by Baker & Taylor Publisher Services